Excel

Get the Results You Want!

SmartStudy 9

Mathematics

Lyn Baker

PASCAL PRESS

Reprinted 2017, 2019, 2020, 2021, 2023, 2025

ISBN 978 1 74125 475 4

Pascal Press
PO Box 250
Glebe NSW 2037
www.pascalpress.com.au

Publisher: Vivienne Joannou
Project editors: May McCool and Rosemary Peers
Edited by May McCool
Answers checked by Peter Little
Indexed by Puddingburn Publishing Services
Cover design by DiZign Pty Ltd
Typeset by Precision Typesetting (Barbara Nilsson)
Printed by Vivar Printing/Green Giant Press

Students
All care has been taken in compiling this book, but please check with your teacher about the exact requirements of the course as these can change from year to year.

TABLE OF CONTENTS

TAKE THESE REVISION STEPS TO SUCCESS!

Step 1 Study Notes

- This section contains valuable **suggestions**, **tips** and **essential points** about the topic.
- When reading this section, **highlight** the points that are new to you or that you consistently forget. You could also **rewrite** these points so that you are making a summary of the Study Notes. Keep this summary in a prominent place at home—you could stick it on a mirror, wall or door.
- Work carefully through the **checklist** at the end of the section to ensure that you have mastered each of the listed skills.
- If there is something that you are **not sure about**, take time out to read through your class notes, textbooks or another relevant ***Excel*** book. (Check our website at www.pascalpress.com.au for more titles.)

Step 2 Skills Check

- Check that you know the **basic skills** you need to successfully complete the topic.
- Once you have completed the check, **mark your work** quickly by looking at the answers at the bottom of the page. This is **instant feedback** for you. The **worked solutions** are located at the back of the book if you want to check the working of any answer.
- If you have **scored less than 50%** in this section then you should revise your basic skills.

Step 3 Intermediate Test

- This test has very similar questions to the ones you will get in your **class test** or **exam**.
- Look carefully for the **Hints** which are provided for the trickiest questions—these appear at the bottom of the page.
- **Marks** are allocated for each question. These are similar to the marks in your tests.
- Time yourself—**check** how much **time** you have got to complete the test. Also look at the **total marks** of the test to calculate approximately how much time you should spend on each question. For example, if there are twenty marks in total and twenty minutes have been allocated for completion of the test, then spend about one minute on each mark. If you cannot complete all the questions within the suggested time, you may need to revise the topic.
- Fill in the **Your Feedback** panel once you have marked your work in order to calculate your percentage mark. Then complete the **Test & Exam Results** on page 198 to keep a running total of all your test marks.

Step 4 Advanced Test

- This test features **extension questions** such as problem–solving questions.
- This test is not like a class test, as all the questions in it are challenging. Mastery of questions in this test, however, will ensure you gain top marks in your class tests and exams, and will also prepare you for the mathematics you will learn next year.
- **Marks** are allocated for each question. These are similar to the marks in your tests.
- Time yourself—**check** how much **time** you have to complete the test. Also look at the **total marks** of the test to calculate approximately how much time you should spend on each question. For example, if there are forty marks in total and forty minutes have been allocated for completion of the test, then spend about one minute on each mark.

- Fill in the **Your Feedback** panel once you have marked your work in order to calculate your percentage mark. Then complete the **Test & Exam Results** on page 198 to keep a running total of all your test marks.

Step 5 Check Your Solutions

- **Worked solutions** to all questions are found at the back of the book. Work through the solutions to any questions that you got wrong.
- **Longer questions** are usually worth two or more marks and will involve some working. You should set out all your working, because in Maths you may get some marks for your working even if your answer is wrong.
- The **ticks** that appear in the worked solutions indicate those parts of the working which receive marks. Therefore, even if your answer is wrong, you may be entitled to some marks for what you have written. Compare the worked solution to your own working to find out whether you are entitled to any marks for the question.
- If you still **cannot understand** how the correct answer was obtained, revise that part of the topic and, if necessary, refer to your class notes or ask your teacher for help. It is important to learn from your mistakes.

Step 6 Test & Exam Results

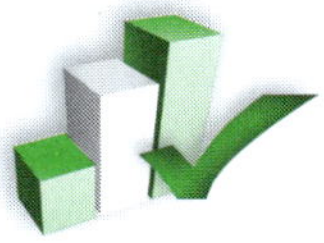

- Go to the **Test & Exam Results** section on page 198 to record your test score as a percentage. When you have completed all topics you will be able to determine your areas of weakness and your areas of strength.
- It is important that you know which **areas need further work**—to 'know what you don't know'. The more you prepare for the topic tests, the more successful you will be and the more you will remember when you sit your end-of-term/semester/year test or exam.

Step 7 Tips for the Sample Exam Papers

- These **useful tips** appear on page 109. Read them before you start one of the Sample Exams (see Step 8 below).

Step 8 Sample Exam Papers

- Three **Sample Exam Papers** are provided at the end of the book. These are of three levels of difficulty: Average, Above Average and Difficult. The Above Average paper will be very similar to your final examination.
- Before attempting the Sample Exam Papers, make sure that you have completed all of the **Tests** and have worked through the solutions to all questions that you answered incorrectly.
- Set aside the **time allowed** for the paper and complete it under **exam conditions**—no sneaking a look at your notes or textbooks! That way you will be better prepared for your final exam.
- **Worked solutions** to the Sample Exam Papers are found at the back of the book. Work through the solutions to any questions that you got wrong. Carefully note the **ticks** in the worked solutions and remember to give yourself marks for correct working. Write down your total marks for each section in the **Your Feedback** boxes at the end of each part of the paper, then add them up to get a total percentage for each test.

HOW TO USE THIS BOOK TO STUDY FOR A CLASS TEST, HALF-YEARLY OR END-OF-YEAR EXAM

Depending on your teacher or school, you will be given a variety of tests and exams each year. There may be a single-topic test, a test that covers a number of topics, a semester test or exam, or even a half-yearly or yearly exam.

Step 1 **Find out** which topics will be covered in the class test.

- To do this, look at your class workbook/textbook, laptop/tablet or online study program, and ask your teacher.
- For example, your class test may be on Statistics.

Step 2 **Match** the topics that your test is on to the topics in this book.

- For example, the three units on pages 89 to 100 of this book cover Statistics.

Step 3 **Use** this book to study the topics being tested.

- For example, the three units on pages 89 to 100 of this book contain the topics you will study for your class test on Statistics!

Note:

- When you are using this book to study for a **half-yearly** test, follow the same steps as above—the only difference being that you will have more topics to revise of course!
- When you are using this book to study for an **end-of-year** test, you will more than likely need to study the whole book!

REAL NUMBERS

Number and Algebra

STUDY NOTES

1 The top part of a fraction is the **numerator** and the bottom part is the **denominator**. An **improper fraction** has its numerator greater than its denominator. A **mixed number** has a whole number part and a fractional part.

2 To **add (or subtract) fractions** there should be a **common denominator**. When the denominators are the same, add the numerators. For example, $\frac{5}{6} + \frac{3}{8} = \frac{20}{24} + \frac{9}{24} = \frac{29}{24} = 1\frac{5}{24}$.

3 To **multiply fractions**, multiply the numerators and multiply the denominators. Mixed numbers must be changed to improper fractions before multiplying. For example, $1\frac{1}{2} \times \frac{3}{5} = \frac{3}{2} \times \frac{3}{5} = \frac{9}{10}$.

4 **Reciprocals** multiply to give one. For example, $2\frac{1}{3} = \frac{7}{3}$. The reciprocal of $2\frac{1}{3}$ is $\frac{3}{7}$ because $\frac{7}{3} \times \frac{3}{7} = 1$.

5 To **divide by a fraction**, multiply by its reciprocal. For example, $\frac{3}{5} \div \frac{2}{3} = \frac{3}{5} \times \frac{3}{2} = \frac{9}{10}$.

6 To **change a fraction to a decimal**, divide the numerator by the denominator. When **changing a decimal to a fraction** the denominator will be a power of ten. $0.06 = \frac{6}{100} = \frac{3}{50}$. You should know the **decimal equivalents** of the **common fractions**.

7 Decimals may either terminate or recur. **Terminating decimals** come to an end. For example, 0.2, 0.125, 0.76543. **Recurring decimals** have digits that repeat. Only one digit might repeat (for example, $0.\dot{1} = 0.111111\ldots$) or a sequence of digits (for example, $0.\overline{285714} = 0.285714285714 28\ldots$).

8 Some numbers cannot be expressed as either terminating or recurring decimals (for example, π). Such numbers are **not rational**.

9 To **add or subtract decimals** keep the decimal points aligned. When **multiplying decimals** there must be the same number of digits after the decimal point in the answer as there were in total in the question. For example, $0.03 \times 0.2 = 0.006$. To **divide by a decimal**, move the decimal point so that you divide by a whole number. Move the decimal point the same number of places to the right in both parts of the question. For example, $0.8 \div 0.02 = 80 \div 2 = 40$.

10 **Directed numbers** include both **positive** and **negative** numbers. When multiplying or dividing **two** directed numbers the answer will be positive if the signs are the same but negative if the signs are different.

11 **Per cent** means '**out of a hundred**'. To find a **percentage of an amount**, change the percentage to a fraction or decimal and multiply. For example, 15% of \$300 = 0.15 × \$300 = \$45. To make **something into a percentage**, multiply by 100%. For example, as a percentage 12 out of 40 is $\frac{12}{40} \times 100\% = 30\%$.

When **finding the whole amount** you are finding 100%. First find 1% then find 100%. For example, if 72% is \$180 then 1% is \$2.50. Multiplying by 100 gives 100% which is \$250.

Checklist

Can you:

1 *Convert fractions to decimals, fractions to percentages, decimals to percentages and vice versa?* ☐

2 *Understand the terms 'numerator', 'denominator', 'improper fractions', 'mixed numbers' and 'reciprocals'?* ☐

3 *Understand terminating and recurring decimals?* ☐

4 *Add, subtract, multiply and divide fractions, decimals and directed numbers?* ☐

5 *Find percentages of amounts, express one amount as a percentage of another, and find the whole amount given a percentage?* ☐

REAL NUMBERS

Number and Algebra

SKILLS CHECK

1 a Simplify $\frac{12}{16}$

b Write $2\frac{5}{7}$ as an improper fraction.

c Write $\frac{15}{6}$ as a mixed number in simplest form.

2 What is the reciprocal of:

a 4? b $\frac{3}{4}$? c $3\frac{2}{5}$?

3 Find:

a $\frac{2}{3}$ of 24 b $1\frac{1}{2} + 2\frac{3}{4}$ c $5 - 2\frac{3}{5}$ d $\frac{3}{7} \times \frac{2}{5}$

e $1\frac{2}{3} \times 2\frac{1}{2}$ f $8 \div \frac{1}{2}$ g $2\frac{3}{5} \div 1\frac{1}{4}$

4 Write each decimal as a fraction (or mixed number) in simplest form:

a 0.8 b 3.5 c 0.35

5 Find:

a $0.7 + 0.12$ b $0.35 - 0.2$ c 0.2×0.4 d 1.6×1000

e $3.5 \div 100$ f $0.06 \div 3$ g $0.08 \div 0.2$

6 Find:

a $-2 - 5$ b $7 - 11$ c $-5 + 8$ d $-7 + 4$

e -2×5 f $-12 \div -3$ g $-4 \times -6 \times -2$

7 Write each fraction as a decimal: a $\frac{3}{4}$ b $\frac{5}{8}$ c $\frac{17}{1000}$

8 Write each fraction as a recurring decimal: a $\frac{1}{3}$ b $\frac{5}{12}$ c $1\frac{1}{6}$

9 State whether each fraction is a terminating or recurring decimal:

a $\frac{2}{5}$ b $\frac{2}{3}$ c $\frac{3}{7}$ d $\frac{7}{8}$

10 Write:

a 60% as a fraction in simplest form b 8% as a decimal

c $\frac{3}{4}$ as a percentage d 0.3 as a percentage

11 a Find 35% of $9000. b What percentage is 18 out of 25?

12 15% of a number is 45. What is the number?

PAGE 123

Answers **1 a** $\frac{3}{4}$ **b** $\frac{19}{7}$ **c** $2\frac{1}{2}$ **2 a** $\frac{1}{4}$ **b** $1\frac{1}{3}$ or $\frac{4}{3}$ **c** $\frac{5}{17}$ **3 a** 16 **b** $4\frac{1}{4}$ **c** $2\frac{2}{5}$ **d** $\frac{6}{35}$ **e** $4\frac{1}{6}$ **f** 16 **g** $2\frac{2}{25}$ **4 a** $\frac{4}{5}$ **b** $3\frac{1}{2}$ **c** $\frac{7}{20}$ **5 a** 0.82 **b** 0.15 **c** 0.08 **d** 1600 **e** 0.035 **f** 0.02 **g** 0.4 **6 a** –7 **b** –4 **c** 3 **d** –3 **e** –10 **f** 4 **g** –48 **7 a** 0.75 **b** 0.625 **c** 0.017 **8 a** $0.\dot{3}$ **b** $0.41\dot{6}$ **c** $1.1\dot{6}$ **9 a** terminating **b** recurring **c** recurring **d** terminating **10 a** $\frac{3}{5}$ **b** 0.08 **c** 75% **d** 30% **11 a** $3150 **b** 72% **12** 300

REAL NUMBERS

Number and Algebra

15 MINUTES

INTERMEDIATE TEST

Part A Multiple Choice

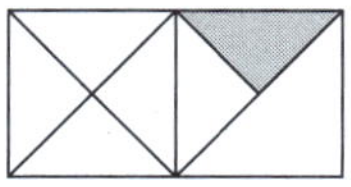

1 A rectangle is twice as long as it is wide. Mikala divided it as shown in the diagram, and shaded one part. What fraction is shaded? *Hint 1*

A $\frac{1}{4}$ **B** $\frac{1}{6}$ **C** $\frac{1}{7}$ **D** $\frac{1}{8}$ (1 mark)

2 Which of the numbers 0.208, 0.28, 0.0208 and 0.028 is the smallest? *Hint 2*

A 0.208 **B** 0.28 **C** 0.0208 **D** 0.028 (1 mark)

3 When 800 is decreased by 8% what is the result?

A 792 **B** 736 **C** 700 **D** 640 (1 mark)

4 Which of 10%, $\frac{1}{10}$ or 0.1 is the largest?

A 10% **B** $\frac{1}{10}$ **C** 0.1 **D** They are all equal. (1 mark)

5 Which fraction lies between $\frac{1}{3}$ and $\frac{1}{4}$? *Hint 3*

A $\frac{3}{10}$ **B** $\frac{3}{7}$ **C** $\frac{2}{9}$ **D** $\frac{5}{12}$ (1 mark)

6 When written as a decimal, correct to one decimal place, $\frac{2}{3} = ?$

A 0.2 **B** 0.3 **C** 0.6 **D** 0.7 (1 mark)

7 Which of these fractions is NOT a recurring decimal?

A $\frac{1}{3}$ **B** $\frac{5}{6}$ **C** $\frac{7}{8}$ **D** $\frac{7}{9}$ (1 mark)

Part B Short Answer

8 On Monday a blizzard occurred and the minimum temperature was –17 °C. The next day conditions improved. The minimum temperature was 9° higher than the previous day. What was the minimum temperature on Tuesday? (1 mark)

9 Evaluate $\frac{2.5 + 1.5}{5 \times 0.4}$ *Hint 4* (1 mark)

10 Alice received 68 marks out of 80 in a test. What was her percentage mark? (1 mark)

11 If $\frac{4}{9}$ of a number is 36, find the number. (1 mark)

12 Evaluate $\left(3\frac{1}{2}\right)^2$ *Hint 5* (1 mark)

13 **a** During a sale, a shop offers a discount of 15% off all marked prices. What would you pay for a shirt marked $60? (1 mark)

b Ian buys the shirt during the sale. As a regular customer, he also receives a discount of 5% applied after the first discount. How much does Ian save altogether? *Hint 6* (2 marks)

Hint 1: Divide the shape into equal parts.
Hint 2: Write each number with 4 digits after the decimal point.
Hint 3: Convert all the fractions to decimals.
Hint 4: Evaluate means 'find the value of'.
Hint 5: Change the mixed number to an improper fraction before squaring.
Hint 6: It is not 20% of the marked price.

Your Feedback

$\frac{\square}{15} \times 100\% = \square\%$

PAGES 123–124

PAGE 198

REAL NUMBERS

Number and Algebra

30 MINUTES

ADVANCED TEST

Note: this test is intended to be done without a calculator.

1 Simplify:

a $2\frac{1}{3}+3\frac{3}{4}$ b $5\frac{2}{7}-2\frac{3}{5}$ c $4\frac{1}{4}\times 3\frac{1}{5}$

d $30\div\frac{1}{3}$ e $\left(2\frac{1}{2}\right)^2$ f $7\frac{1}{2}\div\frac{3}{5}$ (1 mark each)

2 Simplify:

a $\frac{3}{5}+\frac{2}{3}\times\frac{1}{8}$ b $6\frac{4}{5}\times 7\frac{1}{2}\div 5\frac{2}{3}$ (2 marks each)

3 Simplify:

a $-5\times -4\times 7$ b $1.8+0.25$

c 0.2×0.3 d $-2+3\times -5-8\times 2$

e $8-2\times 0.35$ f $(-0.9)^2$ (1 mark each)

4 Evaluate:

a $25\times 37\times 4$ b $39\times 56+39\times 44$ (1 mark each)

5 Find:

a 15% of $8420 b 67% of 3 kg (1 mark each)

6 What percentage is:

a 52 out of 80? b 3 cm out of 6 m? (1 mark each)

7 a Express $\frac{5}{6}$ as a decimal. (1 mark)

b What is the reciprocal of $4\frac{2}{3}$? (1 mark)

c Find $\sqrt{5\frac{4}{9}}$ (2 marks)

d Express $0.2\dot{3}$ as a fraction in simplest form. (2 marks)

8 40% of Year 9 students went on an excursion. There were 240 students altogether on the excursion and one-quarter of those were in Year 9. How many students were in Year 9? (2 marks)

Your Feedback

$\frac{\square}{30}\times 100\% = \square\,\%$

PAGES 124–125

PAGE 198

ALGEBRAIC TECHNIQUES
Number and Algebra

STUDY NOTES

1 **Pronumerals** are symbols, usually letters, used in place of numbers in an expression.
$3x$ means $3 \times x$ (3 lots of x added together $x + x + x$)
x^2 means $x \times x$ (2 factors of x multiplied together) $3x^2$ means $3 \times x^2$ or $3 \times x \times x$
$3x$, x^2 and $3x^2$ are also called algebraic terms.

2 **Like terms** are terms which have exactly the same pronumeral part. $3x^2$ and $7x^2$ are like terms but $4a$ and $4a^3$ are not like terms because the pronumeral parts (a and a^3) are not exactly the same.

3 Algebraic expressions can be simplified by **collecting like terms**. When simplifying algebraic expressions only like terms can be added together or subtracted.
$3x^2 + 7x^2 = 10x^2$ (3 lots of x^2 plus 7 lots of x^2 is 10 lots of x^2)
but $4a + 4a^3 = 4a + 4a^3$ (It cannot be simplified because $4a$ and $4a^3$ are not like terms.)

4 The **operation sign** belonging to any term is immediately in front of it. For example, in the expression $8x^2 - 7x + 3x^2$ the minus belongs to $7x$ and the plus to $3x^2$.
$8x^2 - 7x + 3x^2 = 11x^2 - 7x$
$9a + 5b - 7a + 3b = 2a + 8b$ $[9a - 7a = 2a, 5b + 3b = 8b]$

5 When **multiplying pronumerals** we multiply any numbers and multiply the pronumerals separately. For example, $2x \times 3y = 6xy$ (Note: xy means $x \times y$. The multiplication sign is left out.)

6 **Fractions involving pronumerals** can be simplified by dividing both numerator and denominator by any common factors. For example, $\frac{8ab}{2b} = 4a$ (dividing both numerator and denominator by $2b$)

7 The **order** in which **operations** are performed must be carefully followed. First remove any brackets, then do any multiplications or divisions, working from left to right, and finally any additions or subtractions, again working from left to right.

8 Linear equations are **solved** by performing the same operation on **both sides** of the equation. An equation may be solved by:
- **adding** the same number **to both sides** of the equation
- **subtracting** the same number **from both sides** of the equation
- **multiplying both sides** of the equation by the same amount, or
- **dividing each side** of the equation by the same amount.

9 When **substituting**, a pronumeral is replaced by a number.
$2x$ means $2 \times x$, so when $x = 4$, $2x = 2 \times 4 = 8$
ab means $a \times b$ so that when $a = 2$ and $b = 3$, $ab = 2 \times 3 = 6$

10 By using algebraic symbols, **word problems** can be written as algebraic expressions or equations. A statement should always be written explaining what each symbol represents.

Checklist
Can you:

1 *Understand what we mean by pronumerals, terms and like terms?* ☐
2 *Simplify expressions by collecting like terms?* ☐
3 *Multiply terms together and divide one term by another?* ☐
4 *Simplify an algebraic fraction?* ☐
5 *Understand the correct order of operations to follow when simplifying algebraic expressions?* ☐
6 *Solve simple equations?* ☐
7 *Substitute into algebraic expressions?* ☐
8 *Use symbols to write algebraic expressions?* ☐

ALGEBRAIC TECHNIQUES

Number and Algebra

SKILLS CHECK

1 Find:

a $3x + 5x$ b $2a^2 + 3a^2$ c $7pq - 5pq$

d $8m + m$ e $6e - e$ f $9y - 8y$

g $5a + 7a - 3a$ h $-3p - 7p$ i $6m^3 - 4m^3 - 8m^3$

2 Simplify by collecting like terms:

a $9x + 4y + 2x$ b $7p - 2q - p + 8q$

c $4a + 5b + 2a + 7$ d $6m + 5n - 4m + 2n$

e $-2k + 3 - 4k - 5$ f $4p^2 + 3p + 2p$

g $3p^2 - 4q^2 + 5p - 2q^2 - 3q + 4p$ h $9x^3 - 5x^2 + 8x - 3x^2 + 2x - x^2$

3 Find these products.

a $x \times x$ b $7 \times a$ c $2 \times 4y$

d $k \times 6$ e $3 \times 5x^2$ f $2t \times 4$

g $4m \times 2n$ h $5a \times b$ i $3xy \times 3z$

4 Simplify:

a $7x \div x$ b $8a \div 2a$ c $4g \div 4$

d $12m \div 3$ e $9y^3 \div 3$ f $20xy \div 4y$

g $\dfrac{4p}{2}$ h $\dfrac{6x}{12x}$ i $\dfrac{9}{6b}$

5 If $x = 5$, find the value of:

a $x + 3$ b $4x$ c x^2

d $3x - 1$ e $15 \div x$ f $8 - 2x$

6 If $a = 2$, $b = 3$, $c = 7$ and $d = -2$, find:

a $2a + b$ b ab c $c^2 - a$

d $2b^2 + d$ e ad^2 f $4bc - 2a$

7 Solve:

a $2x + 9 = 17$ b $\dfrac{d-4}{3} = -9$ c $5p + 8 = 29 - 2p$

PAGES 125–126

Answers **1** a $8x$ b $5a^2$ c $2pq$ d $9m$ e $5e$ f y g $9a$ h $-10p$ i $-6m^3$ **2** a $11x + 4y$ b $6p + 6q$ c $6a + 5b + 7$ d $2m + 7n$ e $-6k - 2$ f $4p^2 + 5p$ g $3p^2 - 6q^2 + 9p - 3q$ h $9x^3 - 9x^2 + 10x$ **3** a x^2 b $7a$ c $8y$ d $6k$ e $15x^2$ f $8t$ g $8mn$ h $5ab$ i $9xyz$ **4** a 7 b 4 c g d $4m$ e $3y^3$ f $5x$ g $2p$ h $\frac{1}{2}$ i $\frac{3}{2b}$ **5** a 8 b 20 c 25 d 14 e 3 f -2 **6** a 7 b 6 c 47 d 16 e 8 f 80 **7** a $x = 4$ b $d = -23$ c $p = 3$

ALGEBRAIC TECHNIQUES

Number and Algebra

25 MINUTES

INTERMEDIATE TEST

Part A Multiple Choice

1 $8x + x = ?$

A $8x^2$ **B** $9x$ **C** $9x^2$ **D** none of these (1 mark)

2 $2x \times 3x \times 5x = ?$

A $10x$ **B** $30x$ **C** $10x^3$ **D** $30x^3$ (1 mark)

3 $m^4 + m^4 = ?$

A m^8 **B** m^{16} **C** $2m^4$ **D** $2m^8$ (1 mark)

4 $5a - 2b - 3a + 4b = ?$

A $2a + 2b$ **B** $2a + 6b$ **C** $2a - 6b$ **D** $2a - 2b$ (1 mark)

5 $5pq^2 = ?$

A $5 \times p \times q \times 2$ **B** $5 \times p \times q \times q$ **C** $5 \times p \times p \times q \ \ q$ **D** $5 \times 5 \times p \times p \times q \times q$ (1 mark)

6 $9x^2 - 5x + 2x^2 - x = ?$

A $11x^2 - 6x$ **B** $11x^2 - 4x$ **C** $7x^2 - 4x$ **D** $7x^2 - 6x$ (1 mark)

7 If $k = -2$, $3k^2 = ?$ *Hint 1*

A -36 **B** -12 **C** 12 **D** 36 (1 mark)

8 What is the number 8 more than $2x - 5$? *Hint 2*

A $2x + 3$ **B** $2x + 13$ **C** $10x - 5$ **D** $10x + 13$ (1 mark)

9 $\dfrac{8x^2}{4xy} = ?$

A $2xy$ **B** $2y$ **C** $\dfrac{2}{y}$ **D** $\dfrac{2x}{y}$ (1 mark)

10 If $p = 4$, $q = -3$ and $r = 5$, then $r - pq = ?$

A -7 **B** -2 **C** 17 **D** 48 (1 mark)

Part B Short Answer

11 Simplify $3x^2 - 2x \times 4x$ *Hint 3* (2 marks)

12 Simplify $8a \times 3a + 2 \times 7a^2$ (2 marks)

13 Simplify $\dfrac{3a}{12ab}$ *Hint 4* (2 marks)

14 Solve:

a $8x + 23 = -5$ (2 marks)

b $\dfrac{m}{6} - 32 = 18$ (2 marks)

15 If $p = 5$ and $q = 7$ find the value of $7pq + 8p + 2pq - 5p - 6q + 3p - q$ *Hint 5* (2 marks)

16 Jon charges $5 plus $8 per hour for odd jobs. Write an expression for the amount Jon will charge when he works for n hours on a job. (1 mark)

17 Simplify $(9x - 7x) \div (5x - 3x)$ (2 marks)

Hint 1: Only k is squared.
Hint 2: Add 8 to $2x - 5$.
Hint 3: Take care with the order of operations.
Hint 4: Be careful with the numerator.
Hint 5: Collect like terms before attempting to substitute.

Your Feedback

$\dfrac{\square}{25} \times 100\% = \square\ \%$

PAGE 126

PAGE 198

ALGEBRAIC TECHNIQUES
Number and Algebra

35 MINUTES

ADVANCED TEST

1 Write in simplest form:

a $-9x + x$ b $-9x - x$ c $-9x \times x$ d $-9x \div x$

e $5x + 2y$ f $5x \times 2y$ g $\frac{5ab}{10bc}$ h $\frac{p^2q^2}{(pq)^2}$

(1 mark each)

2 Simplify:

a $2x + 3x \times 4x$ (1 mark)

b $12k \div 3k - 2k$ (1 mark)

c $4n \times -2n \times n \times -3$ (1 mark)

3 Simplify:

a $5x^2 + 2x - x^2 + 7x$ (1 mark)

b $-6a + 8b - 9a - 3b - 4$ (1 mark)

c $4k^3 - 3k^2 + 2k - 5k^2 + 8k$ (1 mark)

d $-3 - 9 - 4m + 7m + 2$ (1 mark)

e $8e + 9f - 4f + 3e + f - 5e$ (1 mark)

f $m^5 - 3m^5 - 4n^4 + 2n^2 - 6n^4$ (1 mark)

4 Simplify:

a $4a \times 5a - 2a \times 8a$ (2 marks)

b $12k^2 \div 6 + 3k \times 4k$ (2 marks)

c $(x^2 + x^2) \times (7x^2 \div x^2)$ (2 marks)

d $(7x - 4x) \div (5x + x)$ (2 marks)

5 If $x = -2$ and $y = 5$ evaluate:

a $y - x$ (1 mark)

b x^2y (1 mark)

c $5x - 2y$ (1 mark)

d $3xy + x$ (1 mark)

6 If an unknown number is given by x, write expressions for each of the following:

a three more than five times the number (1 mark)

b seven less than the product of the number and six (1 mark)

c the sum of twice the number and the square of the number (1 mark)

7 Solve:

a $6a + 9 - a = 8 - 3a + 1$ (2 marks)

b $\frac{2t + 4}{7} - 3 = 5$ (2 marks)

Your Feedback

$\frac{\square}{36} \times 100\% = \square\%$

PAGES 126–127

PAGE 198

INDICES AND SCIENTIFIC NOTATION

Number and Algebra

STUDY NOTES

1 In the expression 6^4 the number 6 is the **base** and 4 is the **index** or **power** or **exponent**.

2 When a number is **raised to a power**, the **index** tells us how many factors of the number are multiplied together. For example, $6^4 = 6 \times 6 \times 6 \times 6$. 4 factors of 6 are multiplied together.

3 **Indices** is the plural of index.

4 **When multiplying**, if the base is the same, **add the indices**. $2^3 \times 2^4 = 2^7$
$(2 \times 2 \times 2) \times (2 \times 2 \times 2 \times 2) = 2^7$. There are 3 factors of 2 and another 4 factors of 2 which gives 7 factors of 2. We don't multiply the base numbers.

5 **When dividing**, if the base is the same, **subtract the indices**. $9^5 \div 9^3 = 9^2$
$\frac{9 \times 9 \times 9 \times 9 \times 9}{9 \times 9 \times 9} = \frac{\cancel{9} \times \cancel{9} \times \cancel{9} \times 9 \times 9}{\cancel{9} \times \cancel{9} \times \cancel{9}} = 9^2$. 5 factors of 9 divided by 3 factors of 9 leaves 2 factors of 9.
We don't divide the base numbers.

6 When **raising a power to a power, multiply the indices.** $(2^3)^2 = 2^6$ because $2^3 \times 2^3 = 2^6$. $(7^4)^3 = 7^{12}$

7 Any number raised to the **zero power** equals one. This is because the answer to a question such as $2^5 \div 2^5$ can be calculated in two ways. Using the index laws $2^5 \div 2^5 = 2^{5-5} = 2^0$. But $2^5 = 32$ and $32 \div 32 = 1$. There are two answers to the same question so the two answers must also be the same; $2^0 = 1$.

8 If a number is raised to a **negative index**, the result is the reciprocal of that number to the equivalent positive index. Using the index laws $2^2 \div 2^3 = 2^{-1}$ but $2^2 = 4$ and $2^3 = 8$ and 4 divided by 8 is a half.
So $2^{-1} = \frac{1}{2}$. Similarly $3^{-2} = \frac{1}{3^2} = \frac{1}{9}$ and $10^{-5} = \frac{1}{10^5} = \frac{1}{100\,000}$.

9 **Scientific notation**, sometimes called standard notation, gives us an easy way of writing very large or very small numbers. In scientific notation, every number can be expressed as the **product** of a number, greater than or equal to 1 and less than 10, with a power of 10.

10 For **large numbers** (numbers greater than one), the power of 10 is **positive**.
For example, $7\,000\,000 = 7 \times 10^6$; $5.24 \times 10^9 = 5\,240\,000\,000$.

11 For **small numbers** (positive numbers less than one), the power of 10 is **negative**.
For example, $0.000\,13 = 1.3 \times 10^{-4}$; $8 \times 10^{-7} = 0.000\,0008$.

12 The **power of 10** does not give the number of zeroes. It tells us the number of places that the decimal point has moved. For example, $7.9 \times 10^8 = 790\,000\,000$ and $8.3 \times 10^{-5} = 0.000\,083$.

1 2 3 4 5 6 7 8   5 4 3 2 1
790 000 000   0.000 083

13 Make sure you understand how to **use your calculator** with scientific notation. Many calculators display scientific notation in a different way to how we write it.

14 Numbers written in scientific notation can be arranged in **ascending (increasing)** or **descending (decreasing) order**. For example, 8×10^5 is smaller than 2×10^6 because the power of 10 is smaller. 1.2×10^{-5} is larger than 2.4×10^{-6} because the power of 10 is larger ($-5 > -6$).

Checklist

Can you:

1 *Understand index notation and the terms base, power, index, exponent and factor?* ☐
2 *Understand and use the zero index?* ☐
3 *Understand and use negative indices?* ☐
4 *Use the index laws when multiplying, dividing and raising a power to a power?* ☐
5 *Understand and use scientific notation?* ☐
6 *Use your calculator with indices and scientific notation?* ☐
7 *Arrange numbers that are in scientific notation in order?* ☐

INDICES AND SCIENTIFIC NOTATION

Number and Algebra

SKILLS CHECK

1 Write $7 \times 7 \times 7 \times 7 \times 7 \times 7$ in index form.

2 Write 3^4 as a product of its factors.

3 Find the value of:

a 2^3 b 3^4 c 5^6 d 10^5
e 6^0 f 9^1 g 37^2 h 7^9

4 Simplify, giving the answer in index form:

a $3^4 \times 3^3$ b $2^5 \times 2$ c $10^2 \times 10^8$

5 Simplify, giving the answer in index form:

a $7^8 \div 7^2$ b $3^6 \div 3^3$ c $2^9 \div 2$

6 Simplify, giving the answer in index form:

a $(5^2)^3$ b $(7^7)^2$ c $(2^4)^4$

7 Find the value as a fraction:

a 2^{-1} b 6^{-1} c 15^{-1}
d 5^{-2} e 2^{-3} f 10^{-6}

8 Write as a decimal:

a 5^{-1} b 10^{-4} c 2^{-2}

9 Write as a power of a prime number:

a $\frac{1}{3}$ b $\frac{1}{49}$ c $\frac{1}{32}$

10 Determine whether the following statements are true or false:

a $3^4 = 64$ b $5^3 = 15$ c $2^5 = 32$
d $7^{-1} = \frac{1}{7}$ e $9^{-2} = \frac{1}{3}$ f $4^{-3} = \frac{1}{81}$
g $10^7 = 1\,000\,000$ h $10^{-4} = 0.000\,01$ i $10^{-2} = 0.01$
j $3 \times 10^2 = 30^2$ k $10 \times 10^{-1} = 1$

11 Write as a power of 10:

a 1000 b 10 000 000 c 0.01 d 0.000 01

12 Write in scientific notation:

a 720 000 b 9000 c 576
d 289 000 000 e 4300 f 26 540 000
g 0.027 h 0.000 05 i 0.001 376

13 Write in decimal form:

a 4×10^2 b 9.1×10^3 c 1.827×10^5
d 3.7×10^{-4} e 8×10^7 f 7×10^{-3}
g 1.2×10^{-6} h 6.45×10^{-5} i 1×10^{-2}

PAGES 127–128

Answers **1** 7^6 **2** $3 \times 3 \times 3 \times 3$ **3** **a** 8 **b** 81 **c** 15 625 **d** 100 000 **e** 1 **f** 9 **g** 1369 **h** 40 353 607 **4** **a** 3^7 **b** 2^6 **c** 10^{10} **5** **a** 7^6 **b** 3^3 **c** 2^8 **6** **a** 5^6 **b** 7^{14} **c** 2^{16} **7** **a** $\frac{1}{2}$ **b** $\frac{1}{6}$ **c** $\frac{1}{15}$ **d** $\frac{1}{25}$ **e** $\frac{1}{8}$ **f** $\frac{1}{1\,000\,000}$ **8** **a** 0.2 **b** 0.0001 **c** 0.25 **9** **a** 3^{-1} **b** 7^{-2} **c** 2^{-5} **10** **a** false **b** false **c** true **d** true **e** false **f** false **g** false **h** false **i** true **j** false **k** true **11** **a** 10^3 **b** 10^7 **c** 10^{-2} **d** 10^{-5} **12** **a** 7.2×10^5 **b** 9×10^3 **c** 5.76×10^2 **d** 2.89×10^8 **e** 4.3×10^3 **f** 2.654×10^7 **g** 2.7×10^{-2} **h** 5×10^{-5} **i** 1.376×10^{-3} **13** **a** 400 **b** 9100 **c** 182 700 **d** 0.000 37 **e** 80 000 000 **f** 0.007 **g** 0.000 0012 **h** 0.000 0645 **i** 0.01

INDICES AND SCIENTIFIC NOTATION

Number and Algebra

20 MINUTES

INTERMEDIATE TEST

Part A Multiple Choice

1 $3^2 \times 3^4 = ?$

A 3^6 B 3^8 C 9^6 D 9^8 (1 mark)

2 $8^{12} \div 8^4 = ?$

A 1^3 B 1^8 C 8^3 D 8^8 (1 mark)

3 $(2^2)^5 = ?$

A 2^7 B 2^{10} C 4^7 D 4^{10} (1 mark)

4 $3^0 + 2^0 = ?$

A 1 B 2 C 3 D 5 (1 mark)

5 $4^{-2} = ?$

A 16 B 2 C $\frac{1}{2}$ D $\frac{1}{16}$ (1 mark)

6 $3^2 \times 2^4 = ?$ *Hint 1*

A 12^2 B 5^6 C 6^5 D 6^6 (1 mark)

7 $8^2 \times 4^3 = ?$ *Hint 2*

A 32^5 B 12^6 C 2^{12} D 2^{10} (1 mark)

8 $2^3 \div 2^{-2} = ?$ *Hint 3*

A 1 B 2 C 2^5 D 2^7 (1 mark)

9 $\dfrac{3^2 \times 2^3}{2^2 \times 3^3} = ?$

A 1 B $\frac{2}{3}$ C $\frac{3}{2}$ D cannot be simplified (1 mark)

10 When written in scientific notation 0.0008 = ?

A 8×10^3 B 8×10^4 C 8×10^{-3} D 8×10^{-4} (1 mark)

11 $2.3 \times 10^6 = ?$

A twenty-three million B two million, three hundred thousand

C two hundred and thirty thousand D none of these (1 mark)

12 What is the sum of 5×10^{-2} and 2.6×10^{-3}? *Hint 4*

A 7.6×10^{-3} B 7.6×10^{-5} C 5.26×10^{-2} D 5.26×10^{-3} (1 mark)

Part B Short Answer

13 Find $\dfrac{32 \times 500}{0.8 \div 0.05}$, giving the answer in scientific notation. (2 marks)

14 Find $\dfrac{(3.2 \times 10^7) \times (5.6 \times 10^3)}{2.8 \times 10^{-5}}$, giving the answer in scientific notation. *Hint 5* (1 mark)

15 Arrange the numbers 2.87×10^{-2}, 6×10^{-2} and 5×10^{-3} in ascending order. *Hint 6* (2 marks)

16 Find $(2^5 \times 2^3)^2 \div (2^4 \div 2)^3$, giving the answer in index form. (3 marks)

Hint 1: The base numbers (3 and 2) are not the same. You can't just add the indices.
Hint 2: Both 8 and 4 are powers of 2.
Hint 3: Take care if subtracting a negative number.
Hint 4: Change the numbers to decimals before adding.
Hint 5: Make sure you divide by the whole denominator.
Hint 6: Ascending order means from lowest to highest. It may help to write each number as a decimal.

Your Feedback

$\dfrac{\square}{20} \times 100\% = \square\%$

PAGE 128

PAGE 198

INDICES AND SCIENTIFIC NOTATION

Number and Algebra

35 MINUTES

ADVANCED TEST

1 Find the value as a decimal:

a 10^{-4} b 2^{-3} c 5^{-2} (1 mark each)

2 Write without indices:

a 7^{-3} b $\left(\frac{1}{2}\right)^{-1}$ c $\left(\frac{2}{3}\right)^{-2}$ (1 mark each)

3 Find the value of x:

a $2^x = 1024$ b $7^x = 1$ c $4^x = 2^{12}$

d $10^x = \frac{1}{1000}$ e $2^x = \frac{1}{64}$ f $x^{-2} = \frac{25}{16}$ (1 mark each)

4 Simplify, leaving answers in simplest index form:

a $5^8 \times 5^{-3}$ b $7^4 \div 7^9$ c $6^{-3} \times 6^{-5}$

d $2^{-4} \div 2$ e $(3^4)^2 \div (3^5)^3$ f $(10^{-3})^{-2} \times (10^{-4})^4$

g $(5^{-5})^2 \div (5^0)^{-2}$ h $2^{-7} \times 4^{-3}$ (1 mark each)

5 Express in correct scientific notation:

a 32×10^7 b 0.35×10^5 c 560×10^{-4}

d 7×100^2 e 0.06×10^{-5} f 97×10^{-9} (1 mark each)

6 If $E = mc^2$ and $c = 3 \times 10^5$, find:

a E when $m = 0.8$ (1 mark)

b m when $E = 5.76 \times 10^{13}$ (1 mark)

7 Find, expressing the answer in scientific notation:

a the product of 7.5×10^8 and 4.8×10^7 (1 mark)

b the square of 3.7×10^6 (1 mark)

c the sum of 6.5×10^{-8} and 5.3×10^{-7} (1 mark)

d the difference between 8.7×10^{-4} and 3.9×10^{-6} (1 mark)

e the square root of 9.604×10^{-5} (1 mark)

8 a The distance from Earth to the sun is 1.49×10^8 km and the distance from the planet Mercury to the sun is 5.8×10^7 km. Assuming that Mercury lies directly between the sun and Earth, what is the distance from Earth to Mercury? (1 mark)

b The planet Venus is 1.08×10^8 km from the sun. When the three planets are all aligned, with Mercury and Venus in a direct line between Earth and the sun, is Venus closer to Mercury or to Earth? Justify your answer. (1 mark)

Your Feedback

$\frac{\square}{35} \times 100\% = \square\,\%$

PAGES 128–129

PAGE 198

PYTHAGORAS' THEOREM

Measurement and Geometry

STUDY NOTES

1 The **hypotenuse** is the longest side of a right-angled triangle. It is the side opposite the right angle.

2 Upper case letters (capitals) are used to name **angles** and lower case letters to name **sides**. The side a is opposite angle A. Angle B is opposite side b.

3 Pythagoras' Theorem states: **In any right-angled triangle the square on the hypotenuse is equal to the sum of the squares on the other two sides.**

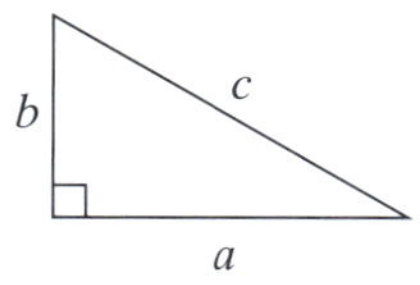

$$c^2 = a^2 + b^2$$

4 Pythagoras' Theorem can be used to find the **length of the hypotenuse** when the lengths of the other two sides are known.

For example,
$$\begin{aligned} x^2 &= 7^2 + 24^2 \\ &= 49 + 576 \\ &= 625 \\ x &= \sqrt{625} \\ &= 25 \end{aligned}$$

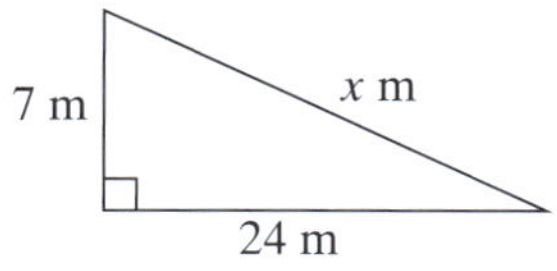

Note: When solving an equation such as $x^2 = 25$, there are two possible answers, one positive and one negative ($5 \times 5 = 25$ and $-5 \times -5 = 25$). But, when finding the side of a triangle, it does not make sense to have a negative value so we don't consider a negative answer.

5 **Pythagoras' Theorem** can also be used to find the **length of one of the other sides** of a triangle.

For example,
$$\begin{aligned} 35^2 &= h^2 + 28^2 \\ h^2 &= 35^2 - 28^2 \\ &= 1225 - 784 \\ &= 441 \\ h &= \sqrt{441} \\ &= 21 \end{aligned}$$

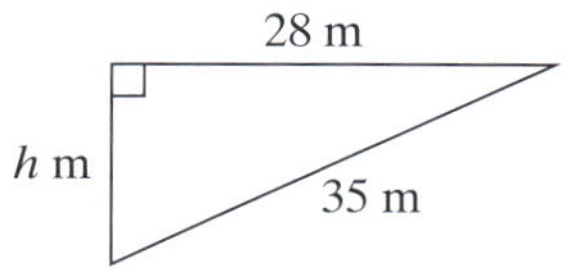

6 A **Pythagorean triad** is a set of three numbers such that the square of the largest one is equal to the sum of the squares of the two smaller ones. For example, the numbers 3, 4 and 5 form a Pythagorean triad. Any multiple of a Pythagorean triad is also a Pythagorean triad. So for example, 6, 8 and 10 and 9, 12 and 15 are also Pythagorean triads.

7 If the sides of a triangle are such that the square of the longest side is equal to the sum of the squares of the other two sides, then the triangle must be **right-angled**.

Checklist

Can you:

1 *Understand what is meant by hypotenuse?* ☐
2 *Name the sides of triangles in relation to their opposite angles?* ☐
3 *Identify the hypotenuse of a right-angled triangle?* ☐
4 *Understand Pythagoras' Theorem?* ☐
5 *Use Pythagoras' Theorem to find the length of the hypotenuse of a right-angled triangle?* ☐
6 *Use Pythagoras' Theorem to find the length of one of the other sides of a right-angled triangle?* ☐
7 *Understand what is meant by a Pythagorean triad?* ☐
8 *Determine whether a triangle is right-angled from the lengths of its sides?* ☐

PYTHAGORAS' THEOREM

Measurement and Geometry

SKILLS CHECK

1 Find: **a** 13^2 **b** $8^2 + 6^2$ **c** $37^2 - 24^2$

2 Find: **a** $\sqrt{5625}$ **b** $\sqrt{25+144}$ **c** $\sqrt{25-9}$

3 Find, to one decimal place: **a** $\sqrt{527}$ **b** $\sqrt{49+16}$ **c** $\sqrt{3.2^2 - 1.7^2}$

4 The hypotenuse of this triangle is BC.

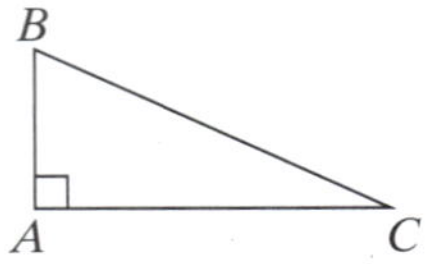

Name the hypotenuse of each of these triangles:

a

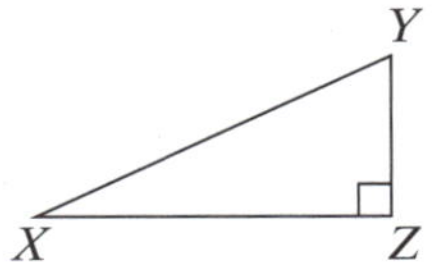

b

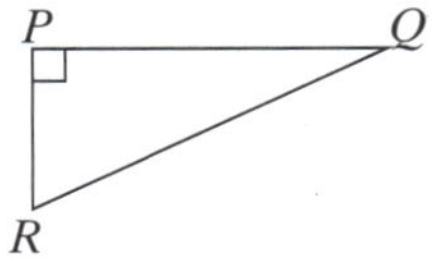

c

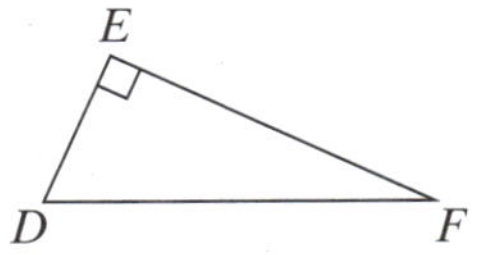

5 Complete for each diagram:

a $a^2 =$ ____________

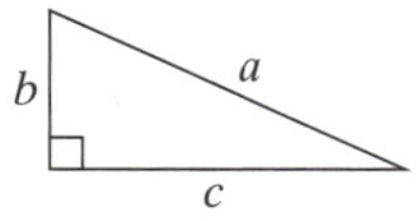

b $p^2 =$ ____________

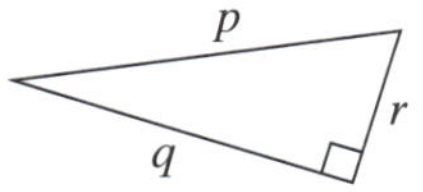

6 Use Pythagoras' Theorem to find the length of the hypotenuse:

a

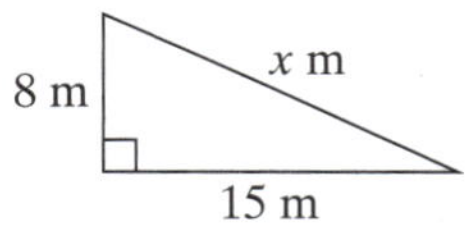

b

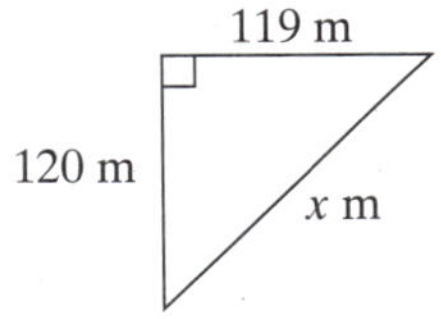

7 Use Pythagoras' Theorem to find the length of the unknown side:

a

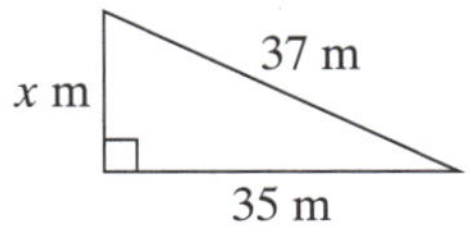

b

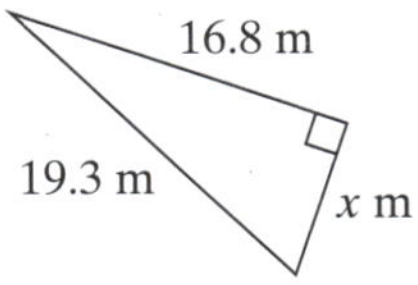

8 Find the length of the side marked x:

a

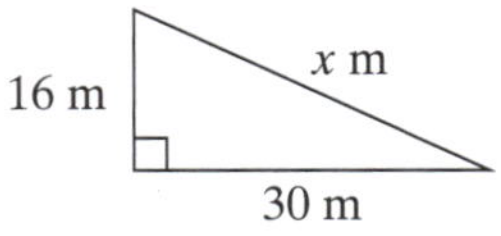

b

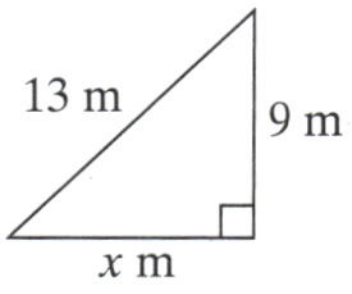

9 Determine whether each statement is true or false:

a $11^2 + 48^2 = 55^2$ **b** $225^2 + 272^2 = 353^2$ **c** $21.9^2 + 29.2^2 = 36.5^2$

PAGES 129–130

Answers **1 a** 169 **b** 100 **c** 793 **2 a** 75 **b** 13 **c** 4 **3 a** 23.0 **b** 8.1 **c** 2.7 **4 a** XY **b** QR **c** DF **5 a** $b^2 + c^2$ **b** $q^2 + r^2$ **6 a** 17 m **b** 169 m **7 a** 12 m **b** 9.5 m **8 a** 34 m **b** 9.4 m (1 d.p.) **9 a** false **b** true **c** true

PYTHAGORAS' THEOREM

Measurement and Geometry

20 MINUTES

INTERMEDIATE TEST

Part A Multiple Choice

1 What is the value of x?

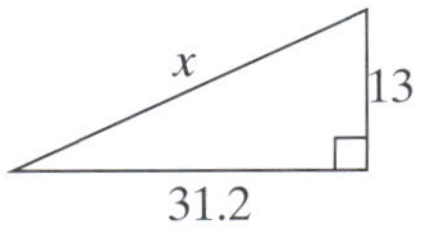

A 31.4 **B** 33.8 **C** 34.1 **D** 35.9 (1 mark)

2 What is the value of z, to one decimal place?

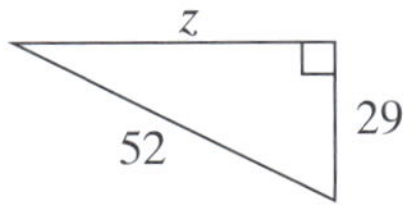

A 43.2 **B** 44.4 **C** 42.8 **D** 59.5 (1 mark)

3 Which triangle is NOT right-angled? (1 mark)

A

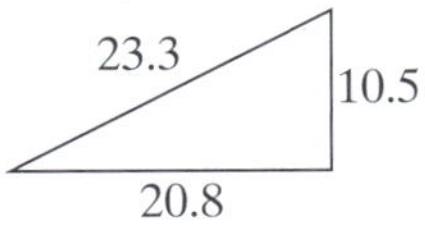

B

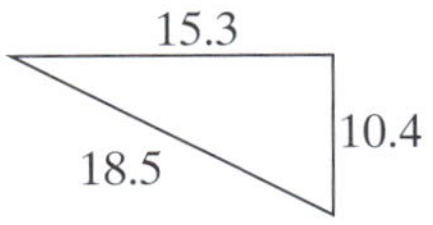

C

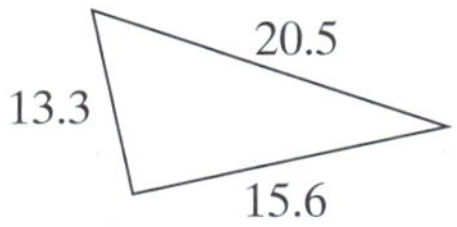

D

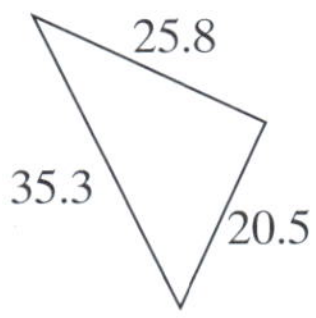

Part B Short Answer

4 A stay is attached to a pole at a height of 2.8 m from the ground. The bottom of the stay is 1.9 m from the base of the pole. How long is the stay? (Give the answer correct to one decimal place.) *Hint 1* (2 marks)

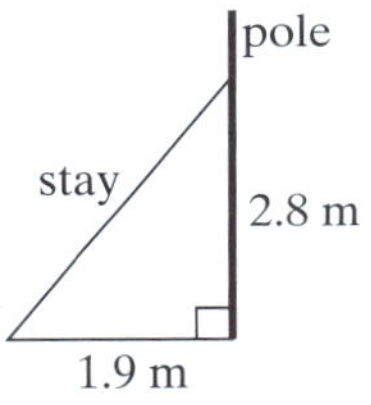

5 A ladder is 2.7 m long. The foot of the ladder must be placed 0.9 m from the wall. How far up the wall will the ladder reach? *Hint 2* (2 marks)

6 A kite is being flown at the end of a 24 m long string. Jay is holding the end of the string at a height of 1.2 m above the ground. Ellen is standing directly under the kite. She is 18 m from Jay on level ground. How high is the kite above the ground? *Hint 3* (3 marks)

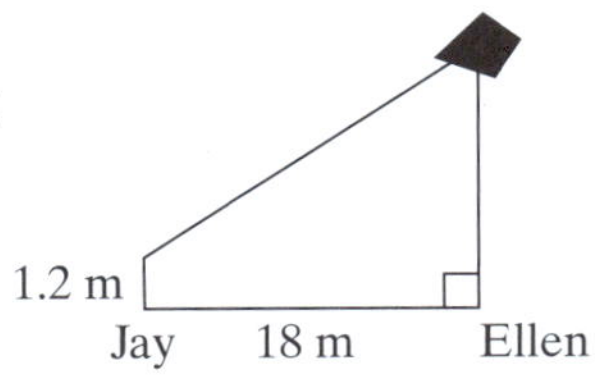

7 A diagonal brace is to be added to a gate that is 5.6 m long and 1.2 m high. How long is the brace (to the nearest centimetre)? *Hint 4* (2 marks)

8 A box, in the shape of a rectangular prism as shown in the diagram, is 24 cm long, 18 cm wide and 16 cm high.

a Find the length of DG. (2 marks)

b Find the length of AG. *Hint 5* (3 marks)

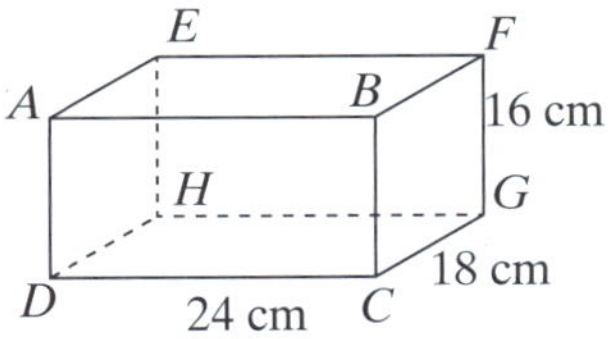

9 Find the area of the triangle. *Hint 6* (3 marks)

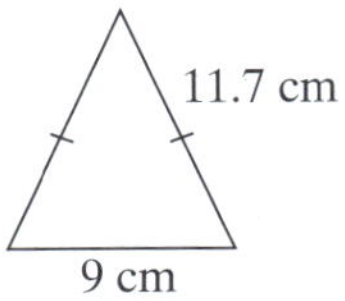

Hint 1: Let x m be the length of the stay.
Hint 2: Draw a diagram.
Hint 3: Find the side of a right-angled triangle using Pythagoras' Theorem. Add the height of Jay's arm.
Hint 4: The gate is a rectangle so opposite sides are equal.
Hint 5: Use the answer to part **a**.
Hint 6: Use Pythagoras' Theorem to find the height. Don't forget to halve the base.

Your Feedback

$\frac{\square}{20} \times 100\% = \square\,\%$

PAGES 130–131

PAGE 198

PYTHAGORAS' THEOREM

Measurement and Geometry

25 MINUTES

ADVANCED TEST

1 Find the length of side AB of each triangle:

a

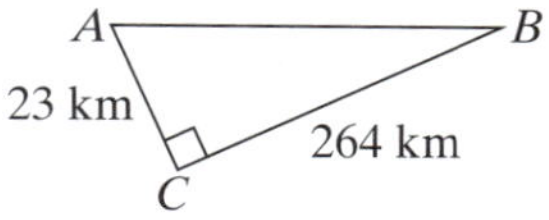

(2 marks)

b

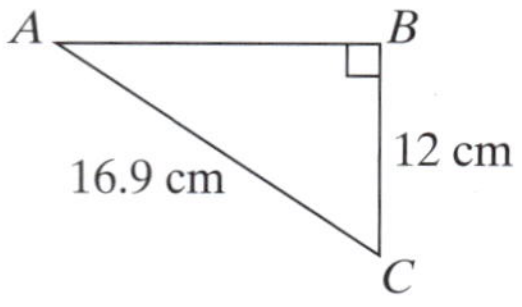

(2 marks)

c

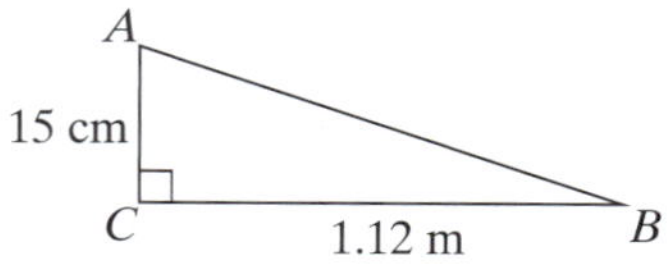

(2 marks)

2 Determine whether each set of numbers forms a Pythagorean triad:

a 48, 55, 73 (1 mark)

b 7.95, 12.92, 15.17 (1 mark)

3 **a** Find the length of the hypotenuse of a right-angled triangle where the other two sides are 11 cm and 25 cm long respectively. Give the answer correct to one decimal place. (2 marks)

b The side that was measured to be 11 cm should actually be 12 cm. How much difference does this make to the length of the hypotenuse? (2 marks)

4 Find the area of this triangle. (3 marks)

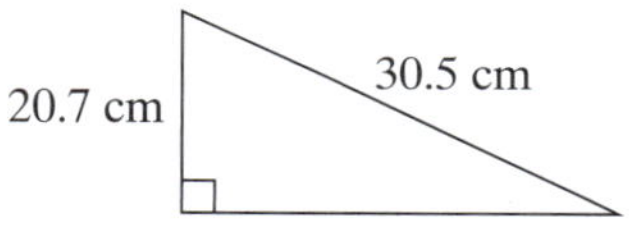

5 The diagonal of a square is 19.8 cm long. What is the length of each side? (3 marks)

6 A pole is to be supported by two wires. One of the wires is 14.9 m long and is attached to the ground at a point 5.1 m from the base of the pole. The second wire is attached to the ground 17.1 m from the base of the pole.

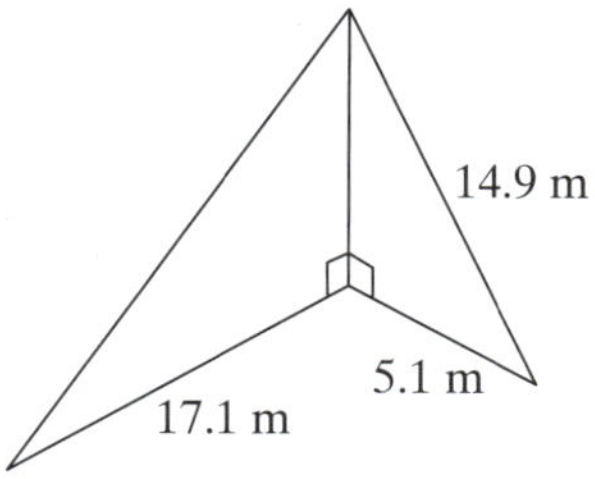

a How tall is the pole? (2 marks)

b How long is the second wire? (2 marks)

7 Ben's javelin is 2.6 m long. He wants to order a case that is 2.5 m long, 70 cm wide and 50 cm high to store some objects in and believes that the javelin will fit in it. Will the javelin fit in the case? Justify your answer. (3 marks)

Your Feedback

$\frac{\square}{25} \times 100\% = \square\%$

PAGES 131–132

PAGE 198

UNITS OF MEASUREMENT AND TIME SCALES

Measurement and Geometry

STUDY NOTES

1 **Units of measurement** in the metric system have **common prefixes**. The most common of these are **milli** (m) meaning one-thousandth, **centi** (c) meaning one-hundredth and **kilo** (k) meaning one thousand. So, for example, 1 mg = 0.001 g, 1 mm = 0.001 m and 1 mL = 0.001 L.

2 **Other prefixes for large amounts** are **mega** (M) meaning one million (10^6), **giga** (G) meaning one thousand million (10^9), and **tera** (T) meaning one million million (10^{12}).

3 **Other prefixes for small amounts** include **micro** (μ) one-millionth (10^{-6}) and **nano** (n) meaning one thousand millionth (10^{-9}).

4 **Digital information** is measured in bytes. 1 kilobyte (kB) is actually 1024 bytes (it has to be a power of 2). Similarly, 1 MB = 1024 kB, 1 GB = 1024 MB, 1 TB = 1024 GB. In many cases we approximate the values, giving the prefixes the same meaning as for other measurements.

5 **Units of long periods of time** include century (100 years) and millennium (1000 years).

6 We often use **approximate values** when dealing with numbers. An approximate value is close to the original value, but not exactly the same. For example, 599 is approximately 600. There are a few different symbols that can be used to mean 'approximately equal to' (for example, $\approx$ or $\doteqdot$).

7 **Numbers** can be **rounded off** to some approximate value. When rounding off to a certain number of **decimal places**, consider the digit after (to the right of) the required number of places. If this digit is lower than 5 the required digits remain unchanged but if the digit is 5 or higher add 1 to the previous digit (e.g. when rounded off to 1 decimal place 0.37 = 0.4, when rounded off to 3 decimal places 0.2397 = 0.240). Never move the decimal point or change the approximate value of the number.

8 If asked for an **exact answer**, give the answer as either a fraction or a terminating or recurring decimal. Do not give a rounded answer.

9 Some thought should always be given to the **accuracy of your answer.** If the question gives units in whole metres, for example, the answer will not be accurate to 7 decimal places.

10 Wherever possible use the **exact value**, not a rounded off answer, in **further calculations**.

11 The **first significant figure** in a number is the first non-zero digit (when reading from left to right). Significant figures count all digits after (and including) the first non-zero digit. For example, 0.006 2085 correct to 4 significant figures is 0.006 209. The first significant figure is 6 so we require 4 figures, the first being 6.The zeroes before (to the left of) the 6 are not significant but they cannot be left out (because if they were the number would be totally different).

12 The **level of accuracy** should be given when answering a question. This means you should state to how many decimal places or to how many significant figures your answer is accurate.

13 **Measuring instruments** are limited as to their accuracy. A measuring instrument is only accurate to half of the smallest unit of which it is capable of measuring. For example, on a ruler that measures in centimetres the accuracy is ±0.5 cm. This means that if a measurement with this ruler is given as 3 cm, say, then the actual measurement is between (3 – 0.5) cm and (3 + 0.5) cm, that is, between 2.5 cm and 3.5 cm.

Checklist

Can you:

1 *Understand and use prefixes with metric units?* ☐
2 *Use units of time and calculate time intervals?* ☐
3 *Understand measurements of digital information?* ☐
4 *Understand approximations and rounding off?* ☐
5 *Write numbers correct to a certain number of significant figures or decimal places?* ☐
6 *Understand the need to consider the accuracy of an answer or of a measuring instrument?* ☐

UNITS OF MEASUREMENT AND TIME SCALES

Measurement and Geometry

SKILLS CHECK

1 Complete:

- **a** 3 m = ______ cm
- **b** 5 kg = ______ g
- **c** 16 L = ______ mL
- **d** 7 cm = ______ mm
- **e** 9000 L = ______ kL
- **f** 23 000 mg = ______ g

2 Convert these times:

- **a** 5 min = ______ s
- **b** 24 h = ______ min
- **c** 480 s = ______ min
- **d** 3 days = ______ h
- **e** 7 decades = ______ years
- **f** 2½ centuries = ______ years

3 What is the time:

- **a** 3 h 15 min after 7:20 am?
- **b** 9 h 48 min after 8:52 pm?

4 How long is it from:

- **a** 6:35 am until 7:45 pm?
- **b** 0850 to 1435?
- **c** 12 noon on 19 May until 12 noon on 5 July?

5 What is the date 45 days after 13 September?

6 Arrange in order from smallest to largest: megalitre, gigalitre, kilolitre.

7 Arrange in order from shortest to longest: microsecond, millisecond, nanosecond.

8 How many years are in a millennium?

9 Round off each number to the nearest one hundred:

- **a** 624
- **b** 578
- **c** 1295
- **d** 45 378 632
- **e** 453 720
- **f** 63 999

10 Write each number correct to one decimal place:

- **a** 0.47
- **b** 2.54
- **c** 8.15
- **d** 24.6378
- **e** 1.0365
- **f** 0.98

11 Write each number correct to two decimal places:

- **a** 0.376
- **b** 1.2345
- **c** 576.399

12 Which is the first significant figure in each number?

- **a** 21 576
- **b** 0.005 78
- **c** 1.027 65
- **d** 0.020 79
- **e** 700 000
- **f** 0.800

13 Write correct to three significant figures:

- **a** 5 876 235
- **b** 562 877
- **c** 9461
- **d** 0.058 76
- **e** 0.003 0854
- **f** 0.000 047 239

PAGES 132–133

Answers **1 a** 300 **b** 5000 **c** 16 000 **d** 70 **e** 9 **f** 23 **2 a** 300 **b** 1440 **c** 8 **d** 72 **e** 70 **f** 250 **3 a** 10:35 am **b** 6:40 am **4 a** 13 h 10 min **b** 5 h 45 min **c** 47 days **5** 28 Oct **6** kilolitre, megalitre, gigalitre **7** nanosecond, microsecond, millisecond **8** 1000 **9 a** 600 **b** 600 **c** 1300 **d** 45 378 600 **e** 453 700 **f** 64 000 **10 a** 0.5 **b** 2.5 **c** 8.2 **d** 24.6 **e** 1.0 **f** 1.0 **11 a** 0.38 **b** 1.23 **c** 576.40 **12 a** 2 **b** 5 **c** 1 **d** 2 **e** 7 **f** 8 **13 a** 5 880 000 **b** 563 000 **c** 9460 **d** 0.0588 **e** 0.003 09 **f** 0.000 0472

UNITS OF MEASUREMENT AND TIME SCALES

Measurement and Geometry

20 MINUTES

INTERMEDIATE TEST

Part A Multiple Choice

1 Sally is going to measure the length of her mother's car. Which would be the most appropriate unit for Sally to use?

A centimetres **B** metres **C** kilometres **D** millimetres (1 mark)

2 Which number is closest to 0.8? *Hint 1*

A 0.7 **B** 0.789 **C** 0.77 **D** 0.83 (1 mark)

3 What fraction of a second is a nanosecond?

A millionth **B** ten millionth **C** hundred millionth **D** thousand millionth (1 mark)

4 Approximately how many kilobytes are in 1 gigabyte?

A 1000 **B** 100 000 **C** 1 000 000 **D** 100 000 000 (1 mark)

5 The value of 5378 correct to two significant figures is:

A 53 **B** 54 **C** 5300 **D** 5400 (1 mark)

6 Michael measured a pencil. The length of the pencil was 7 cm to the nearest centimetre. Which is the best interval to show where the actual length must lie?

A between 65 mm and 75 mm **B** between 6 cm and 8 cm
C between 7 cm and 7.5 cm **D** between 69 mm and 71 mm (1 mark)

7 0.028 19 when written correct to three significant figures is:

A 0.028 **B** 0.029 **C** 0.0281 **D** 0.0282 (1 mark)

8 'When 2.■63 is written correct to one decimal place the answer is 2.6.' This statement is correct but, unfortunately, a black mark from the printer has covered one of the digits in the number 2.■63. What digit has ■ covered?

A 4 **B** 5 **C** 6 **D** 7 (1 mark)

9 1 microgram = ?

A 0.0001 g **B** 0.000 01 g **C** 0.000 001 g **D** 0.000 0001 g (1 mark)

10 The speed of light is 3×10^5 km per second. About how long does it take light to travel 1 km? *Hint 2*

A 3 seconds **B** 3 milliseconds **C** 3 nanoseconds **D** 3 microseconds (1 mark)

Part B Short Answer

11 A bottle holds 700 mL of medicine.

a How many 15 mL doses of medicine does the bottle hold? *Hint 3* (2 marks)

b How many 12 mL doses of medicine does the bottle hold? (1 mark)

c Is the number of 12 mL doses that can be poured from 3 bottles of this medicine the same as three times the answer to part **b**? Explain why or why not. (3 marks)

12 A country has a population of about 20 million people. The country has about 2.6 people per square kilometre. Henry divides 20 million by 2.6 and concludes that the country has an area of 7 692 307.692 km^2. Briefly comment on the accuracy of Henry's answer. *Hint 4* (2 marks)

13 **a** Dinosaurs existed during the Mesozoic era which was between 245 million and 65 million years ago. How long is that time interval? (1 mark)

b The Jurassic period was 208 million years ago. About how many times is this longer than your age? Give your answer to two significant figures. (1 mark)

Hint 1: To easily compare the decimals, write each correct to three decimal places.
Hint 2: Time = distance ÷ speed.
Hint 3: Is it appropriate to round up?
Hint 4: Henry used approximations. Can his answer be accurate to one decimal place?

Your Feedback

$\frac{\square}{20} \times 100\% = \square\%$

PAGE 133
PAGE 198

UNITS OF MEASUREMENT AND TIME SCALES

Measurement and Geometry

25 MINUTES

ADVANCED TEST

1 Find, giving each answer correct to two decimal places:

a 0.065×18.235 (1 mark)

b $\sqrt{912}$ (1 mark)

c $\dfrac{16.35 + 2.97}{4.85 - 3.2}$ (1 mark)

2 Find, giving each answer correct to three significant figures:

a $95 \div 0.875$ (1 mark)

b $(0.246)^2$ (1 mark)

c $563.7 \div 0.006$ (1 mark)

3 Determine the number of:

a micrograms in 1 tonne (1 mark)

b millilitres in 1 gigalitre (1 mark)

c nanoseconds in 1 hour (1 mark)

4 Convert:

a 255 minutes to hours and minutes (1 mark)

b 0.08 grams to milligrams (1 mark)

c 43.5 megalitres to litres (1 mark)

d 720 micrograms to milligrams (1 mark)

e 0.0087 seconds to milliseconds (1 mark)

f 8.5 millennia to centuries (1 mark)

g 12 288 gigabytes to terabytes (1 mark)

h 29 420 nanoseconds to seconds (1 mark)

i 5.6×10^{-7} kilograms to micrograms (1 mark)

j 9.4×10^{8} kilolitres to gigalitres (1 mark)

5 a How many files of size 1240 kB can be stored on a 700-MB disk? (1 mark)

b A 4-GB storage device can hold as much information as how many 700-MB disks? (1 mark)

6 a If light travels at 3×10^5 km/s, how long, in microseconds, will it take light to travel 3 km? (1 mark)

b If sound travels at approximately 1200 km/h, about how long will it take sound to travel 3 km? (1 mark)

c A particular star is 4×10^{13} km from Earth. Light travels at 3×10^5 km/s. About how long, in years and months, will it take light to reach Earth from that star? (2 marks)

Your Feedback

$\dfrac{\square}{25} \times 100\% = \square\%$

PAGES 133–134

PAGE 198

EARNING MONEY

Money and Financial Mathematics

STUDY NOTES

1 A **wage** is money paid for work by the day or week. Wages are usually paid at an hourly rate of pay. For example, if you work 40 hours per week and earn \$20 per hour your weekly wage is \$800 (40 × \$20).

2 A **salary** is a fixed amount earned yearly, and paid in equal amounts by the week or fortnight. For example, if you have an annual salary of \$52 000 and are paid weekly, you will earn \$1000 per week (\$52 000 ÷ 52).

3 When calculating amounts such as the weekly wage from an annual salary we generally use **52 weeks = one year**. A fortnight is two weeks so there are **26 fortnights** in a year. (In actual fact 52 weeks = 364 days not 365). Take care if finding the monthly pay. There are 12 months in a year. One month is not 4 weeks. There are actually 13 4-week periods in 1 year.

4 **Commission** is paid as a percentage of income generated. A commission is often earned by a salesperson as a percentage of the amount of sales made. For example, if a salesperson earns commission of 5% on all sales, then in a week when his or her sales total \$2000 he or she will earn 5% of \$2000.

5 **Piecework** is work done and paid for by the item. For example, fruit pickers are often paid for each bin of picked fruit. Machinists may be paid by the item sewn.

6 **Overtime** is paid for work performed outside the normal working hours. Time-and-a-half means that the worker is paid 1½ times the normal rate of pay. Double time is twice the normal rate and so on. For example, if a woman is paid time-and-a-half for working overtime and her normal hourly rate is \$24 per hour then she will be paid 1.5 × \$24 for every hour of overtime.

7 A **bonus payment** is usually a one-off payment made for a special reason.

8 **Penalty payments** are paid for adverse working conditions. For example, you may be paid at time-and-a-half for work performed in a very cold place, or you may receive an additional \$0.38 per hour for working in dusty conditions.

9 A **holiday loading** is additional money paid to workers while on annual leave. (It was originally intended to compensate for lack of overtime opportunities.) Holiday loading is usually 17½% of four weeks pay. For example, if someone earns \$200 per week, they will receive \$800 for 4 weeks. The holiday loading will be 17.5% of \$800, which is \$140. The total holiday pay will be \$800 + \$140 = \$940.

10 **Gross earnings** are the total earnings. **Deductions** are amounts subtracted from pay. Examples of common deductions are tax, superannuation contributions and union fees. **Net earnings** are sometimes referred to as **take-home pay**. This is because it is the amount that is left to take home after all deductions have been taken from gross earnings.
Net earnings = gross earnings – deductions

Checklist

Can you:

1 *Understand the difference between a wage and a salary?* ☐
2 *Find earnings given hourly rates of pay, weekly wages or salary?* ☐
3 *Understand and find commissions?* ☐
4 *Understand what is meant by piecework?* ☐
5 *Understand the terms 'time-and-a-half', 'double time', 'double-time-and-a-half' and so on?* ☐
6 *Calculate overtime, bonuses and penalty payments?* ☐
7 *Understand what is meant by a holiday loading and calculate holiday pay?* ☐
8 *Understand the difference between gross and net earnings?* ☐

EARNING MONEY

Money and Financial Mathematics

SKILLS CHECK

1 Find the amount earned for working:

- **a** 8 hours at \$37.65 per hour
- **b** 40 hours at \$23.50 per hour
- **c** 35 hours at \$41.70 per hour
- **d** 37½ hours at \$38.76 per hour

2 Find the weekly pay if you earn:

- **a** \$41 028 per year
- **b** \$612.50 per fortnight

3 Find the annual income if you earn:

- **a** \$752 per week
- **b** \$1825 per fortnight
- **c** \$3015 per month

4 A salary of \$54 288 is how much:

- **a** per week?
- **b** per fortnight?
- **c** per month?

5 Find the hourly rate of pay if you earn:

- **a** \$1274.80 for 40 hours
- **b** \$1438.50 for 35 hours
- **c** \$912 for 38 hours

6 Find the total number of hours worked in a week when you work from 9 am to 5 pm on Monday, 9 am to 4 pm on Tuesday, 9 am to 12.30 pm on Wednesday, 9 am to 5 pm on Thursday and 9 am to 6 pm on Friday.

7 If you are paid overtime rates for all hours worked beyond 8 hours in any one day, find the total number of overtime hours if you work 7 hours on Monday, 8.5 hours on Tuesday, 8 hours on Wednesday, 11.5 hours on Thursday and 9.5 hours on Friday.

8 Find a commission of:

- **a** 7% of \$8000
- **b** 2½% of \$15 000

9 Find the earnings if you receive:

- **a** 80 cents per bucket for 56 buckets
- **b** \$5 per shirt for 15 shirts

10 If you earn \$21.60 per hour, what would you earn for:

- **a** 1 hour at time-and-a-half?
- **b** 3 hours at double time?
- **c** 8 hours at double-time-and-a-half?

11 Find:

- **a** the earnings for 4 weeks if you earn \$895 per week
- **b** 17½% of \$2872

12 Find the total deductions: \$209 for tax, \$68 for superannuation contributions, \$29.50 for union fees and \$15 for laundry.

13 Find the net wage if the gross wage is \$876 and deductions total \$277.

PAGES 134–135

Answers **1 a** \$301.20 **b** \$940 **c** \$1459.50 **d** \$1453.50 **2 a** \$789 **b** \$306.25 **3 a** \$39 104 **b** \$47 450 **c** \$36 180 **4 a** \$1044 **b** \$2088 **c** \$4524 **5 a** \$31.87 **b** \$41.10 **c** \$24 **6** 35.5 hours **7** 5.5 hours **8 a** \$560 **b** \$375 **9 a** \$44.80 **b** \$75 **10 a** \$32.40 **b** \$129.60 **c** \$432 **11 a** \$3580 **b** \$502.60 **12** \$321.50 **13** \$599

EARNING MONEY

Money and Financial Mathematics

25 MINUTES

INTERMEDIATE TEST

Part A Multiple Choice

1 What will you be paid for working 12 hours at time-and-a-half if your usual pay rate is $24.38 per hour?

A $731.40 **B** $310.56 **C** $304.75 **D** $438.84 (1 mark)

2 Amanda earns a royalty of 12.5% on sales of her book. If the book sells for $19.95, which is closest to the amount Amanda will receive when 1500 copies are sold?

A $1750 **B** $3030 **C** $2390 **D** $3740 (1 mark)

3 Which results in the highest weekly pay?

A $53 872 p.a. **B** $2052.50 per fortnight

C $30.75 per hour for 35 hours **D** $26.58 per hour for 40 hours (1 mark)

Part B Short Answer

4 Tamsyn earns $28.40 per hour and is paid overtime at time-and-a-half for any hours above the normal 35 hours.

a What is Tamsyn paid for a normal 35-hour week? (1 mark)

b What will Tamsyn earn in a week where she works 42 hours? *Hint 1* (3 marks)

5 Carlo is employed by a finance company on a yearly salary of $78 000.

a What is Carlo's fortnightly gross pay? (1 mark)

b Each fortnight Carlo contributes $215 to the superannuation fund, pays $892 tax, $39.80 health insurance and donates $20 to a charity. What is Carlo's take-home pay? *Hint 2* (2 marks)

6 Eric works for a market research company surveying opinions from the public. He earns $7.50 for every survey he completes.

a How much will Eric earn for a day in which he completes 24 surveys? (1 mark)

b In one particular week, Eric earned $1290. How many surveys did he complete that week? (1 mark)

7 Roger's Real Estate Company charges commission on all property sales. The rate of commission is 4% of the first $100 000 and 3½% of the remainder. What does Roger's Real Estate Company receive for selling a property for $315 000? *Hint 3* (4 marks)

8 Mai works for an international company and earns a wage of $1808.00 for a 40-hour week.

a What is Mai's hourly rate of pay? (1 mark)

b Whenever Mai works on a weekend she earns double time on Saturday and double-time-and-a-half on Sunday. What does Mai receive for a weekend when she works 6 hours on Saturday and 4½ hours on Sunday? *Hint 4* (3 marks)

c Mai is to take her four weeks annual leave and will receive a loading of 17½%. How much holiday pay will Mai receive? *Hint 5* (3 marks)

Hint 1: Multiply the number of overtime hours by 1.5 times the normal rate of pay and add to the normal week's pay.

Hint 2: Take-home pay is net earnings.

Hint 3: 4% of $100 000 plus 3.5% of $(315 000 – 100 000).

Hint 4: Multiply the number of hours on Saturday by two times the normal rate and the number of hours on Sunday by 2.5 times the normal rate.

Hint 5: Don't forget to add the loading to the normal pay for 4 weeks.

Your Feedback

$\frac{\square}{23} \times 100\% = \square\%$

PAGE 135

PAGE 198

EARNING MONEY

Money and Financial Mathematics

25 MINUTES

ADVANCED TEST

1 Sam has been offered two jobs doing similar work. Position A offers a salary of \$87 512 p.a. Position B pays \$3250 per fortnight.

a How much more will Sam receive each year if he takes position A than if he takes position B? (1 mark)

b Give three possible reasons why Sam might be better off taking position B. (3 marks)

2 How much is your monthly income if you earn \$876 per week? (2 marks)

3 The table shows the times Kendall started and finished work last week.

	Monday	Tuesday	Wednesday	Thursday	Friday
Start	9:00 am	8:00 am	8:30 am	11:00 am	11:00 am
Finish	5:00 pm	4:30 pm	5:00 pm	9:00 pm	4:00 pm

a How many hours did Kendall work? (1 mark)

b If Kendall receives \$19.70 per hour, how much did she earn on Monday? (1 mark)

c Kendall receives time-and-a-half for any hours worked after 5:00 pm.
How much did Kendall earn on Thursday? (2 marks)

d How much did Kendall receive last week? (2 marks)

4 Margaret has a gross income of \$92 560. Her allowable deductions for taxation purposes total \$2595.

a What is Margaret's taxable income? (1 mark)

b Use the tax table below to find the tax payable on that income. (2 marks)

Taxable income	Tax on this income
0–\$18 200	Nil
\$18 201–\$37 000	19 cents for each \$1 over \$18 200
\$37 001–\$80 000	\$3572 plus 32.5 cents for each \$1 over \$37 000
\$80 001–\$180 000	\$17 547 plus 37 cents for each \$1 over \$80 000
\$180 001 and over	\$54 547 plus 45 cents for each \$1 over \$180 000

c A Medicare levy of 1.5% of taxable income must also be paid. How much is the Medicare levy that Margaret must pay? (1 mark)

d Margaret paid \$453.50 every week in PAYG tax instalments. Will she receive a tax refund or will she have to pay more tax? How much is the refund or extra amount that she needs to pay? (2 marks)

5 Grant is a salesman and receives a flat amount of \$320 per week plus a commission of 3¼% on that part of sales that exceed \$5000. How much will Grant earn in a week where his sales total \$27 000? (2 marks)

6 Celine is paid \$995.40 for a 36-hour week.

a How much will Celine receive in total for her 4 weeks holiday pay, including a leave loading of 17½%? (2 marks)

b How long, in hours and minutes, would Celine need to work overtime, at time-and-a-half, so that her weekly pay is equivalent to her weekly holiday pay? (3 marks)

Your Feedback

$\frac{\square}{25} \times 100\% = \square\%$

PAGES 135–136

PAGE 198

SIMPLE INTEREST

Money and Financial Mathematics

STUDY NOTES

1 **Interest** is a payment made for the cost of borrowing, or investing, money.

2 When an amount is **borrowed**, interest must be paid as the cost of borrowing that money.

3 When an amount is **invested**, interest is received as payment for investing the money.

4 **Simple interest** is when interest is always paid on the same fixed amount of money. The amount of money is called the **principal**.

5 The **interest rate** is usually given as a **percentage** per annum (**p.a.**) which means per year.

6 Simple interest can be calculated using the formula $\boldsymbol{I = PRN}$ where P is the principal, or amount of money, $R = \frac{r}{100}$ is the interest rate per period of time and N the number of periods. For example, if \$300 is invested at 4% p.a. for 2 years then $P = \$300$, $R = \frac{4}{100}$ (or 0.04) and $N = 2$. (Sometimes T is used in this formula instead of N.)

7 The **number of periods** is often called the **term** of a loan or investment.

8 The **simple interest formula** can also be used when the amount of interest is known, to find the principal or the rate of interest or the term. Substitute the known values into the formula and a simple equation will need to be solved.

9 **Buying on terms** means that a customer agrees to pay off an item usually with equal instalments over a given period of time.

10 If the amount and number of instalments are known, then the **total amount paid** for an item bought on terms can be found. The **difference** between this total and the cash price is the interest.

11 A **deposit** is part payment for an item made at the start. A deposit is often a particular percentage of the total price.

12 The **balance** owing is the price minus the deposit. The balance is then subject to interest.

13 Once the interest has been calculated it is added to the balance to give the **amount to be repaid** (after the deposit).

14 The amount to be repaid is divided by the number of months in the term of the loan to give the amount of each **monthly repayment** or instalment. (Repayments could be weekly, fortnightly, quarterly or yearly instead of monthly.)

15 **Costs** such as **interest** and **fees** apply to many methods of paying for goods. Customers need to consider the costs involved in relation to the convenience offered.

Checklist

Can you: ✓

1 *Understand what is meant by interest?* ☐
2 *Understand what is meant by the terms 'simple interest', 'principal', 'interest rate' and 'p.a.'?* ☐
3 *Know the simple interest formula?* ☐
4 *Calculate simple interest?* ☐
5 *Find the amount of the principal, interest rate or term using the simple interest formula?* ☐
6 *Understand what is meant by buying on terms?* ☐
7 *Understand the terms 'deposit', 'balance', 'amount owing' and 'instalments'?* ☐
8 *Solve problems involving simple interest and buying on terms?* ☐

SIMPLE INTEREST

Money and Financial Mathematics

SKILLS CHECK

1 How many months are in:
- **a** 3 years?
- **b** 8 years?
- **c** 25 years?
- **d** 5½ years?

2 Find:
- **a** 10% of $900
- **b** 5% of $8000
- **c** 3% of $250 000

3 What percentage is:
- **a** 20 of 200?
- **b** 180 of 3000?
- **c** 1380 of 18 400?

4 Find 100% if:
- **a** 5% is $9000
- **b** 12% is $75 000

5 Find the simple interest on:
- **a** $500 at 6% per year for 1 year
- **b** $2000 at 4% per year for 2 years
- **c** $8000 at 5% per year for 6 years
- **d** $950 at 7.5% per year for 4 years

6 $600 is in a bank account. It earns interest at 5% p.a.
- **a** Find the interest earned in 1 year.
- **b** Find the total amount in the account if the interest is added to the $600.
- **c** Find the amount of interest earned on the account in the next year.

7 Find the balance owing on a purchase of $5000 if you pay a deposit of:
- **a** $1500
- **b** 20%

8 Find the total paid if you buy something on these terms:
- **a** 12 monthly payments of $125 per month
- **b** no deposit, monthly payments of $78 for 2 years
- **c** $1000 deposit, 18 equal monthly instalments of $321.70
- **d** $30 000 deposit, monthly payments of $813.47 for 25 years

9 Find the amount of each equal monthly instalment if:
- **a** $2724 is repaid over 12 months
- **b** $11 311.20 is repaid over 3 years

10 Find the total amount of interest charged if:
- **a** the cash price was $2500 and the total paid was $3212.50
- **b** the cash price was $8000 and you paid $220 per month for 4 years

PAGES 136–137

Answers **1 a** 36 **b** 96 **c** 300 **d** 66 **2 a** $90 **b** $400 **c** $7500 **3 a** 10% **b** 6% **c** 7.5% **4 a** $180 000 **b** $625 000 **5 a** $30 **b** $160 **c** $2400 **d** $285 **6 a** $30 **b** $630 **c** $31.50 **7 a** $3500 **b** $4000 **8 a** $1500 **b** $1872 **c** $6790.60 **d** $274 041 **9 a** $227 **b** $314.20 **10 a** $712.50 **b** $2560

SIMPLE INTEREST

Money and Financial Mathematics

25 MINUTES

INTERMEDIATE TEST

Part A Multiple Choice

1 What is the interest paid if \$2750 is borrowed at 4% p.a. simple interest for 3 years?

A \$110 B \$330 C \$440 D \$1100 (1 mark)

2 Wendy borrows \$60 000. Simple interest at 9% p.a. is paid on the loan. How much interest will Wendy pay in total over the 15 years of the loan?

A \$21 000 B \$36 000 C \$72 000 D \$81 000 (1 mark)

3 Ivor invests \$25 000 for 18 months in an account earning 6% p.a. simple interest. How much interest will Ivor receive?

A \$2250 B \$2700 C \$22 500 D \$27 000 (1 mark)

4 An amount of \$500 earns \$100 simple interest in 4 years. What annual rate of interest is paid?

A 2% B 5% C 10% D 20% (1 mark)

Part B Short Answer

5 An amount of \$12 000 is borrowed. Simple interest at 7.5% p.a. is charged on the loan. If the total interest charged is \$5400, for how many years was the loan? (1 mark)

6 Caitlin deposits \$2000 into a bank account that pays 3% p.a. interest.

a Find the amount of interest Caitlin earns in one year. (1 mark)

b The interest is added to the amount in the account. What is the new balance of Caitlin's account? (1 mark)

c How much interest will Caitlin earn in the next year? (1 mark)

d Find the total in the account after three years if interest has just been paid. *Hint 1* (2 marks)

7 Rory decides to buy a motorbike, the cash price being \$7800. He agrees to pay 15% deposit and repay the balance plus interest in equal monthly instalments over 3 years.

a How much is the deposit? (1 mark)

b How much is owing after the deposit has been paid? (1 mark)

c Interest is charged on the balance at 8% p.a. simple interest. Find the total amount of interest. *Hint 2* (2 marks)

d What is the total that Rory must repay? (1 mark)

e How much is each monthly payment? (2 marks)

8 \$2400 interest is earned on an amount of money invested for 4 years. If the simple interest was paid at 5% p.a., what was the amount of money? *Hint 3* (2 marks)

9 Rosa decides to buy a lawn tractor. The cash price is \$5569. Rosa agrees to pay \$700 deposit and \$240 a month for 2 years.

a How much will Rosa pay for the tractor? (2 marks)

b How much interest does she pay? *Hint 4* (1 mark)

c What percentage rate (correct to two decimal places) of interest is Rosa charged? (2 marks)

10 Fiona has a credit card that charges a simple interest rate of 0.0445% per day. Find the amount of interest that Fiona must pay if she uses her credit card to buy furniture for \$1840 and pays the bill after 30 days. *Hint 5* (1 mark)

Hint 1: Caitlin will earn a different amount of interest again in the third year. The interest from the previous year has been added to the account, so the principal is different each year.

Hint 2: Interest is only charged on what is owing.

Hint 3: Substitute the known values into the simple interest formula and solve.

Hint 4: The interest is the extra amount that Rosa pays above the amount borrowed.

Hint 5: The interest rate is given as a percentage.

Your Feedback

$\frac{\square}{25} \times 100\% = \square\%$

PAGES 137–138

PAGE 198

SIMPLE INTEREST

Money and Financial Mathematics

25 MINUTES

ADVANCED TEST

1 Find the simple interest on an investment of \$6400 at 4.8% p.a. for 18 months. (1 mark)

2 Find the simple interest on \$145 000 at 0.8% per month for 25 years. (2 marks)

3 Find the amount invested if it earns \$588 simple interest in 3 years at 7% p.a. (2 marks)

4 Find the number of years that an amount of \$12 500 needs to be invested at 6% p.a. to earn \$6750 in simple interest. (2 marks)

5 Find the annual rate of simple interest if an amount of \$8600 earns \$2494 interest in 4 years. (2 marks)

6 Find the total value of an investment of \$16 000 earning simple interest at 7½% p.a. for 6 years. (2 marks)

7 Simple interest of 0.06% per day is charged on May's credit card purchases from, and including, the date of purchase until, and including, the date of repayment. If May used her credit card to buy a dining table for \$3750 on 18 August and repaid it on 9 October, how much interest did she pay? (2 marks)

8 Find the monthly repayments required to repay a loan for a car if the cash price is \$26 400, a deposit of 15% is paid and the balance is repaid over 5 years with simple interest of 8.5% p.a. charged. (3 marks)

9 Find the annual rate of simple interest charged on a loan of \$16 000 if it is repaid over 4 years with equal monthly instalments of \$450. (3 marks)

10 An amount of \$30 000 is invested and earns interest at 5% p.a. The interest earned each year is added to the investment at the end of the year.

- a Find the amount of interest earned in 1 year. (1 mark)
- b How much interest will be earned in the second year? (1 mark)
- c How much interest in total will have been paid after 3 years? (2 marks)
- d If the interest was not added, a simple interest rate of 5¼% p.a. could be earned. Which option would be better over 3 years? Justify your answer. (2 marks)

Your Feedback

$\frac{\square}{25} \times 100\% = \square\%$

PAGE 138

PAGE 198

RATES AND PROPORTION

Proportion

STUDY NOTES

1 **Rates** compare quantities of different kinds. For example, when expressing a speed in kilometres per hour we are comparing a distance (in km) with a time (in hours). When given in **simplest form**, rates are expressed per 1 unit. For example, a rate of 50 m in 10 seconds is a rate of 5 m in 1 second or 5 m/s.

2 We can **convert a rate** from one set of units to another. For example, if you travel 18 m in 1 second you will travel $18 \times 60 = 1080$ m in 1 minute and $1080 \times 60 = 64\,800$ m in 1 hour. Dividing by 1000 we find that 64 800 m = 64.8 km. So 18 m/s = 64.8 km/h.

3 Items are in **proportion** if they are in the same ratio.

4 Two items are in **direct proportion** if both increase, or both decrease, at the same rate. The graph of one item against the other will have a positive gradient. For example, the total cost of a purchase of tickets is directly proportional to the number bought.

5 If two items, x and y, are in direct proportion then $y = kx$ where k is a constant; it has the same value all the time. To **solve problems** involving direct proportion first use the given information to find the value of k, then find the value that you require.

6 Two items are in **inverse** (or indirect) proportion if one decreases as the other increases. For example, if the cost to hire a bus is fixed, then the cost per person to travel on the bus will decrease the more people who travel on it (up to a certain point). The graph of one item against the other will have a negative gradient.

7 If two items are directly proportional a **conversion graph** can be used to change from one unit to the other.

8 A **constant** relationship is denoted by a straight line segment on a graph.

9 A **variable** relationship is denoted by a curved line on a graph.

10 A line or curve is **increasing** if it leans to the right at the top and **decreasing** if it leans to the left.

11 A **curve** that is increasing can be further **described** according to the **rate of change of the graph**. The steeper the graph the faster it is increasing. If the graph increases quickly at first and then slows, it is increasing at a decreasing rate. If it increases slowly and then more quickly it is increasing at an increasing rate. Similarly, a curve that is decreasing may be decreasing at an increasing rate or decreasing at a decreasing rate.

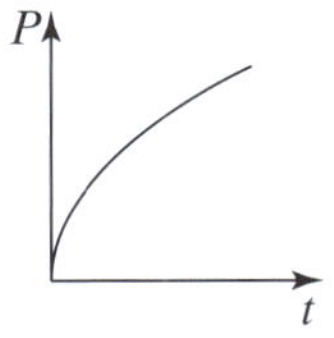

Increasing at a decreasing rate

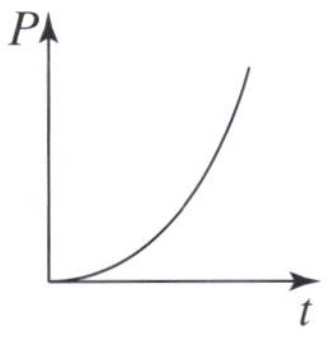

Increasing at an increasing rate

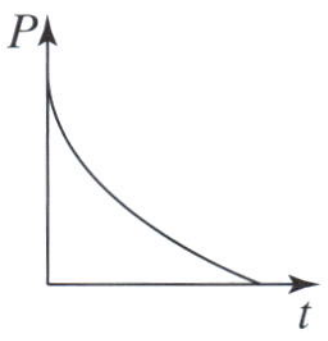

Decreasing at a decreasing rate

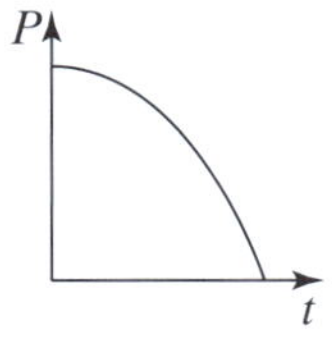

Decreasing at an increasing rate

Checklist

Can you:

1 *Understand what is meant by a rate and solve problems involving rates?* ☐
2 *Convert a rate from one set of units to another?* ☐
3 *Understand what it means for items to be in proportion?* ☐
4 *Understand the difference between direct proportion and inverse proportion?* ☐
5 *Use graphs with proportions, including graphing software, and conversion graphs?* ☐
6 *Use an equation and given information to solve problems involving direct proportions?* ☐
7 *Understand the difference between constant and variable relationships?* ☐
8 *Understand what it means for a curve to be increasing or decreasing and describe rates of change?* ☐

RATES AND PROPORTION
Proportion

SKILLS CHECK

1 Express as a rate in simplest form:
- **a** 80 km in 2 hours
- **b** 200 m in 10 seconds
- **c** $8.00 for 4 litres
- **d** 189 km in 3 hours
- **e** $86 for 4 people
- **f** $120 for 5 hours work

2 Find the average speed (in km/h) if a car travels:
- **a** 140 km in 2 hours
- **b** 720 km in 9 hours
- **c** 420 km in 5 hours
- **d** 348 km in 6 hours
- **e** 180 km in 2½ hours
- **f** 27 km in 20 minutes

3 Find the distance travelled if a car travels for 3 hours at an average speed of 75 km/h.

4
- **a** 12 m/s = ______ metres per minute
- **b** 60 km/h = ______ km per minute
- **c** 16 litres per day = ______ mL per day
- **d** 25 000 g per minute = ______ kg per minute

5 Determine whether these are examples of direct or inverse proportion.
- **a** As the radius increases the circumference increases.
- **b** As the age of the car increases the value decreases.
- **c** As the length of the span decreases the strength increases.
- **d** As the speed decreases the distance covered decreases.

6 This graph is used to convert litres to gallons.
- **a** Approximately how many gallons is 400 litres?
- **b** Approximately how many litres is 70 gallons?

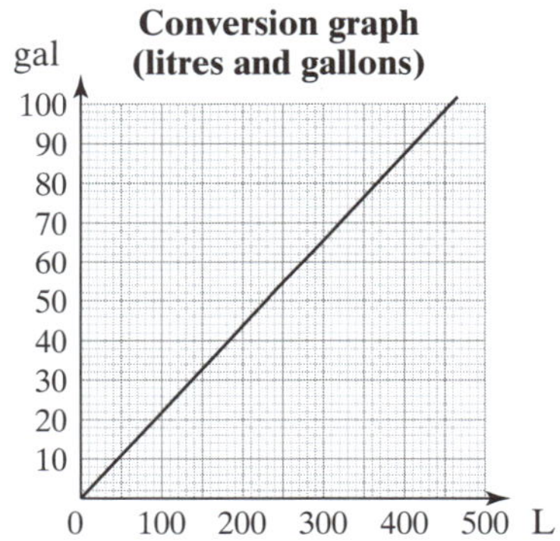

7 If $y = 700x$ find:
- **a** y when $x = 6$
- **b** x when $y = 3150$

8 Given $y = kx$, and $y = 72$ when $x = 15$, find:
- **a** k
- **b** y when $x = 25$
- **c** x when $y = 96$

9 Each graph shows the depth of fluid in a container over time.
- **i** State whether the depth is increasing or decreasing.

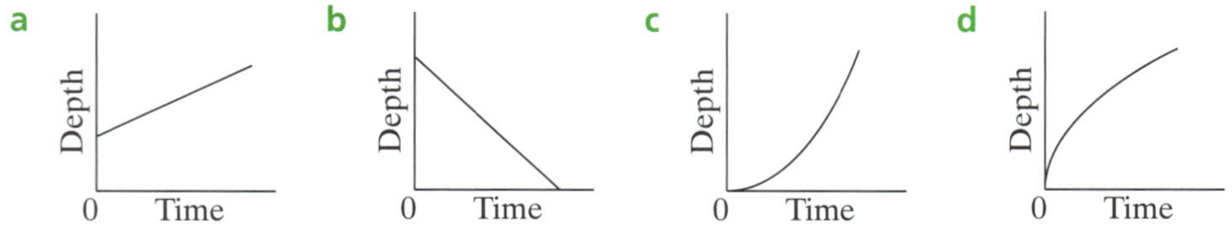

- **ii** For each of the graphs in part **i**, state whether the rate of change of the depth is constant or variable.

PAGES 138–139

Answers **1 a** 40 km/h **b** 20 m/s **c** $2/L **d** 63 km/h **e** $21.50 per person **f** $24 per hour **2 a** 70 km/h **b** 80 km/h **c** 84 km/h **d** 58 km/h **e** 72 km/h **f** 81 km/h **3** 225 km **4 a** 720 **b** 1 **c** 16000 **d** 25 **5 a** direct **b** inverse **c** inverse **d** direct **6 a** 88 **b** 320 **7 a** 4200 **b** 4.5 **8 a** 4.8 **b** 120 **c** 20 **9 i a** increasing **b** decreasing **c** increasing **d** increasing **ii a** constant **b** constant **c** variable **d** variable

RATES AND PROPORTION

Proportion

Part A Multiple Choice

1 Water is flowing through a pipe at the rate of 40 litres every 5 minutes. How much water flows through the pipe in 12 minutes? *Hint 1*

A 48 L **B** 88 L **C** 96 L **D** 480 L (1 mark)

2 A car travelled 378 km at an average speed of 72 km/h. How long was the journey? *Hint 2*

A 5 h 15 min **B** 5 h 25 min **C** 5 h 35 min **D** 5 h 45 min (1 mark)

3 Mitchell scored 75 runs in 1 hour and 20 minutes. What is his run rate, in runs per minute?

A 0.625 **B** 0.9375 **C** $1.0\dot{6}$ **D** 1.6 (1 mark)

4 Water is poured into this container at a constant rate.
Which diagram shows the height of the water level as time passes? *Hint 3*

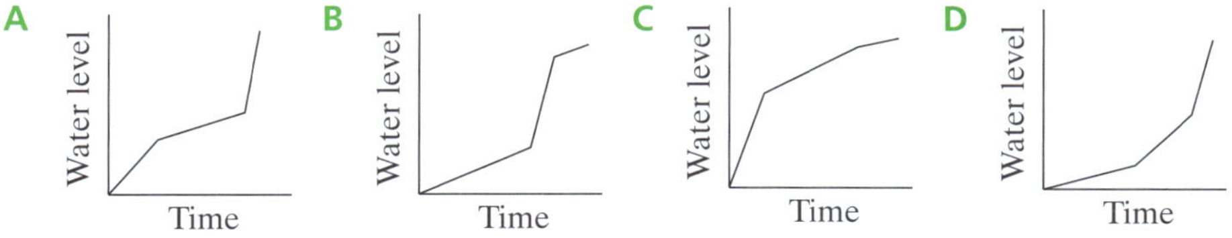

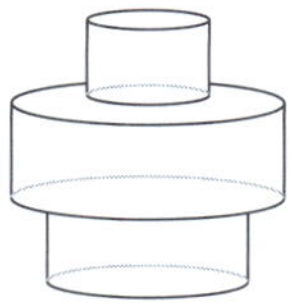

(1 mark)

Part B Short Answer

5 The sides of two rectangles are in proportion. One rectangle is 69 mm long and 39 mm wide. The other rectangle is 91 mm wide. How long is it? (1 mark)

6 Shelley can type 50 words a minute.

a How many words can Shelley type in 5 minutes? (1 mark)

b How long will it take Shelley to type a 2000-word essay? (1 mark)

c Jeremy can type 60 words a minute. How much more quickly than Shelley can Jeremy type a 3000-word essay? (2 marks)

7 Patrick runs 100 metres in 10 seconds. What is his speed in kilometres per hour? (2 marks)

8 To control weeds, 5 mL of weedkiller is to be used in every 10 litres of water.

a How much weedkiller should be used in a 500-litre spray tank? (1 mark)

b The full spray tank controls weeds in an area of 2 ha. How many litres of weedkiller will be needed to spray 15 ha? (3 marks)

9 Convert 78 km/h to m/s. *Hint 4* (2 marks)

10 The graph shows the water level in a washing machine as it goes through part of its cycle.

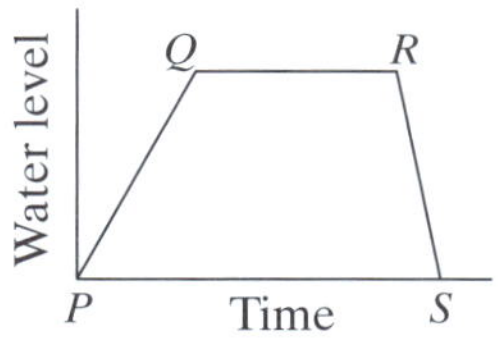

a What can you say about the level of water in the machine between: **i** P and Q? **ii** Q and R? **iii** R and S? (3 marks)

b Does the washing machine fill or empty quicker? How can you tell? *Hint 5* (2 marks)

11 In a racing car trial, the distance travelled in kilometres, d, is directly proportional to the speed, S km/h, so $d = kS$. At 80 km/h, the distance travelled is 12 km. Find the distance travelled at 100 km/h. *Hint 6* (3 marks)

Hint 1: First find the rate at which the water is flowing.
Hint 2: Be careful changing a time expressed as a decimal to a time in hours and minutes.
Hint 3: When will the water level rise most quickly? When will it rise the most slowly?
Hint 4: First convert to metres per hour, then metres per minute, then metres per second.
Hint 5: Consider the steepness of the lines.
Hint 6: Use the given information to find k first.

Your Feedback

$\frac{\square}{25} \times 100\% = \square\%$

PAGES 139–140
PAGE 198

RATES AND PROPORTION

Proportion

ADVANCED TEST

1 When Jim filled his car with petrol it cost a total of \$65.52. The price of the petrol was \$1.56 per litre.

- **a** How many litres did Jim put in his car? (1 mark)
- **b** Jim had travelled 450 km since last filling his fuel tank. What was his car's petrol consumption in litres per 100 km? (1 mark)
- **c** The petrol tank on Jim's car holds 60 litres altogether. Jim believes he could take a trip of 625 km without needing to refill the tank. Do you agree? Discuss, justifying your answer. (2 marks)

2 Over time, production has decreased at an increasing rate. Sketch a possible graph of production over time. (1 mark)

3 The diagram shows the side wall (cross-section) of a swimming pool. Draw a graph to illustrate the depth of water in the pool as it is filled? (2 marks)

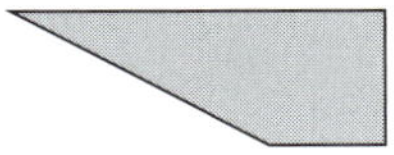

4 Grain is being augured into a silo at a constant rate. Sketch a graph showing the height of grain in the silo over the time that it is being filled. (2 marks)

5 Oliver leaves home and walks at a constant rate for 5 minutes, reaching a point 400 m from his home. He then begins to run slowly, gradually increasing his speed until reaching a point 1000 m from home after a further 5 minutes. He stops for 10 minutes then returns home, running quickly at first but gradually decreasing his pace. He is away from home for 30 minutes. Draw a graph to show Oliver's journey. (3 marks)

6 The graph shows a bus journey.

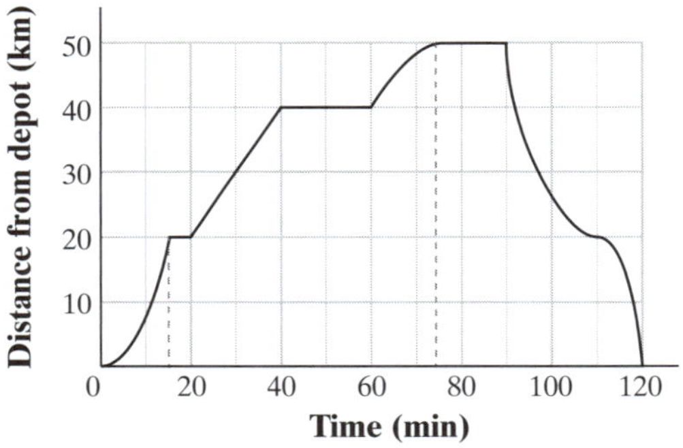

- **a** How long did the journey take? (1 mark)
- **b** Briefly describe what happens in the first 40 minutes of the journey. (2 marks)
- **c** What is the average speed on the return leg of the journey? (2 marks)
- **d** How far did the bus travel between the second and third stops? (1 mark)
- **e** 'The last part of the graph can't be right,' said Ruby. 'The bus should only travel at a maximum of 100 km/h and it couldn't possibly reach more than 120 km/h.' Assuming Ruby is correct, explain why the graph must be wrong. (2 marks)

Your Feedback

$\frac{\square}{20} \times 100\% = \square\%$

PAGES 140–141

PAGE 198

THE INDEX LAWS

Algebra

STUDY NOTES

1 x^4 is written using **index notation**. x is the **base**. 4 is the **index** or **power**. The plural of index is **indices**. The index tells us how many times the base is multiplied by itself.
For example, $x^4 = x \times x \times x \times x$ (There are four factors of x.)
$p^2q^3 = p \times p \times q \times q \times q$ (There are 2 factors of p and 3 factors of q.)

2 A symbol without an index actually has a **power of one**. $x = x^1$ We don't usually write a power of one.
For example, $a^1 = a$ $e = e^1$ $2y^1 = 2y$

3 When **multiplying**, if the base is the same, **add the indices**. $x^m \times x^n = x^{m+n}$
For example, $x^2 \times x^3 = x^{2+3} = x^5$ $a^5 \times a^4 = a^{5+4} = a^9$ $2^3 \times 2^8 = 2^{3+8} = 2^{11}$

4 When multiplying, if the **base is not the same**, we cannot add the indices. For example, $x^2 \times y^3 = x^2y^3$

5 If a term involves a number and a pronumeral, or more than one pronumeral, we **multiply each separately**.
For example, $2x^5 \times 3x^2 = 6x^7$ $[2 \times 3 = 6, x^5 \times x^2 = x^{5+2}]$
$a^2b^3 \times a^4b^7 = a^6b^{10}$ $[a^2 \times a^4 = a^{2+4}, b^3 \times b^7 = b^{3+7}]$
$4p^3q^4 \times 5pq^2 = 20p^4q^6$ $[4 \times 5 = 20, p^3 \times p = p^{3+1}, q^4 \times q^2 = q^{4+2}]$

6 When **dividing**, if the base is the same, **subtract the indices**. $x^m \div x^n = x^{m-n}$
For example, $a^8 \div a^2 = a^{8-2} = a^6$ $e^9 \div e^2 = e^{9-2} = e^7$ $6^5 \div 6^3 = 6^{5-3} = 6^2$

7 When dividing, if the **base is not the same**, we do not subtract the indices.
For example, $m^5 \div n^3 = \frac{m^5}{n^3}$. It cannot be simplified.

8 If a term involves a **number** and a **pronumeral**, or **more than one pronumeral**, we **divide each separately**.
For example, $8a^8 \div 2a^2 = 4a^6$ $[8 \div 2 = 4, a^8 \div a^2 = a^{8-2}]$
$12y^4z^9 \div 4y^2z^8 = 3y^2z$ $[12 \div 4 = 3, y^4 \div y^2 = y^{4-2}, z^9 \div z^8 = z^{9-8}]$

9 When **simplifying a fraction** involving indices, divide both the numerator and the denominator by a **common factor**. For example, $\frac{10n^5}{2n^8} = \frac{5}{n^3}$ [dividing both numerator and denominator by $2n^5$]

10 When a **power is raised to a power**, we **multiply the indices**. $(x^m)^n = x^{mn}$
For example, $(a^2)^4 = a^{2\times4} = a^8$ $(m^5)^3 = m^{5\times3} = m^{15}$

11 Only those **terms inside the brackets** are raised to the power but **every part of a term** inside the brackets must be raised to the power. So $(7x^3)^2 = 49x^6$ but $7(x^3)^2 = 7x^6$
For example, $(2a^3b^4)^3 = 8a^9b^{12}$ $[2^3 = 8, (a^3)^3 = a^{3\times3}, (b^4)^3 = b^{4\times3}]$

12 Any number raised to the **zero power** equals **one**. $x^0 = 1$
For example, $5^0 = 1$ $37^0 = 1$ $h^0 = 1$ $d^0 = 1$
The answers to questions involving the **zero index** are not always one.
For example, $p^0 + q^0 = 1 + 1 = 2$ $3x^0 = 3 \times x^0 = 3 \times 1 = 3$

13 If an expression is raised to a negative index, the result is the reciprocal of that expression raised to the equivalent positive index. $a^{-1} = \frac{1}{a}$; $3a^{-2} = \frac{3}{a^2}$; $(5a)^{-3} = \frac{1}{(5a)^3}$

Checklist

Can you:

1. *Understand the terms 'base', 'power', 'index' and 'indices'?* ☐
2. *Multiply expressions involving indices?* ☐
3. *Divide expressions involving indices?* ☐
4. *Simplify fractions involving indices?* ☐
5. *Raise a power to a power?* ☐
6. *Understand and use the zero index?* ☐

THE INDEX LAWS

Algebra

SKILLS CHECK

1 Write in index form:

a $2^4 \times 2^5$ b $3^2 \times 3^3$ c $7^{12} \times 7^2$

2 Simplify:

a $x^4 \times x^5$ b $a^2 \times a^3$ c $t^{12} \times t^2$

d $2e^7 \times e$ e $8m^4 \times 2m^2$ f $4p^7 \times 3p^7$

g $b^4 \times 5b^2$ h $4a^2 \times 3a^7 \times 2a^3$ i $6x^2y^3 \times 2x^3y^5$

j $2a^4 \times 3a^2b$ k $4t^3u^7 \times 4tu^2$ l $3p^2qr^3 \times 5pqr^4$

3 Write in index form:

a $2^{12} \div 2^2$ b $5^8 \div 5^3$ c $6^9 \div 6^6$

4 Simplify:

a $x^{12} \div x^2$ b $a^8 \div a^3$ c $h^9 \div h^6$

d $12m^{12} \div 2m^2$ e $16x^2 \div 4x$ f $15p^7 \div 5p^6$

g $x^4y^6 \div x^2y^5$ h $9m^4n^3 \div 3m^2n^2$ i $8a^3b^2 \div 2ab^2$

j $\dfrac{8x^2}{12x}$ k $\dfrac{7n^6}{14n^3}$ l $\dfrac{9a^2b^7}{6a^2b^3}$

5 Simplify:

a $(x^4)^2$ b $(a^2)^3$ c $(g^7)^4$ d $(m^3)^3$

e $(2n)^5$ f $(4k^2)^2$ g $(3t^6)^3$ h $5(p^4)^9$

i $(x^2y^3)^2$ j $(a^3b^2)^4$ k $(pq^2)^3$ l $(2m^2n^4)^4$

6 Evaluate:

a 7^0 b 9^0 c x^0 d k^0

e $2a^0$ f $(2a)^0$ g $5e^0$ h $m^0 + n^0$

i $7x^0 + 2$ j $4m^0 + 2m^0$ k $9e^0 + 3e$ l $b^0 - c^0$

PAGE 141

Answers **1** a 2^9 b 3^5 c 7^{14} **2** a x^9 b a^5 c t^{14} d $2e^8$ e $16m^6$ f $12p^{14}$ g $5b^6$ h $24a^{12}$ i $12x^5y^8$ j $6a^6b$ k $16t^4u^9$ l $15p^3q^2r^7$ **3** a 2^{10} b 5^5 c 6^3 **4** a x^{10} b a^5 c h^3 d $6m^{10}$ e $4x$ f $3p$ g x^2y h $3m^2n$ i $4a^2$ j $\frac{2x}{3}$ k $\frac{n^3}{2}$ l $\frac{3b^4}{2}$ **5** a x^8 b a^6 c g^{28} d m^9 e $32n^5$ f $16k^4$ g $27t^{18}$ h $5p^{36}$ i x^4y^6 j $a^{12}b^8$ k p^3q^6 l $16m^8n^{16}$ **6** a 1 b 1 c 1 d 1 e 2 f 1 g 5 h 2 i 9 j 6 k $9 + 3e$ l 0

THE INDEX LAWS

Algebra

20 MINUTES

INTERMEDIATE TEST

Part A Multiple Choice

1 $3a^3 \times 5a^5 = ?$

A $8a^8$ B $8a^{15}$ C $15a^8$ D $15a^{15}$ (1 mark)

2 $4n^9 \div 4n^3 = ?$

A n^3 B n^6 C $4n^3$ D $4n^6$ (1 mark)

3 Which is NOT equal to $4x^2$?

A $2x \times 2x$ B $8x^2 \div 2x^2$ C $4x \times x$ D $3x^2 + x^2$ (1 mark)

4 $(2x^3)^3 = ?$

A $2x^6$ B $2x^9$ C $8x^6$ D $8x^9$ (1 mark)

5 $7x^0 + 7^0 = ?$

A 1 B 2 C 7 D 8 (1 mark)

6 $8x^2 \times 3x = ?$

A $24x^2$ B $24x^3$ C $24x^4$ D none of these (1 mark)

7 $2 + 4p^0 = ?$

A 0 B 1 C 2 D 6 (1 mark)

8 $\frac{12m^8}{16m^9} = ?$

A $\frac{3}{4m}$ B $\frac{3m}{4}$ C $\frac{4}{3m}$ D $\frac{4m}{3}$ (1 mark)

9 $2n^5 + 3n^2 \times n^3 = ?$ *Hint 1*

A $5n^5$ B $5n^{10}$ C $5n^{11}$ D $2n^5 + 3n^6$ (1 mark)

10 $(x^2y^3)^4 = ?$

A x^6y^7 B x^8y^{12} C $x^{16}y^{81}$ D $(xy)^{24}$ (1 mark)

11 $3(m^5)^2 = ?$

A $9m^7$ B $9m^{10}$ C $3m^7$ D $3m^{10}$ (1 mark)

12 $a^2b^3c^4 \times ab^2c^3 = ?$ *Hint 2*

A $a^2b^5c^7$ B $a^3b^5c^7$ C $a^2b^6c^{12}$ D $a^3b^6c^{12}$ (1 mark)

Part B Short Answer

13 $a^{12} \div a^2 \times a^6 = ?$ (1 mark)

14 Simplify $12x^8 - 3x^3 \times 5x^5$ *Hint 3* (2 marks)

15 Simplify $\frac{6x^7 \div 3x^3}{5x^2 \times 2x^4}$ *Hint 4* (3 marks)

16 Simplify $(12p^5 \div 6p^2)^2$ *Hint 5* (2 marks)

Hint 1: Do any multiplication before addition.
Hint 2: Add the indices separately for each pronumeral.
Hint 3: Take care to perform the operations in the correct order.
Hint 4: Simplify the numerator and simplify the denominator. Then simplify the fraction.
Hint 5: Work out the part in the bracket first.

Your Feedback

$\frac{\square}{20} \times 100\% = \square\%$

PAGES 141–142

PAGE 198

THE INDEX LAWS

Algebra

35 MINUTES

ADVANCED TEST

1 Simplify:

a $\dfrac{12a^3b^5c^2}{20ab^3c^2}$

b $-2x^3y^5z^8 \times 5xy^3z^7$

c $(3x^2y^3)^3$

d $(5x^2y)^0$

e $18p^9q^7r \div -3p^5qr$

f $4k^3mn^4 \times 3kmn^2 \times 2m^6n$ (1 mark each)

2 Simplify:

a $5a^3b^2 \times 3a^2b^6 + 9ab^5 \times 2a^2b^4$

b $\dfrac{2m^5 \times 3m^4}{12m^{12} \div 2m^2}$ (1 mark each)

3 Simplify:

a $36a^8b^7c^6 \div 3ab^3c^3 \div 3abc^3$ (2 marks)

b $(-3x^4y^5 \times 2x^3y^2)^2$ (2 marks)

c $\dfrac{9p^5q^6 \times 8p^3q^3}{6pq^4 \times 6p^4q^5}$ (2 marks)

4 Simplify, where possible:

a $3x^2 \times 4y^2$ (1 mark)

b $((x^2y^3)^2)^2$ (1 mark)

c $a^m \times a^n$ (1 mark)

d $b^m \div c^n$ (1 mark)

5 Write with negative indices:

a $\dfrac{2}{b^2}$

b $\dfrac{1}{5x}$

c $\dfrac{2}{(3a)^3}$ (1 mark each)

6 Write as fractions:

a $5a^{-1}$ (1 mark)

b $(3t)^{-1}$ (1 mark)

c $6p^{-2}$ (1 mark)

7 Determine whether the statement is true or false. If false, give the correct answer.

a $20a^7 \div 4a^7 = 5a$

b $(4x^3)^2 = 4x^6$

c $3^0 + 3x^0 = 2$

d $\left(\dfrac{2}{x}\right)^{-1} = \dfrac{x}{2}$

e $6m^4 \div 2m^7 = 3m^3$

f $\dfrac{1}{2}n^{-2} = \dfrac{1}{2n^2}$ (1 mark each)

8 Simplify:

a $5x^{-2} \times 3x^{-5}$ (1 mark)

b $2(x^{-4})^{-3}$ (1 mark)

c $(4k^3)^{-1}$ (1 mark)

d $8n^5 \div 4n^{-3}$ (1 mark)

e $12a^{-2} \div -3a^{-9}$ (1 mark)

Your Feedback

$\dfrac{\square}{35} \times 100\% = \square\%$

PAGE 142

PAGE 198

THE DISTRIBUTIVE LAW

Algebra

STUDY NOTES

1 **When expanding**, every term inside the brackets must be multiplied by the number or term outside. For example, $a(b + c) = a \times b + a \times c = ab + ac$. This is an example of the **distributive law**.

2 **After expanding**, it may be possible to **simplify** an expression **by collecting like terms**.
For example, $x(x + 4) + 5(x + 4) = x^2 + 4x + 5x + 20$
$= x^2 + 9x + 20$

3 When simplifying, remember that the **operation sign** before a term is the one belonging to that term. For example, $4x^2 - 3x + 2x^2 - 7x = 6x^2 - 10x$ because $4x^2 + 2x^2$ is $6x^2$ and $-3x - 7x$ is $-10x$

4 We should always follow the correct **order of operations**, which generally means doing what is inside the brackets first but, in most cases when we need to expand, the expression inside the brackets cannot be simplified.

5 Remember, when **multiplying two terms**, if the signs are the same (both positive or both negative) the answer is **positive** but if the signs are different the answer is **negative**.
$-2(3x + 4) = -6x - 8$ $\quad$ $-2(3x - 4) = -6x + 8$

6 If there is **no number or term outside the brackets**, we are really multiplying by 1.
$2x + (3x + 5) = 2x + 3x + 5$ $\quad$ $2x - (3x + 5) = 2x - 3x - 5$

7 When **expanding expressions** like $3(x + 5) - 2(x - 1)$ take extra care to get the **signs correct**.
$3(x + 5) - 2(x - 1) = 3x + 15 - 2x \mathbf{+} 2$

8 **Factorising** is the opposite operation to expanding. When factorising we look for a **common factor** and place it outside the brackets. For example, 5 is a common factor of $5x$ and 15 so $5x - 15 = 5(x - 3)$.

9 You can always **check a factorisation by expanding** the answer.

10 A **binomial** is an algebraic expression involving two terms added together or subtracted. Examples of binomials involving the sum of two terms are $3x + 7$, $x^2 + 8x$ and $9ab^2 + 7abc$. Some binomials involving the difference of two terms are $5x - 1$, $m^2 - 6m$ and $3p^2q - 4qr$.

11 A **diagram** can be used to illustrate the process that occurs when we **expand a binomial**.

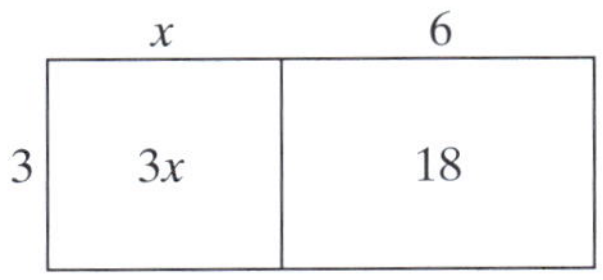

$3(x + 6) = 3x + 18$

Note: the diagram is only to illustrate the result. You are not expected to draw one in order to expand.

12 **Equations** involving grouping symbols are solved in the usual way after first expanding.

Checklist

Can you:

1 *Expand expressions?* ☐
2 *Expand and simplify?* ☐
3 *Understand how the signs work and get them correct?* ☐
4 *Understand what is meant by factorisation?* ☐
5 *Check a factorisation by expanding?* ☐
6 *Understand what is meant by a binomial?* ☐
7 *Solve equations involving grouping symbols?* ☐

THE DISTRIBUTIVE LAW

Algebra

SKILLS CHECK

1 Expand:

a $2(x + y)$ b $5(p - q)$ c $8(a + b + c)$

d $3(x + 2)$ e $6(a - 4)$ f $9(t + 1)$

g $2(3e + 4)$ h $4(5p - 2)$ i $7(2a + 3b)$

2 Factorise:

a $5a + 10$ b $2e - 8$ c $6t + 8u$

3 Expand:

a $4(a + 3)$ b $x(x - 2)$ c $2m(3m - 1)$

d $4(x - 2y + 5)$ e $3ab(a^2 - 5b)$ f $x^2y(xy - y^4)$

4 Expand:

a $-2(3x + 5)$ b $-5(6 - 5e)$ c $-4(x + y - z)$

d $-3(4a - 5b)$ e $-x(x + 8)$ f $-7e(f - 2g)$

5 Expand:

a $-(3x - 4)$ b $-(a + 2b)$ c $-(4x^2 - 1)$

d $-(2x - 3y + z)$ e $-(-2x + 7)$ f $-(-3ab - 4c)$

6 Simplify:

a $6x + 5 + 3x + 8$ b $8a - 4 + 3a - 6$

c $7p - 2 - 3p + 5$ d $x^2 + 3x + 5x + 15$

e $a^2 - 2a + 3a - 6$ f $n^2 - 5n - n + 5$

7 Expand and simplify:

a $12 + 3(4t + 1)$ b $6(x - 2) + 5$

c $4(2p - 3) + 3p - 2$ d $8x - (x - 3)$

8 Expand and simplify:

a $8(a + 2) + 6(b + 3)$ b $2(3x + 5) + 4(2x + 7)$

c $5(x + 4) + 3(x - 7)$ d $7(5x - 4) - (3x + 2)$

e $4(3p + 5q) - 3(2p + q)$ f $6(2a + b) - 5(3a - b)$

g $-4(5a + 2b) - 3(6a + 5)$ h $-2(8x - y) - (3x + y)$

PAGES 142–143

Answers **1** a $2x + 2y$ b $5p - 5q$ c $8a + 8b + 8c$ d $3x + 6$ e $6a - 24$ f $9t + 9$ g $6e + 8$ h $20p - 8$ i $14a + 21b$ **2** a $5(a + 2)$ b $2(e - 4)$ c $2(3t + 4u)$ **3** a $4a + 12$ b $x^2 - 2x$ c $6m^2 - 2m$ d $4x - 8y + 20$ e $3a^3b - 15ab^2$ f $x^3y^2 - x^2y^5$ **4** a $-6x - 10$ b $-30 + 25e$ c $-4x - 4y + 4z$ d $-12a + 15b$ e $-x^2 - 8x$ f $-7ef + 14eg$ **5** a $-3x + 4$ b $-a - 2b$ c $-4x^2 + 1$ d $-2x + 3y - z$ e $2x - 7$ f $3ab + 4c$ **6** a $9x + 13$ b $11a - 10$ c $4p + 3$ d $x^2 + 8x + 15$ e $a^2 + a - 6$ f $n^2 - 6n + 5$ **7** a $12t + 15$ b $6x - 7$ c $11p - 14$ d $7x + 3$ **8** a $8a + 6b + 34$ b $14x + 38$ c $8x - 1$ d $32x - 30$ e $6p + 17q$ f $-3a + 11b$ g $-38a - 8b - 15$ h $-19x + y$

THE DISTRIBUTIVE LAW

Algebra

20 MINUTES

INTERMEDIATE TEST

Part A Multiple Choice

1 $-2(3 - 7x) = ?$
A $-6 - 7x$ B $-6 + 7x$ C $14x + 6$ D $14x - 6$ (1 mark)

2 $3x - (2x - 1) = ?$ *Hint 1*
A $x + 1$ B $x - 1$ C $2x$ D none of these (1 mark)

3 $7(x + 2) + 3 = ?$
A $7x + 5$ B $7x + 11$ C $7x + 17$ D $7x + 35$ (1 mark)

4 $5a - 2(3 + a) = ?$
A $2a - 6$ B $3a - 6$ C $6a - 6$ D $7a - 6$ (1 mark)

5 $3x^2(4x^3 + 2xy) = ?$ *Hint 2*
A $12x^5 + 6x^3y$ B $12x^6 + 6x^3y$ C $14x^5y$ D $18x^6y$ (1 mark)

6 Which factorisation is NOT correct? *Hint 3*
A $3x^2 - 6 = 3(x^2 - 2)$ B $8a - 4b = 4(2a - b)$
C $12xy + 8y + 4 = 4(3xy + 2y + 1)$ D $15k - 27m = 3(5k - 8m)$ (1 mark)

7 $x(x - 1) - 5(x - 1) = ?$
A $x^2 - 6x + 5$ B $x^2 - 6x - 5$ C $x^2 - 4x + 5$ D $x^2 - 4x - 5$ (1 mark)

8 $2p^3(p^2 - q) - 2p^3q = ?$
A $2p^5$ B $2p^6$ C $2p^5 - 4p^3q$ D $2p^6 - 4p^3q$ (1 mark)

9 $2a^2(3ab^2 - 5a^4) = ?$
A $6a^2b^2 - 10a^6$ B $6a^2b^2 - 10a^8$ C $6a^3b^2 - 10a^6$ D $6a^3b^2 - 10a^8$ (1 mark)

10 $a(a + b) + b(a + b) = ?$ *Hint 4*
A $a^2 + ab + b^2$ B $a^2 + 2ab + b^2$ C $2a + ab + 2b$ D none of these (1 mark)

Part B Short Answer

11 Expand $4u^3(u^3 - 1)$ (1 mark)

12 Expand $3x^2y^3(x^4 - 5xy^2)$ (1 mark)

13 Expand and simplify $6(3x - 4) - 3(2x + 7)$ *Hint 5* (2 marks)

14 Solve $5(3a - 7) = 2(a + 2)$ (2 marks)

15 Expand and simplify $a(3a + 2b) - 4(2a^2 - 3b)$ *Hint 6* (2 marks)

16 Expand and simplify $2(3x - 1) - (x - 2)$ *Hint 7* (2 marks)

Hint 1: Everything inside the brackets must be subtracted.
Hint 2: Use the Index Laws.
Hint 3: Check by expanding.
Hint 4: $b \times a = a \times b = ab$
Hint 5: Take care that the operation signs are correct.
Hint 6: A negative multiplied by a negative is a positive.
Hint 7: There is a 1 outside the second brackets.

Your Feedback

$\frac{\square}{20} \times 100\% = \square\ \%$

PAGE 143
PAGE 198

THE DISTRIBUTIVE LAW

Algebra

40 MINUTES

ADVANCED TEST

1 Expand:

a $-3x^2(2x^2 - 5y)$ (1 mark)

b $7a^2b(2ab^2 - 3b^3)$ (1 mark)

c $2xy^2z^3(-4xy + 11xz^2)$ (1 mark)

d $9p^4q^3(7p^5r^2 - 6q^7)$ (1 mark)

e $11t^5u^9(3t^3u^5 - 8v^6)$ (1 mark)

f $-5a^3b^7(2a^4 + 9b^5)$ (1 mark)

g $p^{-3}(p^4 - p^{-2})$ (1 mark)

h $5n^4(2n^4 + n^{-4})$ (1 mark)

i $6xy\left(\dfrac{1}{x} + \dfrac{1}{y}\right)$ (1 mark)

2 Expand and simplify:

a $2 - (5 - x)$ (1 mark)

b $3x^2 + x(x - 7)$ (1 mark)

c $(p + q) - (p - q)$ (1 mark)

d $9x(3x - 5) - 17x + 2x^2$ (1 mark)

e $a(a - b) + b(a - b)$ (1 mark)

3 Expand and simplify:

a $3p^2(q - 4) + 5pq(p - 1)$ (2 marks)

b $3x^2y^3(5x^3 - 2y) - 7x^4y^2(6xy + 1)$ (2 marks)

c $12(9 - 4a) + 8(6a - 17)$ (2 marks)

d $-6a^2bc^5(2a + 7bc^2) - (6a^3bc^5 - 7a^2b^2c^7)$ (2 marks)

4 Expand and simplify where possible:

a $7p^3q^5(4p^5q^6 + 3p^6q^7) - 5p^8q^{10}(2q + 5p - 3pq)$ (2 marks)

b $9a^4(3a - 5b^3c^2) + 2abc^2(7a^3b^2 - 6b) - a(2bc - 12a^4)$ (2 marks)

5 Simplify:

a $(a^2 \times b^2)(a^2 + b^2)$ (2 marks)

b $(x^3y^2 \times xy^4)(5x + 3y)$ (2 marks)

c $(2xy^2)^2(3x^2 - 1)$ (2 marks)

d $(12k^2 \div 3k)(2k + m)$ (2 marks)

6 Solve:

a $4(3x - 2) + 3(2x + 1) = 7$ (2 marks)

b $6(4m - 5) - 5(2m + 3) = 9(m - 5)$ (2 marks)

c $7(5k - 1) - 9(3k - 4) = 2(6 - k) - 23$ (2 marks)

Your Feedback

$\dfrac{\square}{40} \times 100\% = \square\,\%$

PAGE 144

PAGE 198

DISTANCE BETWEEN TWO POINTS

Linear Relationships

STUDY NOTES

1 A **number plane** is formed when two straight lines, or **axes**, meet at right angles. The horizontal (going straight across) line is the ***x*-axis** and the vertical (up and down) line is the ***y*-axis**. Any point on the number plane can be named by an **ordered pair** of numbers. The first number in the pair, the ***x*-coordinate**, is found on the *x*-axis. The second number, the ***y*-coordinate** is found on the *y*-axis. When naming a point, the *x*-coordinate is always given first. The **origin** is the point **(0, 0)**.

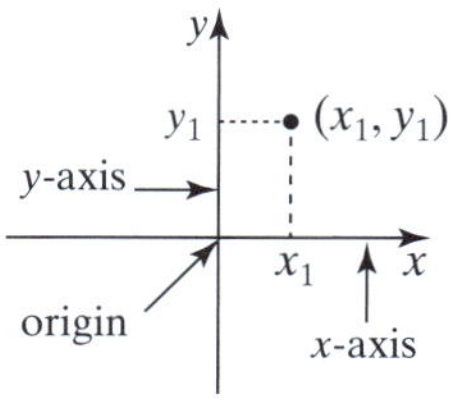

2 The **distance between two points** is the **length** of the interval joining the two points. The distance between two points on the same horizontal line, or on the same vertical line, can be found by counting the units between the points.

3 The **distance** between two points, not on the same horizontal or vertical line, can be found by forming a right-angled triangle (by drawing a vertical line down from the higher point and a horizontal line across from the lower point), and using **Pythagoras' Theorem**.
For example, $d^2 = 3^2 + 4^2 = 25$. So $d = \sqrt{25} = 5$
$\therefore$ The distance between the two points is 5 units.

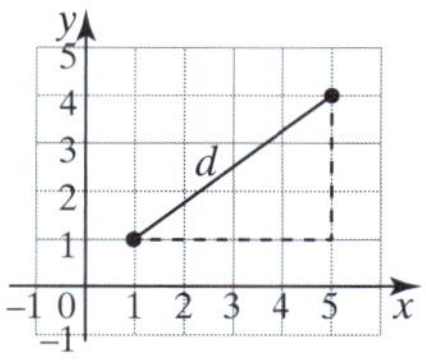

4 (x_1, y_1) and (x_2, y_2) can be any two **points**. x_1 is the *x*-coordinate of the first point and y_1 is the *y*-coordinate of the first point. x_2 and y_2 are the *x*- and *y*-coordinates, respectively, of the second point. For example, if the points are (3, –2) and (–4, 1) then $x_1 = 3$, $y_1 = -2$, $x_2 = -4$ and $y_2 = 1$.

5 If two points lie on the same horizontal line the *y*-coordinates will be the same. If two points lie on the same vertical line the *x*-coordinates will be the same. The distance between two points on the same **horizontal or vertical line** can be found by subtracting the coordinates that are not the same. For example, the distance between the points (2, 8) and (2, –3) is 8 – (–3) = 11 units.

6 The **distance formula** is $d = \sqrt{(x_2 - x_1)^2 + (y_2 - y_1)^2}$, where *d* is the length of the interval joining (x_1, y_1) to (x_2, y_2).

7 The number under the square root will not always be a perfect square. When it is not a perfect square we usually leave the answer as a surd (square root) rather than as a decimal approximation.

8 **Graphing software** applications can be used to find the distance between points.

9 The length of an interval can be used to make judgements about **triangles** and **quadrilaterals**. For example, we can find the lengths of the sides of a triangle to determine whether the triangle is scalene, isosceles or equilateral.

Checklist

Can you:

1 *Understand the terms 'number plane', 'coordinates', 'axes', 'ordered pairs' and 'origin'?* ☐
2 *Understand that the distance between two points is the length of the interval joining them?* ☐
3 *Find the length of the interval joining two points?* ☐
4 *Use graphing software to find the distance between two points?* ☐
5 *Use the distance formula?* ☐
6 *Make judgements about types of triangles and quadrilaterals?* ☐

DISTANCE BETWEEN TWO POINTS

Linear Relationships

SKILLS CHECK

1 Write down the coordinates of each point.

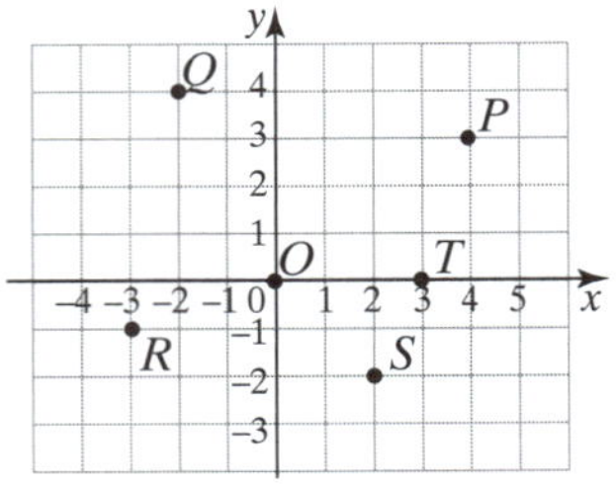

a P
b Q
c R
d S
e T
f O

2 Use Pythagoras' Theorem to find the length of the interval joining A to B.

a

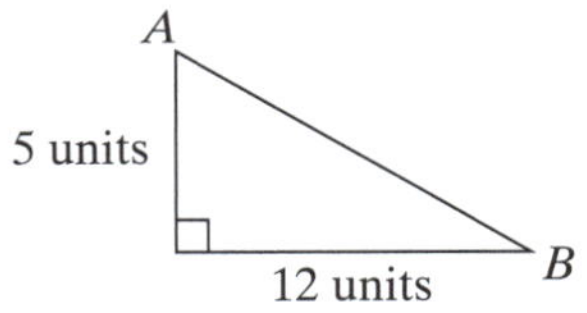

b

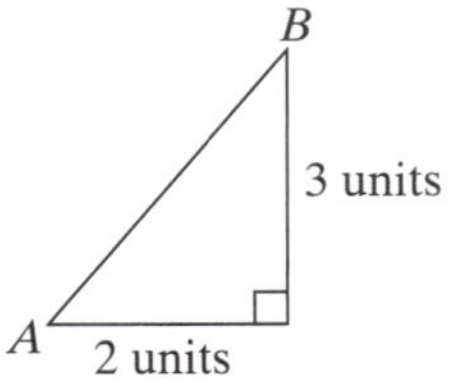

3 Using the points in the diagram, find the distance between:

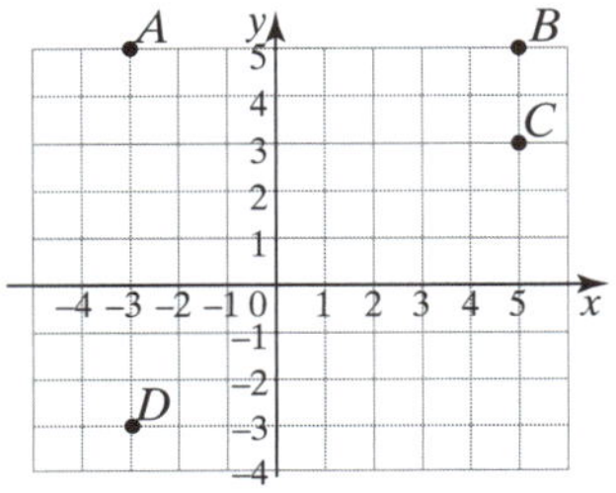

a A and B
b B and C
c A and D
d C and D

4 Find the distance between these pairs of points:

a (3, 2) and (3, 7)
b (4, –8) and (4, 1)
c (–5, –3) and (–5, –9)
d (5, 0) and (11, 0)
e (–6, –2) and (12, –2)
f (–1, 7) and (–3, 7)

5 If $d = \sqrt{(x_2 - x_1)^2 + (y_2 - y_1)^2}$, find d when:

a $x_1 = -3, x_2 = 5, y_1 = 1$ and $y_2 = 7$
b $x_1 = -9, x_2 = 6, y_1 = -2$ and $y_2 = 6$

6 Given $d = \sqrt{(x_2 - x_1)^2 + (y_2 - y_1)^2}$ and $x_1 = 2, x_2 = 3, y_1 = -4$ and $y_2 = -2$, find d leaving the answer in square root form.

7 Find the distance between the points:

a (4, –5) and (–1, 7)
b (6, 2) and (4, 5)

PAGES 144–145

Answers **1 a** (4, 3) **b** (–2, 4) **c** (–3, –1) **d** (2, –2) **e** (3, 0) **f** (0, 0) **2 a** 13 units **b** $\sqrt{13}$ units **3 a** 8 units **b** 2 units **c** 8 units **d** 10 units **4 a** 5 units **b** 9 units **c** 6 units **d** 6 units **e** 18 units **f** 2 units **5 a** 10 **b** 17 **6** $\sqrt{5}$ **7 a** 13 units **b** $\sqrt{13}$ units

DISTANCE BETWEEN TWO POINTS

Linear Relationships

20 MINUTES

INTERMEDIATE TEST

Part A Multiple Choice

1 What is the distance from (–2, 5) to (–2, –8)?

A 13 units **B** 3 units **C** 5 units **D** $\sqrt{173}$ units (1 mark)

2 Which point is NOT 5 units from the origin?

A (0, –5) **B** (–5, 0) **C** (3, 4) **D** (5, 5) (1 mark)

3 What is the distance from (–1, –3) to (5, 2)?

A $\sqrt{17}$ units **B** $\sqrt{37}$ units **C** $\sqrt{41}$ units **D** $\sqrt{61}$ units (1 mark)

Part B Short Answer

4 **a** Draw a diagram and plot the points $A(3, -2)$ and $B(-5, 4)$. (1 mark)

b Find the length of the interval AB. (2 marks)

5 The points $P(1,4)$, $Q(5, 1)$ and $R(1, -1)$ form a triangle.

a What is the length of the side PR? *Hint 1* (1 mark)

b What is the length of the side PQ? (2 marks)

c What type of triangle is PQR? Justify your answer. *Hint 2* (2 marks)

d What is the area of the triangle? *Hint 3* (2 marks)

6 $P(-11, 7)$, $Q(14, 7)$ and $R(-2, -5)$ are the vertices of a triangle.

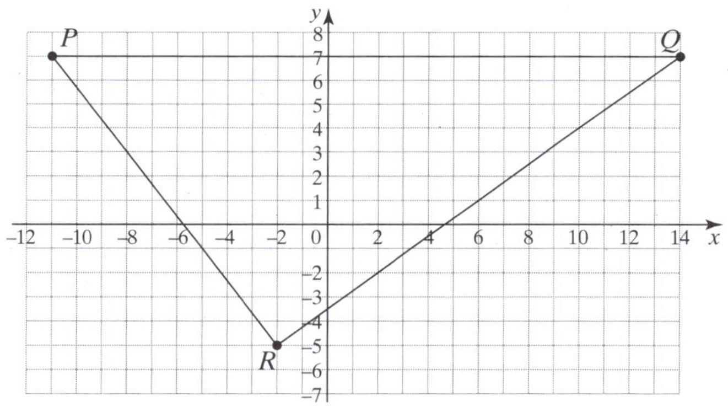

a What is the length of PQ? (1 mark)

b Find the length of PR. (2 marks)

c Find the length of QR. (2 marks)

d Is ΔPQR a right-angled triangle? Justify your answer. *Hint 4* (2 marks)

Hint 1: There is no need to use Pythagoras' Theorem.
Hint 2: What do you know about the sides of the triangle?
Hint 3: Find the distance between the points *P* and *Q* that you found in parts **b** and **c**.
Hint 4: Do the sides form a Pythagorean triad? (Does Pythagoras' Theorem work?)

Your Feedback

$\frac{\square}{20} \times 100\% = \square\%$

PAGES 145–146

PAGE 198

DISTANCE BETWEEN TWO POINTS

Linear Relationships

30 MINUTES

ADVANCED TEST

1 Find the distance between each pair of points:

a (3, 7) and (11, 1) (2 marks)

b (–2, 8) and (–7, –4) (2 marks)

c (5, –9) and (–7, 7) (2 marks)

2 Find the distance between each pair of points. Leave the answer in surd (square root) form.

a (5, 8) and (1, 7) (2 marks)

b (3, –6) and (8, 0) (2 marks)

c (–4, –5) and (1, –9) (2 marks)

3 $A(0, 2)$, $B(4, -1)$ and $C(9, -1)$ are the vertices of a triangle.

a Show that ΔABC is isosceles. (2 marks)

b Find the area of the triangle. (2 marks)

4 $P(7, 7)$, $Q(8,0)$, $R(3, -5)$ are three vertices of a quadrilateral.

a Find the length of PQ. (2 marks)

b Show that $QR = PQ$. (2 marks)

c Draw a diagram and mark the position of S so that $PQRS$ is a rhombus. (1 mark)

d Find the length of diagonal PR. (2 marks)

e Find the area of the rhombus. (3 marks)

5 P is the point (0, 4) and Q is the point (8, 10). R lies on the x-axis.

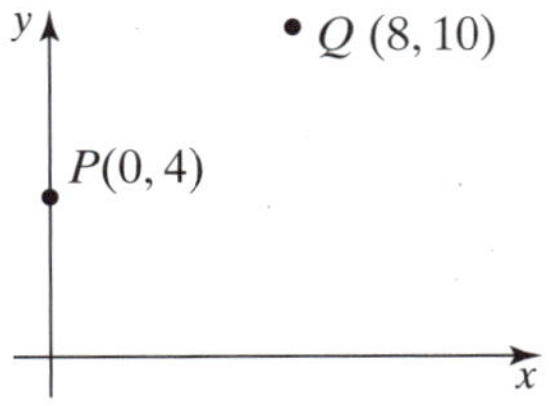

a Find the distance from P to Q. (2 marks)

b The distance from Q to R is equal to the distance from P to Q. Find the coordinates of R. (2 marks)

Your Feedback

$\frac{\square}{30} \times 100\% = \square\%$

PAGE 146

PAGE 198

MIDPOINT AND GRADIENT

Linear Relationships

STUDY NOTES

1 To find the **average** of a set of numbers add them up and divide by how many there are. So the average of two numbers is the sum of those numbers divided by two.

2 The **midpoint** is the point halfway between two given points. The ***x*-coordinate of the midpoint** is the average of the *x*-coordinates of the two points and the ***y*-coordinate of the midpoint** is the average of the *y*-coordinates of the two points.

3 The **midpoint formula** is $M(x, y) = \left(\frac{x_1 + x_2}{2}, \frac{y_1 + y_2}{2}\right)$, where M is the midpoint of (x_1, y_1) and (x_2, y_2).

Note: we write $M(x, y)$ to show that M is a point, not just a single variable.

4 The **gradient** is the measure of the steepness of a line. (Another name for gradient is slope.) A line that leans to the right has a **positive gradient** while a line that leans to the left has a **negative gradient**.

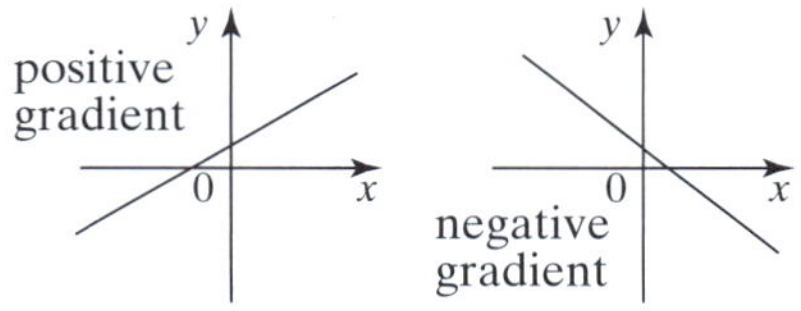

5 The gradient of a line joining two points can be found by forming a right-angled triangle. **Gradient =** $\frac{\textbf{rise}}{\textbf{run}}$,

e.g. In diagram I you move across 4 units and rise up 3 units when moving from A to B so the gradient is $\frac{3}{4}$. In diagram II you move across 4 units and move down 2 units when moving from A to B. The 'rise' is actually a fall, so it is negative. The gradient is $\frac{-2}{4} = -\frac{1}{2}$.

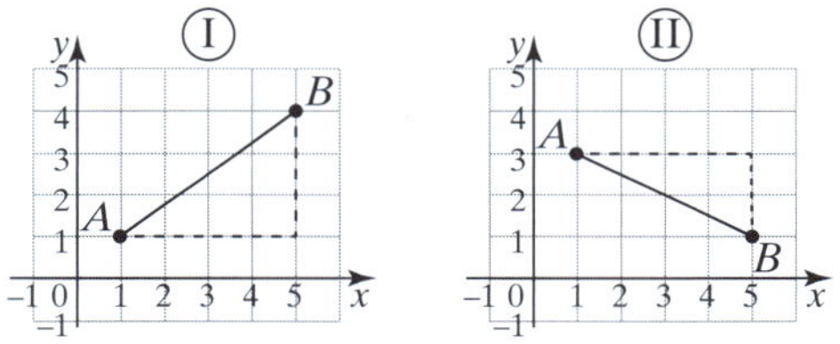

6 The **gradient**, m, of the line joining (x_1, y_1) to (x_2, y_2) is given by the **formula** $m = \frac{y_2 - y_1}{x_2 - x_1}$ $(x_2 \neq x_1)$.

7 Two lines are **parallel** if they have the same gradient.

8 Applications of **graphing software** can be used to find the midpoint or gradient.

9 The gradient of a line or midpoint of an interval can be used to make **judgements** about triangles and quadrilaterals. For example, we can find the gradients of the sides of a quadrilateral to see if it is a parallelogram, or we can find the midpoint of the diagonals of a quadrilateral to see if they bisect each other.

Checklist

Can you:

1 *Understand what is meant by the midpoint and find the midpoint of two points?* ☐
2 *Use the midpoint formula?* ☐
3 *Determine whether a line has positive or negative gradient?* ☐
4 *Find the gradient of a line?* ☐
5 *Use the gradient formula?* ☐
6 *Use graphing software?* ☐
7 *Know what it means for two lines to be parallel?* ☐
8 *Make judgements about types of triangles and quadrilaterals?* ☐

MIDPOINT AND GRADIENT

Linear Relationships

SKILLS CHECK

1 Find the average of:

a 3 and 9   b −3 and 7   c −5 and −4

2 Write down the coordinates of the midpoint of:

a A and B

b B and C

c A and D

d C and D

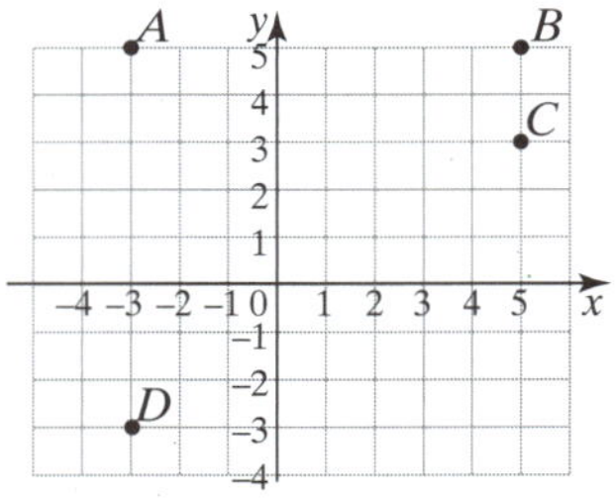

3 State whether the gradient of each line is positive or negative.

a
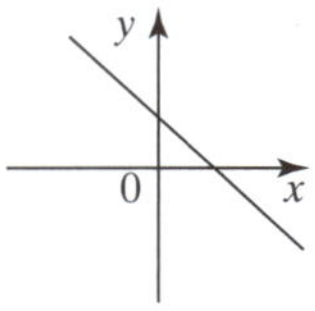

b
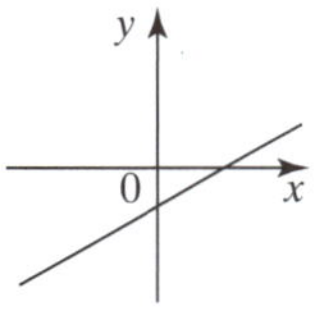

c
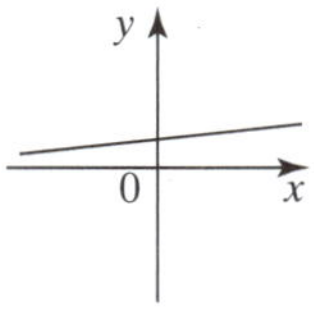

d
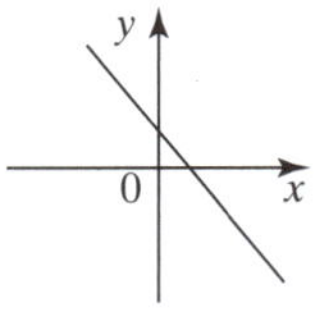

4 Find the gradient of each line:

a
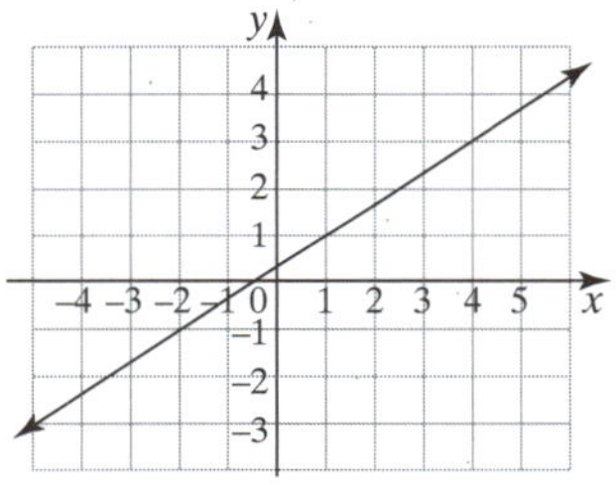

b
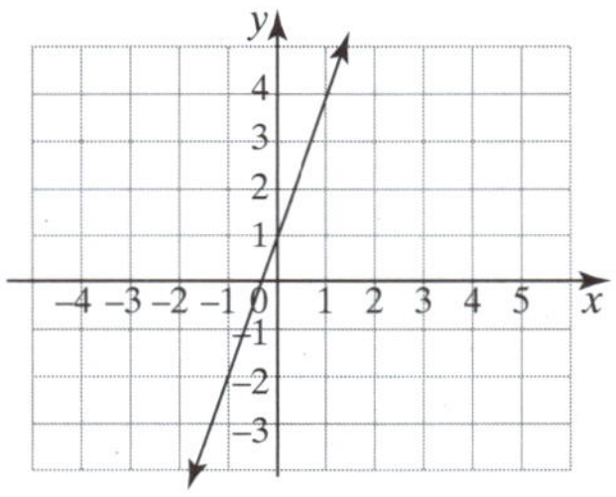

c
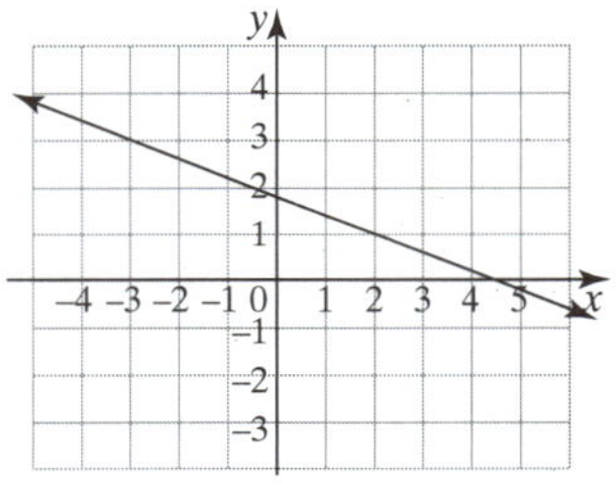

5 If $M(x, y) = \left(\dfrac{x_1 + x_2}{2}, \dfrac{y_1 + y_2}{2}\right)$, find $M(x, y)$ when:

a $x_1 = 5, x_2 = 9, y_1 = -2$ and $y_2 = 4$ b $x_1 = -3, x_2 = 8, y_1 = 3$ and $y_2 = -4$

6 Use the midpoint formula to find the midpoint of:

a (3, 7) and (9, 1)   b (4, 0) and (−2, 7)

7 If $m = \dfrac{y_2 - y_1}{x_2 - x_1}$, find m when:

a $x_1 = -7, x_2 = -1, y_1 = 5$ and $y_2 = 8$ b $x_1 = 2, x_2 = 3, y_1 = 1$ and $y_2 = -5$

8 Find the gradient of the line joining the points:

a (3, 7) and (9, −1)   b (−8, −2) and (6, 5)

9 Complete:

a If the gradients of two lines are equal then the lines are ________________.

b If two lines are parallel then their gradients are ________________.

PAGE 147

Answers **1 a** 6 **b** 2 **c** $-4\frac{1}{2}$ **2 a** (1, 5) **b** (5, 4) **c** (−3, 1) **d** (1, 0) **3 a** negative **b** positive **c** positive **d** negative **4 a** $\frac{2}{3}$ **b** 3 **c** $-\frac{2}{5}$ **5 a** (7, 1) **b** $(2\frac{1}{2}, -\frac{1}{2})$ **6 a** (6, 4) **b** $(1, 3\frac{1}{2})$ **7 a** $\frac{1}{2}$ **b** −6 **8 a** $-\frac{4}{3}$ **b** $\frac{1}{2}$ **9 a** parallel **b** equal

MIDPOINT AND GRADIENT

Linear Relationships

25 MINUTES

INTERMEDIATE TEST

Part A Multiple Choice

1 Which is the midpoint of (3, 7) and (7, 3)?

A (3, 3) **B** (0, 0) **C** (5, 5) **D** (7, 7) (1 mark)

2 $M(3, 8)$ is the midpoint of $P(-1, 4)$ and which other point? *Hint 1*

A (1, 6) **B** (5, 6) **C** (5, 12) **D** (7, 12) (1 mark)

3 Which line has gradient $-\frac{1}{3}$?

A l_1
B l_2
C l_3
D l_4

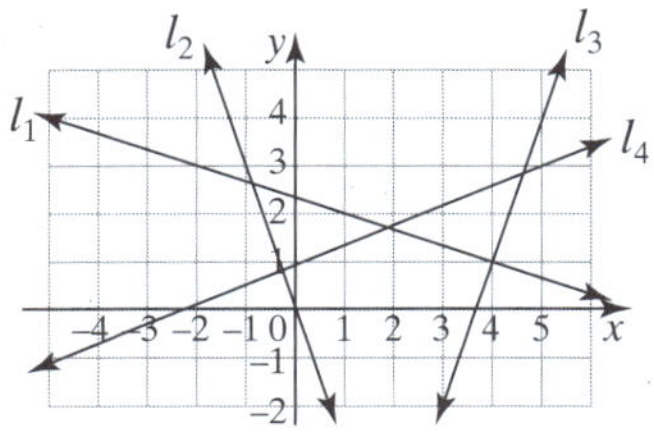

(1 mark)

4 Which point is the midpoint of (3, 7) and (8, –5)?

A (5.5, –6) **B** (5.5, 1) **C** (2.5, –6) **D** (2.5, 1) (1 mark)

5 What is the gradient of the line joining (–2, 4) to (5, –10)?

A $\frac{1}{2}$ **B** $-\frac{1}{2}$ **C** 2 **D** –2 (1 mark)

Part B Short Answer

6

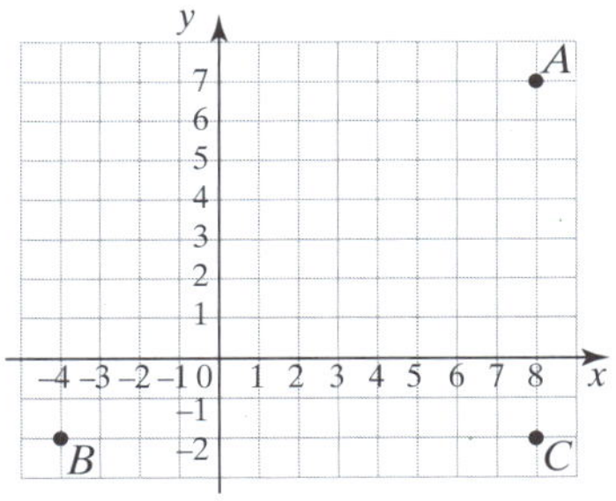

The diagram shows the points $A(8, 7)$, $B(-4, -2)$ and $C(8, -2)$.

a What is the midpoint of AB? (2 marks)

b What is the gradient of the line joining A and B? *Hint 2* (2 marks)

c What is the gradient of BC? (1 mark)

7 **a** Draw a diagram and plot the points $P(-2, 5)$ and $Q(6, 1)$. (1 mark)

b Find the gradient of the line joining P and Q. *Hint 3* (2 marks)

8 $A(-3, -6)$, $B(1, 4)$ and $C(7, 2)$ are the vertices of a triangle.

a Find the gradients of the sides AB, AC and BC. (3 marks)

b Find the coordinates of the midpoints, P, Q and R of sides AB, AC and BC respectively. *Hint 4* (3 marks)

c Find the gradients of QR, PR and PQ. (3 marks)

d What conclusions can be drawn about the lines joining the midpoints? (1 mark)

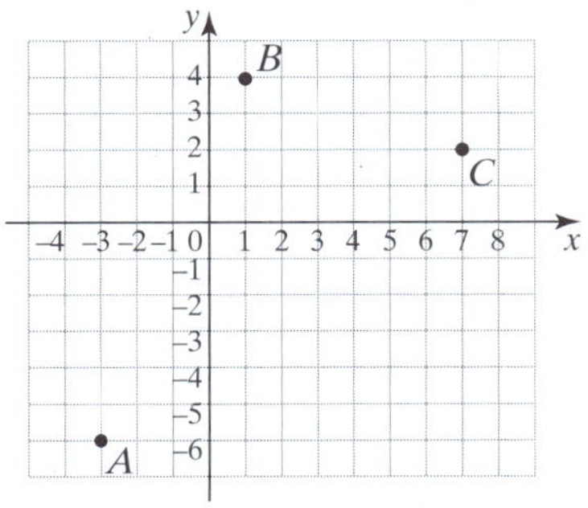

Hint 1: Try each of the choices.
Hint 2: You can use any two points on the line but the line must pass through the corners of the grid.
Hint 3: The line leans to the left.
Hint 4: P is the midpoint of AB, Q is the midpoint of AC and R is the midpoint of BC.

Your Feedback

$\frac{\square}{23} \times 100\% = \square\,\%$

PAGE 148
PAGE 198

MIDPOINT AND GRADIENT

Linear Relationships

30 MINUTES

ADVANCED TEST

1 a Briefly explain what it means if the gradient of a line is zero. (1 mark)
b What is the gradient of a vertical line? (1 mark)

2 A, B, C and D are the points (–2, 1), (5, 2), (10, –3) and (3, –4) respectively.
a Find the gradient of AB. (1 mark)
b Is DC parallel to AB? Justify your answer. (2 marks)
c Find the gradient of BC. (1 mark)
d Is AD parallel to BC? Justify your answer. (2 marks)
e What type of quadrilateral is $ABCD$? (1 mark)
f Find the midpoint of AC. (1 mark)
g Find the midpoint of BD. (1 mark)
h What properties have been illustrated in this question? (1 mark)

3 The midpoint of $P(3, -7)$ and Q is $M(5, 2)$. Find the coordinates of Q. (2 marks)

4 G is the point (–6, 5) and H lies on the y-axis. The gradient of GH is $-\frac{2}{3}$.
What are the coordinates of H? (3 marks)

5 A, B, C and D are the points (3, 5), (7, –4), (–2, –3) and (–6, 6) respectively.
a Find the gradient of the line joining B to D. (1 mark)
b Find the coordinates of the midpoint, M, of AC. (1 mark)
c Find the gradient of BM. (1 mark)
d What conclusion can be made about point M? Justify your answer. (2 marks)

6 A is the point (4, 1), B is the point (–2, –5) and C is the point (p, q).
a If the gradient of AC is –1, show that $q = -p + 5$. (2 marks)
b If the gradient of BC is 3, show that $q = 3p + 1$. (2 marks)
c Briefly explain why $3p + 1 = -p + 5$. (1 mark)
d Solve $3p + 1 = -p + 5$ (1 mark)
e What are the coordinates of C? (2 marks)

Your Feedback

$\frac{\square}{30} \times 100\% = \square\%$

PAGES 148–149

PAGE 198

SKETCHING LINEAR GRAPHS

Linear Relationships

STUDY NOTES

1 An **equation** such as $y = x + 4$ gives a **relationship** between two **variables**, x and y. Variables are symbols which can have different values at different times. For example in the above equation we can let x have any value we like and for any of these values we can find a matching value of y which makes the equation true.

2 A **table of values** is used to find points that lie on a line. By substituting a value for one variable into an equation, we find the value of the other variable.

3 Any **equation** of the form $\boldsymbol{x = k}$ (where k is a number) is a **vertical line**. It is a line parallel to the y-axis.

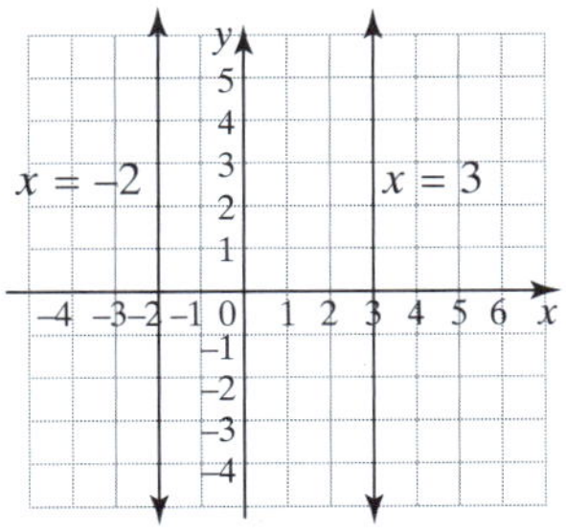

4 Any **equation** of the form $\boldsymbol{y = k}$ (where k is a number) is a **horizontal line**. It is a line parallel to the x-axis.

5 The ***x*-axis** is the line $y = 0$.

6 The ***y*-axis** is the line $x = 0$.

7 The ***x*-intercept** is where the line meets the ***x*-axis**.

8 The ***y*-intercept** is where the line meets the ***y*-axis**.

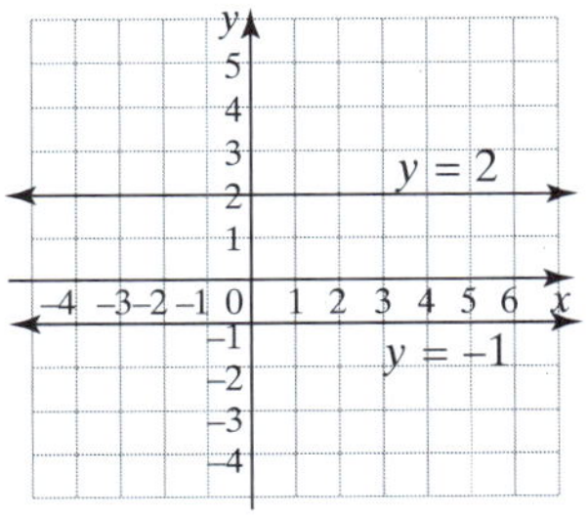

9 When drawing a graph, always remember to **label the axes**.

10 The **graph of a linear relationship** is a **straight line**. By constructing a table of values and plotting the coordinates we can then graph the line. The graph of a line extends across the whole number plane. We don't just 'join the dots'. Always write the name of the line on the graph.

11 To **graph a straight line** we only need **two points**, but it is a good idea to use at least one more as a check point.

12 Given the **gradient and *y*-intercept**, any **line can be graphed**. The y-intercept gives us one point on the line. Use the gradient, $\frac{\text{rise}}{\text{run}}$, to find a second point. For example, if $y = \frac{2}{3}x - 4$, the y-intercept is –4 so one point on the line is (0, –4). The gradient is $\frac{2}{3}$ so run across 3 units and rise up 2 units to find another point on the line.

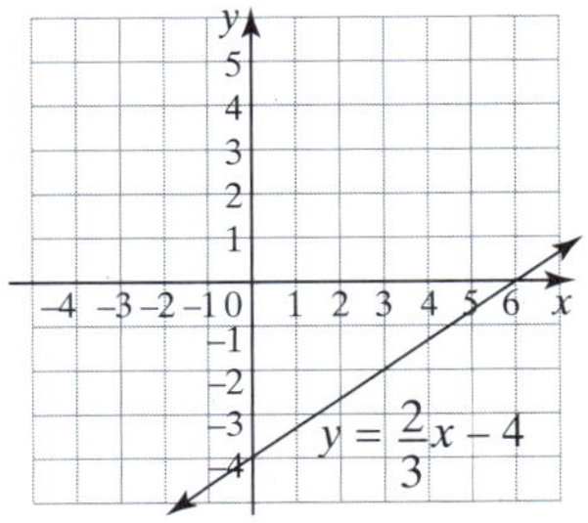

Checklist

Can you:

1 *Complete tables of values?* ☐
2 *Plot points and draw straight lines?* ☐
3 *Know the equations of horizontal and vertical lines?* ☐
4 *Find x- and y-intercepts?* ☐
5 *Know what we mean by a linear relationship?* ☐
6 *Graph lines using the gradient and y-intercept?* ☐

SKETCHING LINEAR GRAPHS

Linear Relationships

SKILLS CHECK

1 Copy and complete the table of values for the given equation.

a $y = x + 1$

x	0	1	2
y			

b $y = 4x$

x	1	2	3
y			

c $y = 6 - x$

x	–1	0	1
y			

d $y = 2x + 5$

x	2	3	4
y			

e $y = 7$

x	0	1	2
y			

f $x = 5$

x			
y	1	2	3

2 Find an equation describing the relationship for the points in each of the tables of values.

a

x	0	1	2	3	4
y	2	1	0	–1	–2

b

x	–2	–1	0	1	2
y	–3	–2	–1	0	1

c

x	–1	0	1	2	3
y	4	2	0	–2	–4

d

x	–2	–1	0	1	2
y	2	2	2	2	2

3 On separate number planes, plot the points in each table of values from question 2 and draw the line on which those points lie.

4 Find the gradient for each of the lines draw in question 3.

5 Write down the y-intercept of each of the lines drawn in question 3.

6 Write down the x- and y-intercepts for each line.

a

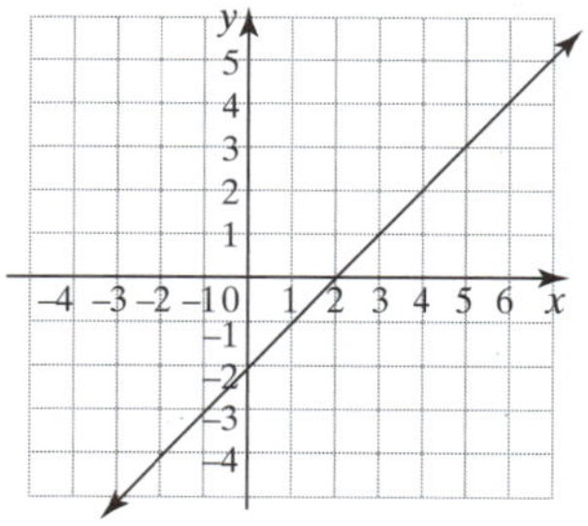

b

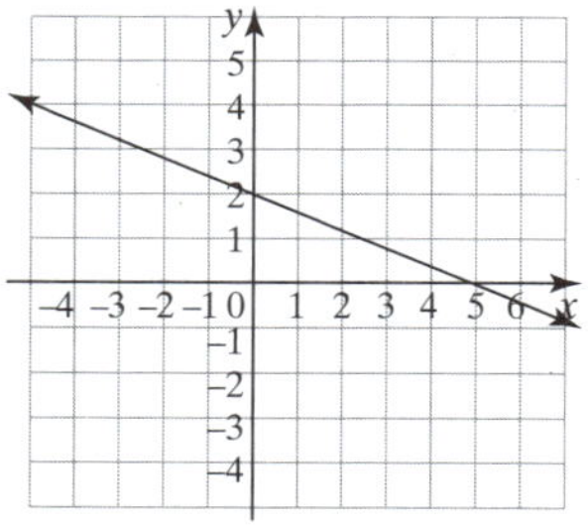

7 Graph each of the straight lines:

a $y = 3$ **b** $x = -3$

8 Graph the lines with:

a gradient 2, y-intercept 1 **b** gradient $-\frac{1}{2}$, y-intercept 3

PAGE 150

Answers **1 a** 1, 2, 3 **b** 4, 8, 12 **c** 7, 6, 5 **d** 9, 11, 13 **e** 7, 7, 7 **f** 5, 5, 5 **2 a** $x + y = 2$ **b** $y = x - 1$ **c** $y = -2x + 2$ **d** $y = 2$
3 see page 150 **4 a** –1 **b** 1 **c** –2 **d** 0 **5 a** 2 **b** –1 **c** 2 **d** 2 **6 a** $x = 2$, $y = -2$ **b** $x = 5$, $y = 2$ **7** see page 150
8 see page 150

SKETCHING LINEAR GRAPHS

Linear Relationships

25 MINUTES

INTERMEDIATE TEST

Part A Multiple Choice

1 The graph of which of these lines will pass through the origin?

A $y = 4$ B $y = 2x - 2$ C $y = -x$ D $y = -3x + 3$ (1 mark)

2 Which is the graph of $y = 2 - x$?

A
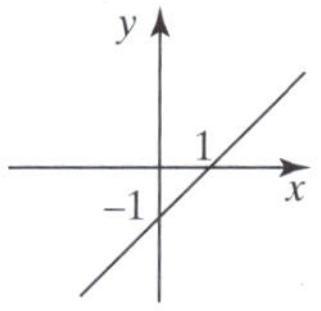

B
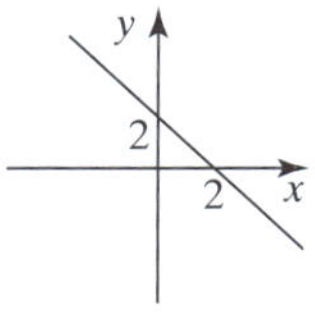

C
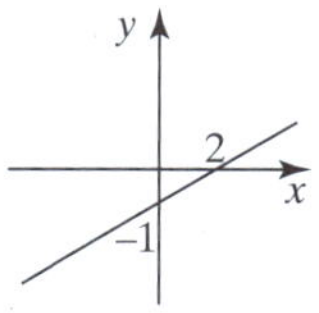

D
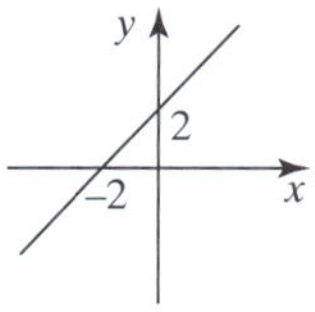

(1 mark)

3 Which equation does this table of values represent?

A $y = 3x - 2$ B $y = 2x + 3$

C $y = 3 - 2x$ D $y = x + 6$

x	–1	0	2	3
y	5	3	–1	–3

(1 mark)

Part B Short Answer

4 a Draw up a table of values, using $x = 1, 2, 3$ and 4, for the equation $y = 2x - 3$. (2 marks)

b Graph the line $y = 2x - 3$ on a number plane. *Hint 1* (2 marks)

c What is the y-intercept? (1 mark)

5 a Copy and complete the table of values for $y = \frac{1}{2}x - 2$ (2 marks)

x	–1	0	1
y			

b Graph the line $y = \frac{1}{2}x - 2$ (2 marks)

c What is the x-intercept of the line $y = \frac{1}{2}x - 2$? (1 mark)

d What is the gradient of the line $y = \frac{1}{2}x - 2$? (1 mark)

6 a If $x = 2$, for what value of y will $x + y = 5$? (1 mark)

b Solve the equation $-2 + y = 5$ (1 mark)

c Graph the line $x + y = 5$ on a number plane. *Hint 2* (2 marks)

7 a On the same diagram graph the lines $x = -2$, $x = 5$, $y = -1$ and $y = 6$. (2 marks)

b What is the area of the rectangle formed by the four lines in part **a**? (2 marks)

8 a On the same diagram sketch the lines $y = 3x - 2$ and $y = 4 - x$. *Hint 3* (2 marks)

b Write down the coordinates of the point where the two lines in part **a** intersect. (1 mark)

Hint 1: Plot each of the points from the table of values. The line should pass through all four points.

Hint 2: From parts **a** and **b** you know two points on the line. Plot those points and draw the line that passes through those two points.

Hint 3: Find two points that lie on each line. For example, what will be the y-values for each line when $x = 0$ and when $x = 1$.

Your Feedback

$\frac{\square}{25} \times 100\% = \square\%$

PAGES 150–151

PAGE 198

SKETCHING LINEAR GRAPHS

Linear Relationships

30 MINUTES

ADVANCED TEST

1 Determine whether it is true or false that each equation represents a straight line:

a $y = 8 - 3x$ (1 mark)

b $y = 5$ (1 mark)

c $2x + 9y - 7 = 0$ (1 mark)

d $5x = 2y + 1$ (1 mark)

e $y = \frac{x}{4}$ (1 mark)

f $x = 0$ (1 mark)

2 a On the same diagram, graph the lines $y = \frac{1}{2}x - 2$, $y = \frac{1}{2}x$, $y = \frac{1}{2}x + 1$ and $y = \frac{1}{2}x + 3$. (4 marks)

b Briefly comment on any similarities between the four lines drawn in part **a**. (1 mark)

c What is the effect on the graph of adding a number to the equation of a line? (1 mark)

3 Consider the equation $3x + 2y - 6 = 0$.

a Rearrange the equation $3x + 2y - 6 = 0$ into gradient/intercept form ($y = mx + b$). (1 mark)

b Write down the gradient (m) and y-intercept (b). (2 marks)

c Graph the line $3x + 2y - 6 = 0$ on a number plane. (2 marks)

d The line $3x + 2y - 6 = 0$ cuts the x-axis at Q. Write down the coordinates of Q. (1 mark)

4 For the line $3x - 2y + 12 = 0$:

a what is the x-intercept? (1 mark)

b what is the y-intercept? (1 mark)

c graph the line. (2 marks)

5 a Sketch the graphs of $y = 3x - 4$ and $y = -2x + 1$ on the same diagram. (2 marks)

b What are the coordinates of A, the point of intersection of the two lines? (1 mark)

c On the diagram sketch the line l that passes through A and also through the point (5, 3). (1 mark)

d What is the gradient of line l? (1 mark)

6 P and Q are the points (–3, 4) and (5, –2) respectively.

a Show the points P and Q on a diagram and the line on which they both lie. (1 mark)

b On the same diagram graph the line $4x - 3y - 1 = 0$. (1 mark)

c From the diagram, what is the geometrical significance of the line $4x - 3y - 1 = 0$? (1 mark)

Your Feedback

$\frac{\square}{30} \times 100\% = \square\%$

PAGES 151–152

PAGE 198

THE EQUATION OF A LINE

Linear Relationships

STUDY NOTES

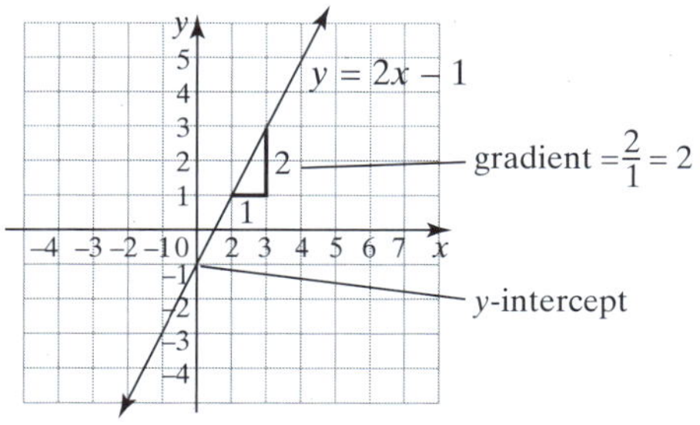

1 Any equation of the form $\boldsymbol{y = mx + b}$ is the **equation of a straight line** and all straight lines have an equation that can be expressed in that form. m will be the gradient of the line and b will be the y-intercept.

2 We say $\boldsymbol{y = mx + b}$ is the **gradient/intercept form** of the equation of a straight line.

3 The equation $\boldsymbol{Ax + By + C = 0}$ is the equation of a straight line in **general form**. This equation can be rearranged into the form $y = mx + b$.

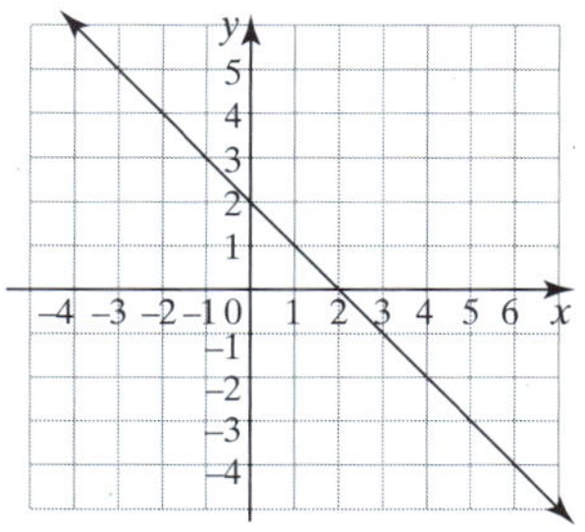

4 The gradient and y-intercept can be **read from a graph**. The **equation of the line** can then be formed. For example, the gradient of this line is –1. The y-intercept is 2. So the equation of the line is $y = -x + 2$ ($y = -1x + 2$).

5 To determine whether a **point lies on a line**, substitute the coordinates of the point into the equation of the line. If the coordinates **satisfy the equation**, the point does lie on the line.

6 The equation of a line can be found whenever we know its **gradient** and any **point** on the line. This can be done by substituting the gradient and the coordinates of the point into the equation $y = mx + b$ and solving to find the value of b.

7 The **point/gradient formula** is another way to find the equation of a line given its gradient and a point on the line. The equation of the line with gradient m and passing through the point (x_1, y_1) is given by the formula $y - y_1 = m(x - x_1)$.

8 The equation of a line can also be found whenever **two points** on the line are known. One way of finding this equation is to find the gradient and then use the point/gradient formula.

9 The **two-point formula** is another way of finding the equation of the line when two points on the line are known. The equation of the line passing through the points (x_1, y_1) and (x_2, y_2) is given by

$$\frac{y - y_1}{x - x_1} = \frac{y_2 - y_1}{x_2 - x_1}.$$

Checklist

Can you:

1 *Understand that y = mx + b is the equation of a straight line in gradient/intercept form?* ☐
2 *Understand that, in the equation y = mx + b, m is the gradient and b is the y-intercept?* ☐
3 *Understand what is meant by the general form of an equation of a line?* ☐
4 *Change the equation of a line in general form to the equation in gradient/intercept form?* ☐
5 *Change the equation of a line in gradient/intercept form to general form?* ☐
6 *Find the gradient, y-intercept and equation of a line from a graph?* ☐
7 *Determine whether a point lies on a line?* ☐
8 *Find the equation of a line given the gradient and a point on the line?* ☐
9 *Find the equation of a line given two points on the line?* ☐

THE EQUATION OF A LINE

Linear Relationships

SKILLS CHECK

1 Write down the gradient and y-intercept for each line:

a $y = 2x + 5$ **b** $y = 3x - 7$ **c** $y = -4x + 1$ **d** $y = x - 3$

e $y = -x$ **f** $y = 4 - 3x$ **g** $y = -\frac{3}{4}x - \frac{1}{2}$ **h** $y = \frac{x}{2} + 2$

2 Write down the equation of the line with gradient m and y-intercept b if:

a $m = 5, b = 1$ **b** $m = -2, b = 0$ **c** $m = \frac{2}{3}, b = -\frac{1}{3}$

3 For each line:

i

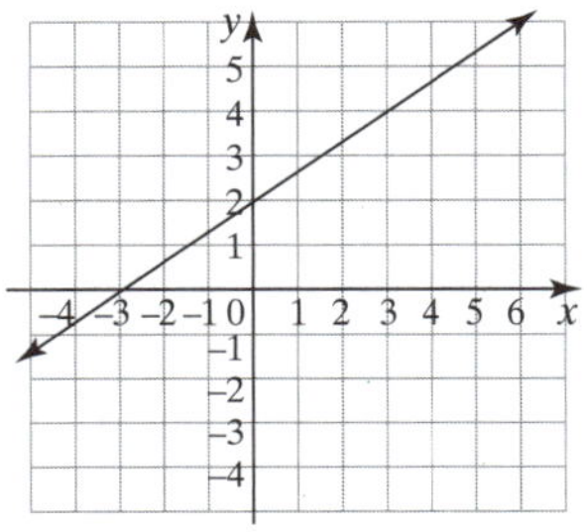

ii

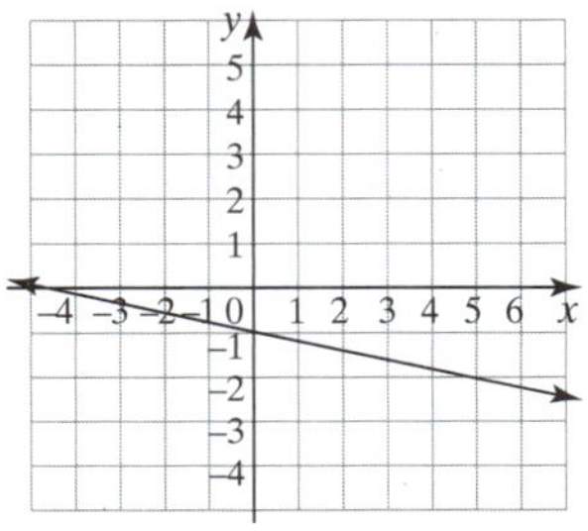

a find the gradient;

b what is the y-intercept?

c what is the equation of the line?

4 Rearrange each equation to the form $y = mx + b$:

a $x + y = 3$ **b** $x - y = 1$ **c** $x + y + 4 = 0$

d $3x - y + 5 = 0$ **e** $4x - 2y = 1$ **f** $x + 4y + 2 = 0$

5 Each point lies on the line $2x + y = 5$. True or false?

a $(0, 5)$ **b** $(-2, 1)$ **c** $(3, -1)$

6 Rearrange each equation into general form:

a $y = 4x + 1$ **b** $y = -3x + 5$ **c** $y = \frac{1}{2}x - 3$

d $y = \frac{2}{3}x + 4$ **e** $y = -\frac{4}{3}x + 2$ **f** $y = \frac{-x}{5} - 8$

7 Find b if $y = mx + b$ and:

a $x = 1, y = 5$ and $m = 3$ **b** $x = 6, y = -7$ and $m = -\frac{1}{2}$

PAGES 152–153

Answers **1 a** 2, 5 **b** 3, −7 **c** −4, 1 **d** 1, −3 **e** −1, 0 **f** −3, 4 **g** $-\frac{3}{4}, -\frac{1}{2}$ **h** $\frac{1}{2}$, 2 **2 a** $y = 5x + 1$ **b** $y = -2x$ **c** $y = \frac{2}{3}x - \frac{1}{3}$ **3 i a** $\frac{2}{3}$ **b** 2 **c** $y = \frac{2}{3}x + 2$ **ii a** $-\frac{1}{5}$ **b** −1 **c** $y = -\frac{1}{5}x - 1$ **4 a** $y = -x + 3$ **b** $y = x - 1$ **c** $y = -x - 4$ **d** $y = 3x + 5$ **e** $y = 2x - \frac{1}{2}$ **f** $y = -\frac{1}{4}x - \frac{1}{2}$ **5 a** true **b** false **c** true **6 a** $4x - y + 1 = 0$ **b** $3x + y - 5 = 0$ **c** $x - 2y - 6 = 0$ **d** $2x - 3y + 12 = 0$ **e** $4x + 3y - 6 = 0$ **f** $x + 5y + 40 = 0$ **7 a** 2 **b** −4

THE EQUATION OF A LINE

Linear Relationships

20 MINUTES

INTERMEDIATE TEST

Part A Multiple Choice

1 The x-intercept of the line $y = -3x + 6$ is:

A 6 B –3 C 2 D 0 (1 mark)

2 What is the equation of the line $y = \frac{2}{3}x - \frac{1}{2}$ in general form?

A $4x - 6y - 3 = 0$ B $4x - 6y + 3 = 0$ C $4x + 6y - 3 = 0$ D $4x + 6y + 3 = 0$ (1 mark)

3 What is the gradient of the line $2x + y - 3 = 0$? *Hint 1*

A $\frac{1}{2}$ B $-\frac{1}{2}$ C 2 D –2 (1 mark)

Part B Short Answer

4 Determine whether the line $y = -3x + 1$ passes through these points.

a (3, –10) (1 mark)
b (–2, 7) (1 mark)
c (5, –14) (1 mark)

5 Determine whether the point (–2, 5) lies on these lines.

a $y = 2x + 1$ (1 mark)
b $y = 3x - 1$ (1 mark)
c $y = -x - 7$ (1 mark)

6 The diagram shows the graph of two lines, k and l.

a Find the equation of line k. *Hint 2* (2 marks)
b Find the equation of line l. (2 marks)

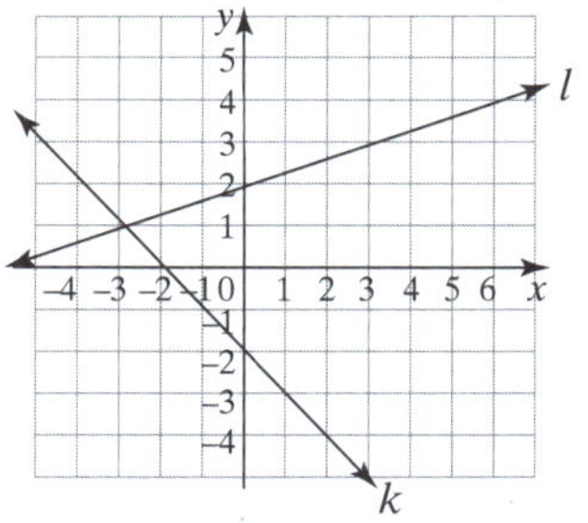

7 A line passes through the points $A(-2, 5)$ and $B(4, -1)$.

a Draw a set of axes, plot the points A and B and show the line that passes through these points. (2 marks)
b Find the gradient of line joining A to B. (1 mark)
c Where does this line meet the y-axis? (1 mark)
d Write down the equation of the line AB. *Hint 3* (1 mark)

8 Find the equation of the line passing through the given point with the given gradient.

a (4, 3); gradient –2 (1 mark)
b (–1, –2); gradient $\frac{2}{3}$ (1 mark)

Hint 1: Rearrange the equation into the form $y = mx + b$.
Hint 2: What is the y-intercept? What is the gradient?
Hint 3: Use the gradient from part **b** and the y-intercept from part **c**.

Your Feedback

$\frac{\square}{20} \times 100\% = \square\%$

PAGE 153

PAGE 198

THE EQUATION OF A LINE

Linear Relationships

30 MINUTES

ADVANCED TEST

1 a What is the y-intercept of the line $2x + y - 8 = 0$? (1 mark)

b What is the x-intercept of the line $2x + y - 8 = 0$? (1 mark)

c What is the area of the triangle bounded by the line $2x + y - 8 = 0$ and the x and y axes? (1 mark)

2 a What is the equation of the line with gradient –2 that passes through the point (0, 5)? (1 mark)

b What is the point of intersection of the line in part **a** and the line $x = 3$? (1 mark)

c What is the point of intersection of the line in part **a** and the line $y = -7$? (1 mark)

3 What is the equation of the line, in gradient/intercept form, passing through (–2, 7) and (5, –14)? (2 marks)

4 a What is the equation of the line, in general form, with gradient $\frac{2}{3}$ and passing through the point (3, –1)? (2 marks)

b Does the point (–4, –5) lie on the line in part **a**? Justify your answer. (2 marks)

5 What is the equation, in general form, of the line with x-intercept 3 and y-intercept 5? (2 marks)

6 a Write the equation $5x - 2y + 3 = 0$ in gradient/intercept form. (1 mark)

b What is the gradient of any line parallel to the line $5x - 2y + 3 = 0$? (1 mark)

c What is the equation, in general form, of the line through (–1, 5) parallel to $5x - 2y + 3 = 0$? (2 marks)

d What is the equation, in general form, of the line through (7, 1) parallel to $5x - 2y + 3 = 0$? (2 marks)

e Briefly comment on the similarities and differences between the equations of the three lines. (1 mark)

7 a Find the equation, in general form, of the line passing through the points (1, 5) and (–3, –1). (2 marks)

b Find the equation, in general form, of the line passing through the points (–1, 2) and (5, 11). (2 marks)

c What conclusion can be drawn about the points (1, 5), (–3, –1), (–1, 2) and (5, 11)? (1 mark)

8 $P(-1, 5)$, $Q(4, 5)$, $R(1, 1)$ and S(–4, 1) are the vertices of a rhombus.

a Find the equation, in gradient/intercept form, of diagonal PR. (1 mark)

b Find the equation, in gradient/intercept form, of diagonal QS. (1 mark)

c What is the y-intercept of each diagonal? What is the geometrical significance of this? (2 marks)

Your Feedback

$\frac{\square}{30} \times 100\% = \square\%$

PAGES 154–155

PAGE 198

NON-LINEAR GRAPHS

Non-linear Relationships

STUDY NOTES

1 A **parabola** is a special type of curve. Parabolic shapes can be seen in the environment, such as the arch of a bridge or the path of a ball.

2 The graph of $\mathbf{y = ax^2 + c}$ is a **parabola**. If $\mathbf{a > 0}$, the parabola **opens upwards** ($\cup$) but if $\mathbf{a < 0}$ the parabola **opens downwards** ($\cap$).

3 The **turning point** of a parabola is called the **vertex**.

4 The parabola $\mathbf{y = ax^2}$ has its **vertex at the origin**. The value of a determines the width of the parabola. The larger the value of a, the thinner the parabola.

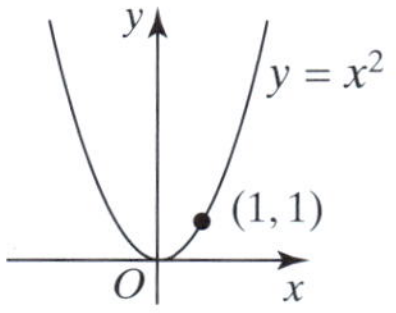

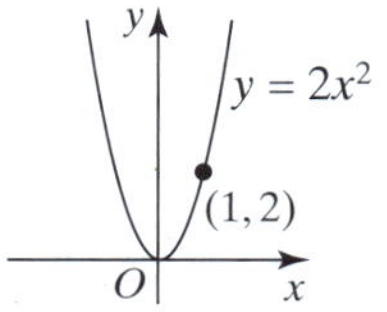

5 The **constant, *c*,** in the equation $y = ax^2 + c$ has the effect of moving the parabola up or down the y-axis. The **vertex** of the parabola $y = ax^2 + c$ is at the point $(0, c)$.

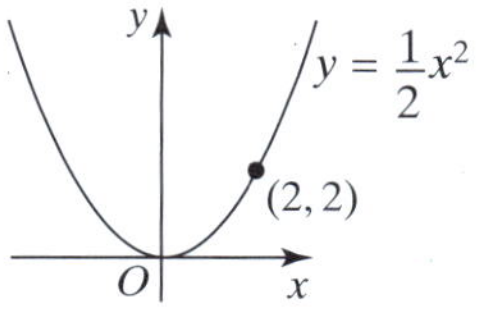

6 The ***x*-intercepts** of a parabola are the points where the curve meets the x-axis. A parabola may meet the x-axis in 0, 1 or 2 points. The x-intercepts are found when $y = 0$. The y-intercept is the point where the curve meets the y-axis. The ***y*-intercept** occurs when $x = 0$. If the equation of the parabola is of the form $y = ax^2 + c$ then the y-intercept will be c.

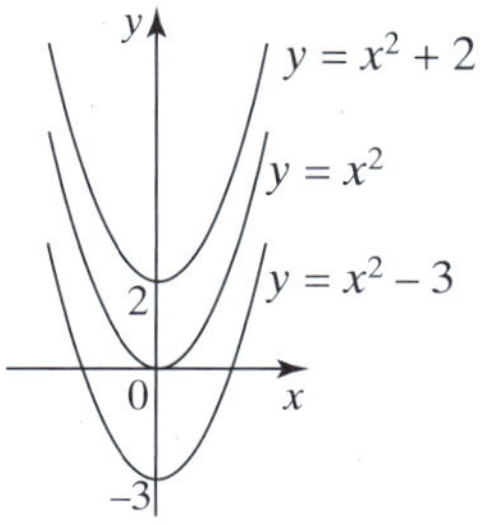

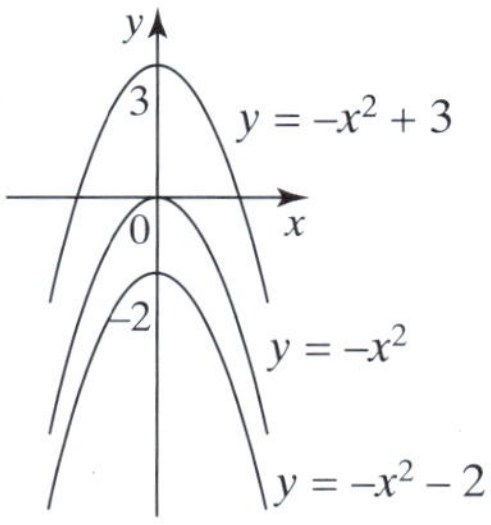

7 Any **equation** of the form $\mathbf{y = a^x}$ is an **exponential curve**. $y = a^x$ has y-intercept 1 for all positive values of a (because $a^0 = 1$). This type of exponential curve has no value of x that will make $y = 0$, so there is no x-intercept.

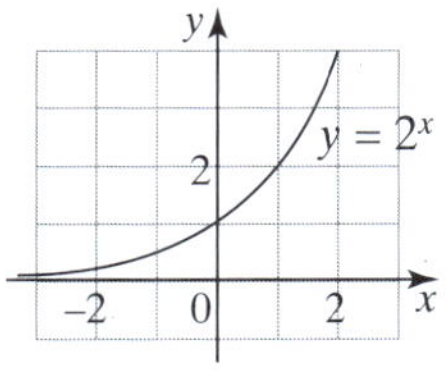

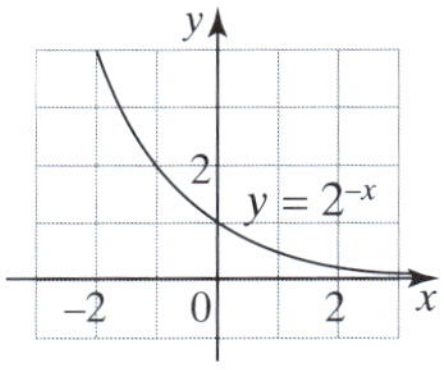

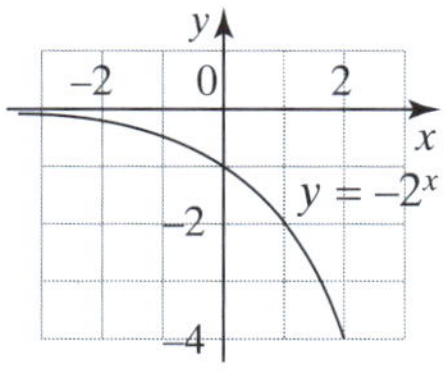

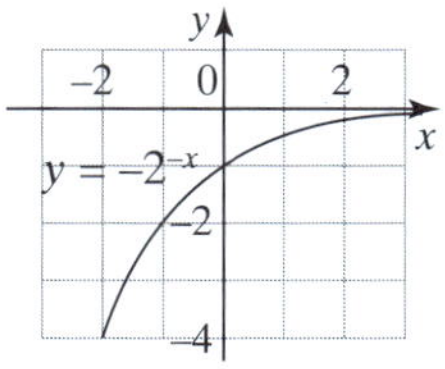

8 A **circle**, **centre the origin**, and **radius** of length r units has equation $x^2 + y^2 = r^2$.

9 Graphs of non-linear relations such as parabolas and exponential curves can be drawn after drawing up a **table of values**. **Digital technologies** such as graphics calculators and graphing software can be used to show the properties of these types of relationships.

Checklist

Can you:

1 *Recognise the graphs of parabolas?* ☐
2 *Draw the graphs of parabolas?* ☐
3 *Understand the significance of a and c in the equation $y = ax^2 + c$?* ☐
4 *Identify the type of graph from an equation?* ☐
5 *Graph parabolas, exponential curves and circles?* ☐
6 *Find the radius of a circle (centre the origin) given its equation?* ☐
7 *Use digital technologies to draw graphs?* ☐

NON-LINEAR GRAPHS

Non-linear Relationships

SKILLS CHECK

1 What name is given to the type of graph shown?

a
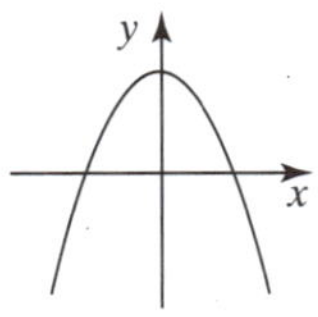

b
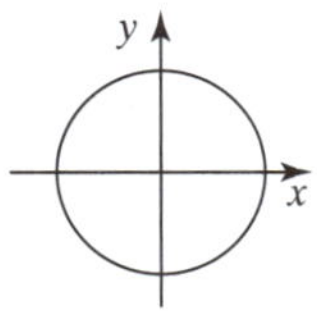

c
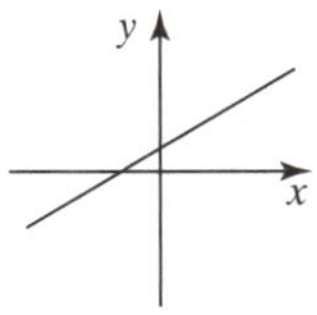

d
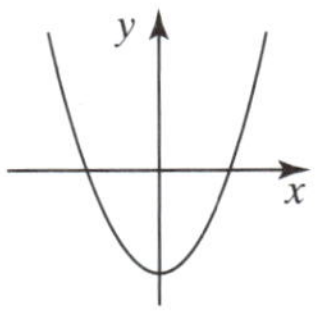

e
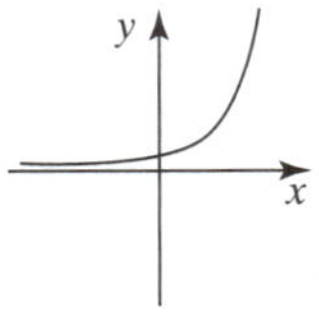

2 **a** Complete the table of values for $y = x^2 + 3$

x	−3	−2.5	−2	−1.5	−1	−0.5	0	0.5	1	1.5	2	2.5	3
y													

b Sketch a graph of $y = x^2 + 3$

3 Name the type of graph (parabola, exponential curve, straight line or circle) with equation:

a $y = 2 - x$ **b** $x^2 + y^2 = 9$ **c** $y = 2^x$ **d** $y = x^2$

e $y = \frac{x}{2}$ **f** $y = -x^2 + 2$ **g** $x + y = 9$ **h** $x^2 + y^2 = 2$

4 Choose the correct equation, from the list below, for each graph.

$y = 4x$, $y = 4x^2$, $x^2 + y^2 = 16$, $y = \frac{x}{4}$, $y = 4 - x$, $y = x^2 - 4$, $x^2 + y^2 = 8$, $y = \frac{-4}{x}$, $y = 4 - x^2$

a
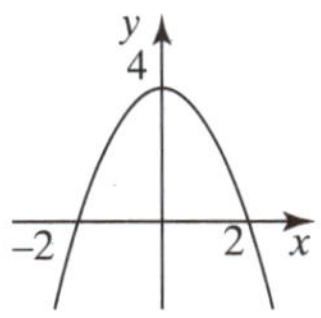

b
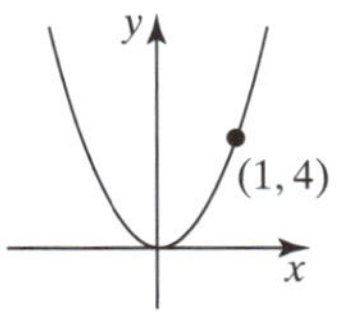

c
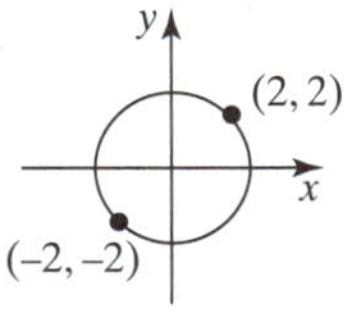

d
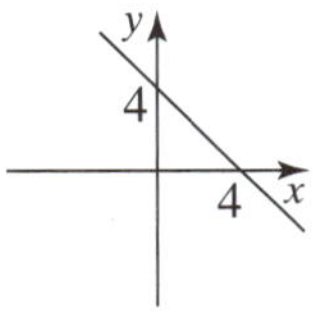

e
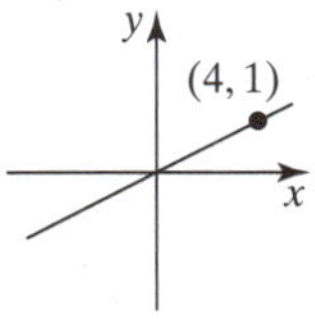

f
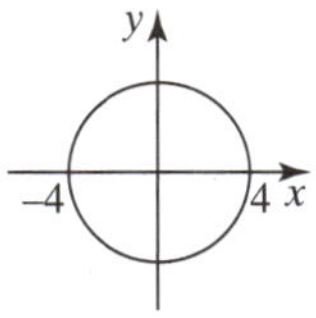

g
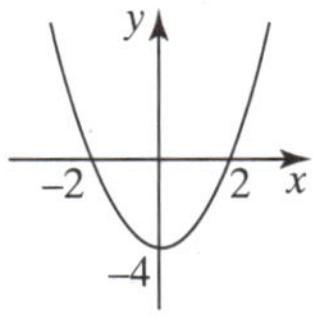

h
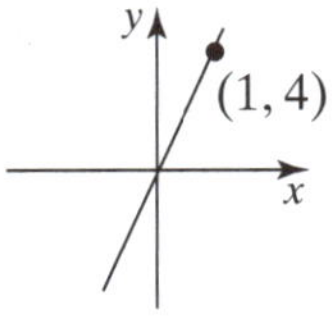

5 Sketch the graph of:

a $y = 2^{-x}$ **b** $y = x^2 - 1$ **c** $y = 2 - x^2$ **d** $x^2 + y^2 = 4$

PAGES 155–156

Answers **1 a** parabola **b** circle **c** straight line **d** parabola **e** exponential curve **2 a** 12, 9.25, 7, 5.25, 4, 3.25, 3, 3.25, 4, 5.25, 7, 9.25, 12 **b** see page 155 **3 a** straight line **b** circle **c** exponential curve **d** parabola **e** straight line **f** parabola **g** straight line **h** circle **4 a** $y = 4 - x^2$ **b** $y = 4x^2$ **c** $x^2 + y^2 = 8$ **d** $y = 4 - x$ **e** $y = \frac{x}{4}$ **f** $x^2 + y^2 = 16$ **g** $y = x^2 - 4$ **h** $y = 4x$ **5** see page 156

Part A Multiple Choice

1 Which curve does NOT pass through the point (3, 4)?

A $x^2 + y^2 = 25$ **B** $y = 3x - 5$ **C** $y = (x - 1)^2$ **D** $y = 2^x$ (1 mark)

2 Which diagram could show the graph of $y = -2^x$? (1 mark)

A
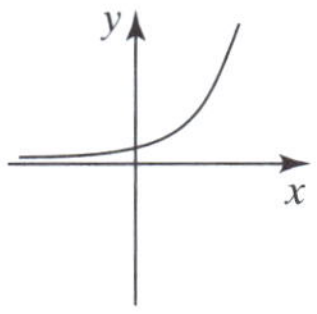

B
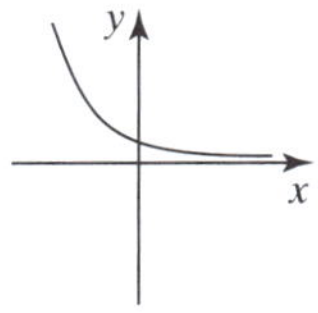

C
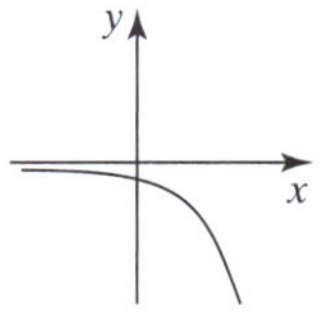

D
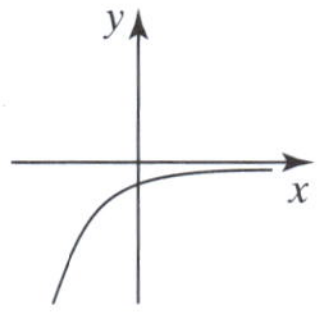

3 Which point lies on the parabola $y = 2x^2 - 5$?

A (0, –3) **B** (–1, –7) **C** (2, 11) **D** (5, 45) (1 mark)

4 Which point does NOT lie on the circle $x^2 + y^2 = 25$?

A (0, –5) **B** (–3, –4) **C** (5, 0) **D** (–2, 3) (1 mark)

Part B Short Answer

5 **a** On the same set of axes sketch the graphs of $y = x^2$, $y = x^2 - 3$, $y = x^2 + 1$ and $y = x^2 + 2$ for values of x between –4 and 4. (4 marks)

b Briefly explain the effect of c on the graph of $y = x^2 + c$. *Hint 1* (2 marks)

6 The diagram shows a sketch of a parabola.
The equation of the parabola is of the form $y = ax^2 + c$.

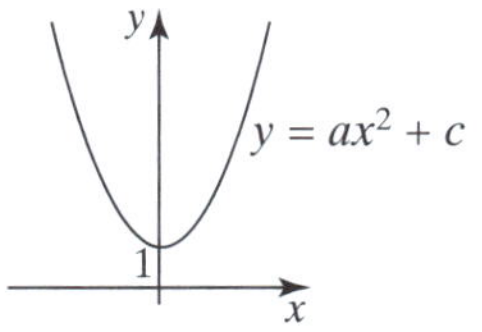

a What is the value of the constant c? (1 mark)

b The graph passes through the point (2, 3).
What is the value of a? *Hint 2* (2 marks)

c What is the value of y when $x = -4$? *Hint 3* (2 marks)

7 For the parabola $y = 9 - x^2$:

a find the values of x for which $y = 0$. *Hint 4* (2 marks)

b What is the largest possible value of y? Briefly justify your answer. *Hint 5* (2 marks)

c Draw a sketch of the graph of $y = 9 - x^2$. *Hint 6* (3 marks)

8 The diagram shows the points $O(0, 0)$ and $P(-8, 6)$.

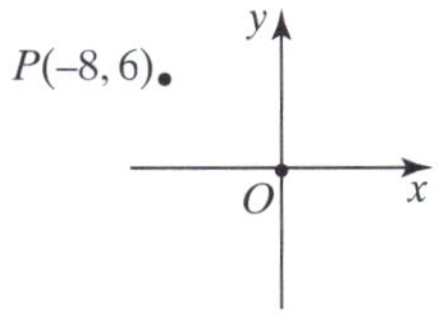

a Use Pythagoras' Theorem to find the length of the interval joining O and P. (1 mark)

b Write down the equation of the circle which has its centre at O and which passes through P. *Hint 7* (2 marks)

Hint 1: Look at the graphs in part **a**. What was the effect of the added constant in each equation?
Hint 2: Put the value of c into the equation of the parabola. Substitute $x = 2$ and $y = 3$ and solve an equation.
Hint 3: Put the values of both a and c into the equation of the parabola.
Hint 4: What values of x, when squared, equal 9. There will be two answers.
Hint 5: x^2 is never negative.
Hint 6: Only a sketch is required. There is no need to draw up a table of values. Use your answers from parts **a** and **b**.
Hint 7: The radius of the circle will equal the length of the interval joining O to P.

Your Feedback

$\frac{\square}{25} \times 100\% = \square\%$

PAGES 156–157

PAGE 198

NON-LINEAR GRAPHS

Non-linear Relationships

30 MINUTES

ADVANCED TEST

1 Consider the parabola $y = 2x^2 - 3$.

- **a** What is the y-intercept? (1 mark)
- **b** For what value(s) of y does $x = -2$? (1 mark)
- **c** For what value(s) of x does $y = 15$? (2 marks)
- **d** Briefly explain the main feature that is different between the graph of $y = x^2 - 3$ and $y = 2x^2 - 3$. (1 mark)

2 Without sketching, describe the curve represented by the equation $x^2 + y^2 = 12\frac{1}{4}$. (2 marks)

3 Consider the curve $y = 3^x$.

- **a** What is the value of y when $x = 0$? (1 mark)
- **b** What happens to the y-values as x gets large? (1 mark)
- **c** On separate diagrams sketch the graphs of $y = 3^x$, $y = 3^{-x}$, $y = -3^x$ and $y = -3^{-x}$. (4 marks)
- **d** Briefly describe the similarities and differences between the curves. (1 mark)

4 A circle, centre the origin, passes through the point $P(3, -5)$.

- **a** Find the equation of the circle. (2 marks)
- **b** Q is the point (6, 0). Does Q lie inside, on, or outside the circle? Justify your answer. (2 marks)

5 The diagram shows the graph of a parabola.

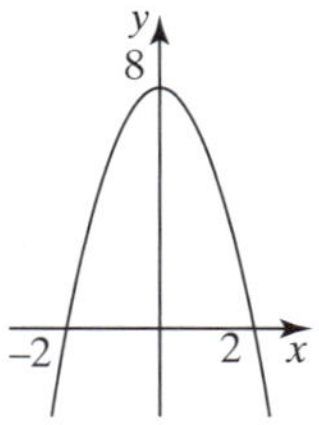

- **a** What is the equation of the parabola? (3 marks)
- **b** Copy the diagram and sketch the graph of $y = x^2 - 4$ on the same diagram. (2 marks)
- **c** Briefly comment on any similarities and differences between the two curves. (2 marks)

6

- **a** Does the parabola $y = x^2 + 5$ open upwards or downwards? (1 mark)
- **b** What, if any, is the solution to the equation $x^2 + 5 = 2$? (1 mark)
- **c** By considering the curve $y = x^2 + 5$ explain the significance of the answer to part **b**. (1 mark)

7 Without sketching, briefly explain the effect of adding a constant (positive or negative) to the equation of a curve. (2 marks)

Your Feedback

$\frac{\square}{30} \times 100\% = \square\,\%$

PAGE 157

PAGE 198

AREA OF COMPOSITE SHAPES

Area and Surface Area

STUDY NOTES

1 **Area** is the measurement of the extent of a plane surface. Area is measured in **square units** (m^2, cm^2, mm^2, km^2, ha). One hectare = 10 000 m^2.

2 The **area of a rectangle** is found by multiplying its length by its breadth ($A = lb$). A **square** is a special rectangle where the length and breadth are equal. To find the area of a square we can use the formula for the area of a rectangle or use $A = s^2$ where s is the length of the side.

3 The **area of a triangle** is half the area of a rectangle with the same base and height. The area is found by multiplying half the base by the perpendicular height ($A = \frac{1}{2}bh$).

4 The **area of a parallelogram** can be found by multiplying the base by the perpendicular height ($A = bh$).

5 The **area of a rhombus** (or kite) can be found by halving the product of the diagonals ($A = \frac{1}{2}xy$). This is because the diagonals of a rhombus meet at right angles. Don't forget that a rhombus is a parallelogram. The area can also be found by using the formula for the area of a parallelogram.

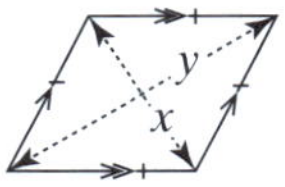

6 To find the **area of a trapezium**, multiply half the sum of the parallel sides by the perpendicular height ($A = \frac{1}{2}h(a + b)$). h must be the perpendicular height, the distance between the parallel sides.

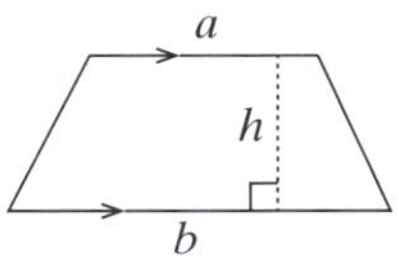

7 The **area of a circle** is given by $A = \pi r^2$, where r is the radius.

8 A **sector** is part of a **circle**. It is a piece bounded by two radii and part of the circumference. (Radii is the plural of radius.) The **area of a sector** of a circle is just a fraction of the area of the whole circle. For example, if the angle at the centre of the sector is 50° then the area will be $\frac{50}{360} \times \pi r^2$ where r is the radius.

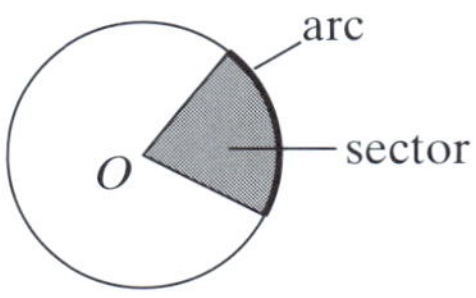

A **quadrant** is one-quarter of a circle. It is a sector with the angle at the centre equal to 90°.

9 A **composite shape** is made up of two or more other regular shapes. For example, figure I is made up of two rectangles. Figure II is a rectangle plus two semicircles.

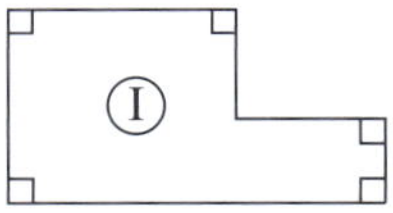

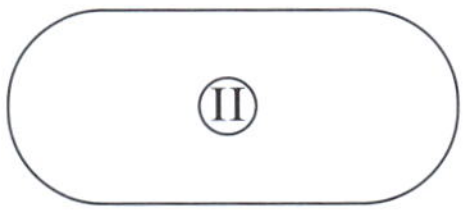

10 The **area of a composite shape** can be found by **adding** the areas of each of the regular figures that make up the shape. For example, the area of this shape is found by adding the area of a triangle to the area of a semicircle.

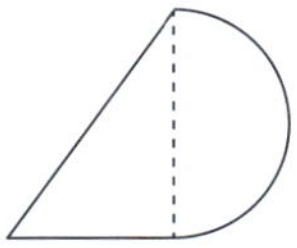

11 An **area between two shapes** can be found by **subtracting** the smaller area from the larger area. For example, in this diagram the shaded area is equal to the area of a circle minus the area of a rectangle.

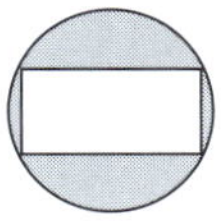

12 An **annulus** is the shape of a ring. It is the space between two circles with common centre. To find the area of an annulus subtract the area of the smaller circle from the larger.

Checklist

Can you:

1 *Understand what is meant by area?* ☐
2 *Find areas of triangles, quadrilaterals, circles and parts of circles?* ☐
3 *Understand what is meant by a composite shape and find the area of composite shapes?* ☐
4 *Find the area between two shapes by subtracting one area from another?* ☐

AREA OF COMPOSITE SHAPES

Area and Surface Area

SKILLS CHECK

1 Find the area of each figure:

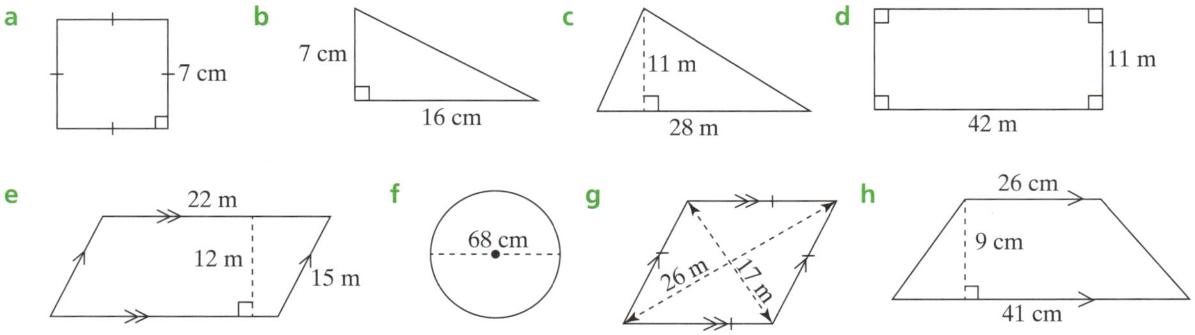

2 Find the area:

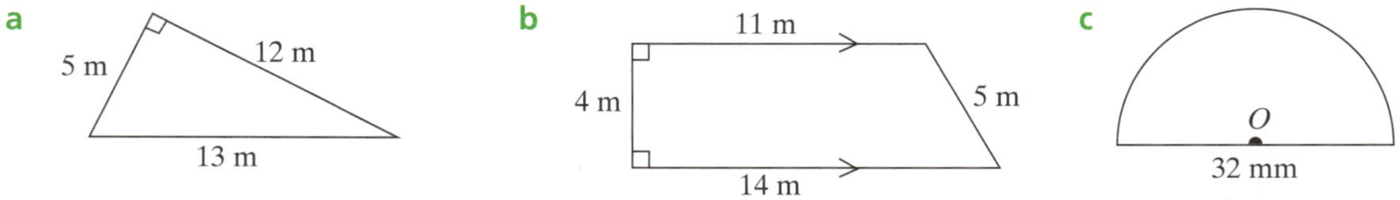

3 Find the area by dividing into two regular shapes along the dotted line.

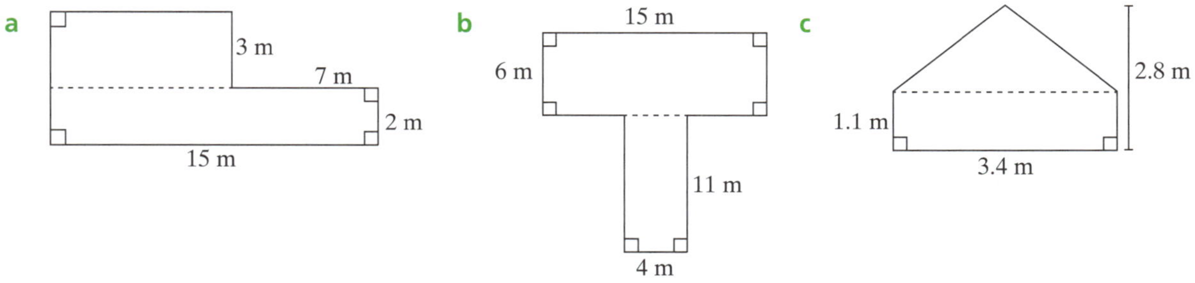

4 Find the shaded area by subtracting the area of one regular shape from the area of another:

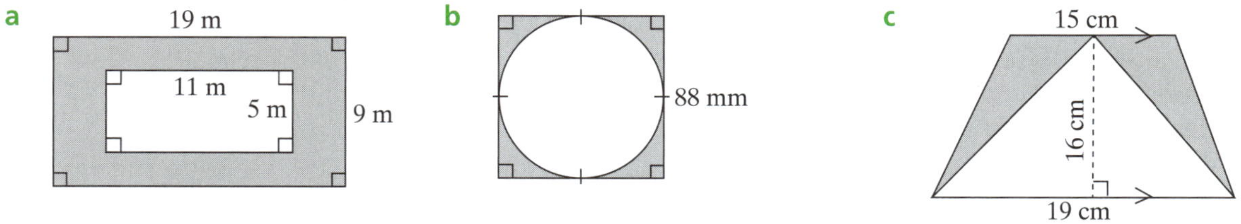

5 **a** What fraction of this circle is shaded? (*O* is the centre.)

b Find the area of the shaded sector, given that the radius is 9 cm. (Give the answer correct to one decimal place.)

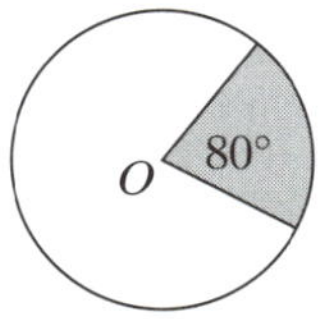

PAGES 157–158

Answers **1 a** 49 cm² **b** 56 cm² **c** 154 m² **d** 462 m² **e** 264 m² **f** 3631.7 cm² (1 d.p.) **g** 221 m² **h** 301.5 cm² **2 a** 30 m² **b** 50 m² **c** 402.1 mm² (1 d.p.) **3 a** 54 m² **b** 134 m² **c** 6.63 m² **4 a** 116 m² **b** 1661.9 mm² (1 d.p.) **c** 120 cm² **5 a** $\frac{2}{9}$ **b** 56.5 cm²

AREA OF COMPOSITE SHAPES

Area and Surface Area

20 MINUTES

INTERMEDIATE TEST

Part A Multiple Choice

1 Find the area of this figure.

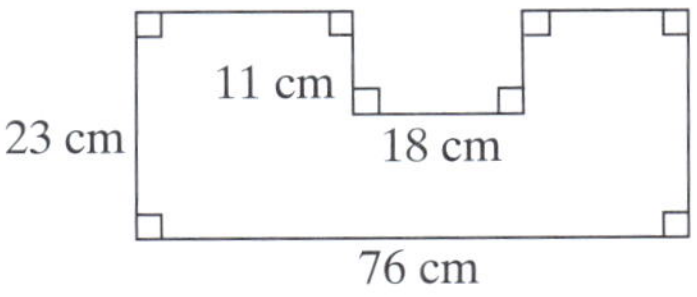

A 1115 cm^2 **B** 1478 cm^2
C 1550 cm^2 **D** 1874 cm^2 (1 mark)

2 What is the area of this paddock in hectares?

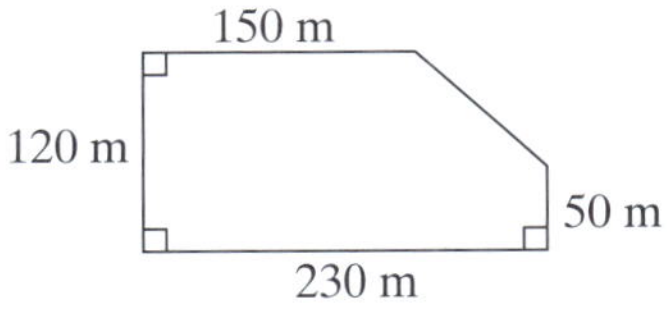

A 2.48 ha **B** 2.2 ha
C 2.07 ha **D** 2.385 ha (1 mark)

3 The area of the shaded section of this circle is 33.5 cm^2.

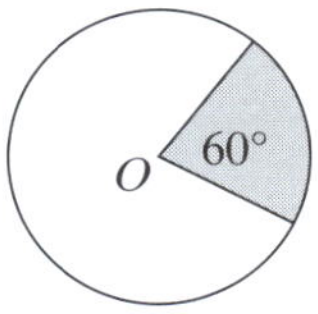

What is the radius?

A 11 cm **B** 3 cm
C 6 cm **D** 8 cm (1 mark)

Part B Short Answer

4 A swimming pool is 8 m long and 3 m wide. A fence is to be built around the pool, 1 m from the pool on three sides and 2 m on the fourth side, as shown in the diagram. The area between the pool and the fence is to be tiled. Find the cost of tiles for this area if they cost \$23.00 per square metre. *Hint 1*

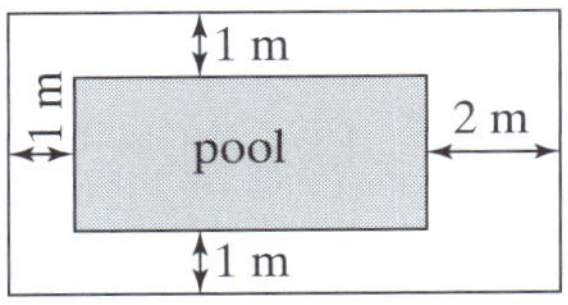

(4 marks)

5 This shape is made up of a quadrant (a quarter of a circle) and a triangle. Find its area, to the nearest square metre. *Hint 2*

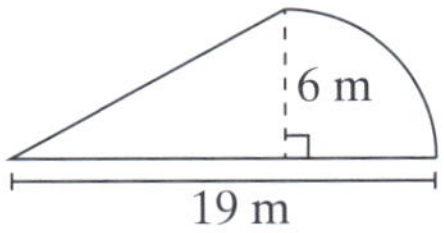

(3 marks)

6 The diagram shows the end wall of a shed. What is its area? *Hint 3*

(3 marks)

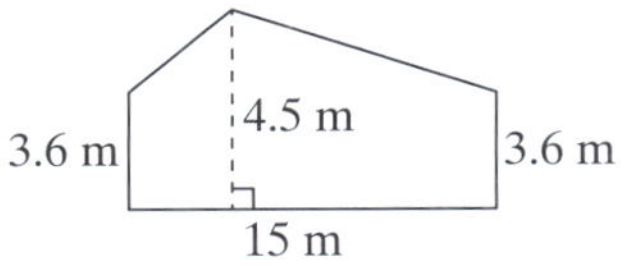

7 Find the area of this annulus. Give the answer to two decimal places.

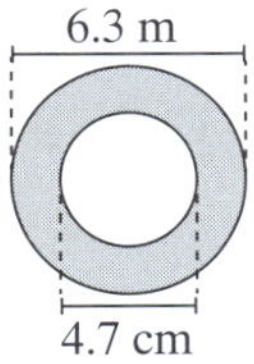

(3 marks)

8 The area of this shape can be found by each of these calculations. Briefly explain why.

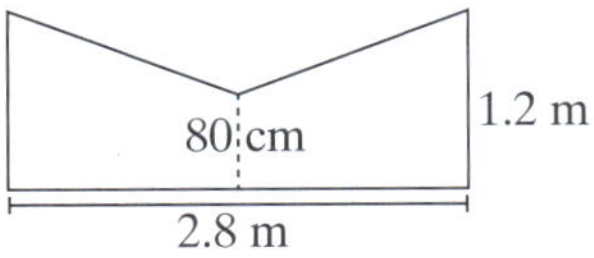

a $2.8 \times 1.2 - \frac{1}{2} \times 2.8 \times 0.4$ (2 marks)

b $2 \times \frac{1}{2} \times 1.4 \times (1.2 + 0.8)$ (2 marks)

Hint 1: Add the length of the appropriate gaps to the length and breadth of the pool to get the length and breadth of the fence.
Hint 2: Because you know the length of the radius you can find the length of the base of the triangle.
Hint 3: Divide into a rectangle and triangle.

Your Feedback

$\frac{\square}{20} \times 100\% = \square\%$

PAGES 158–159
PAGE 198

AREA OF COMPOSITE SHAPES

Area and Surface Area

25 MINUTES

ADVANCED TEST

1 Find the area of this kite.

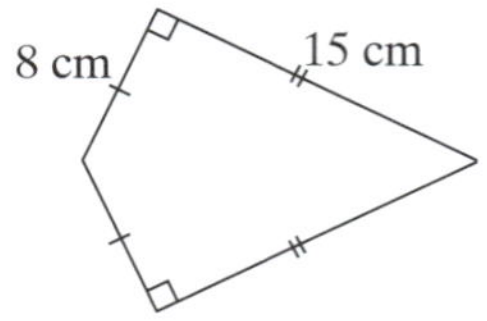

(2 marks)

2 The diagram shows a quadrant.

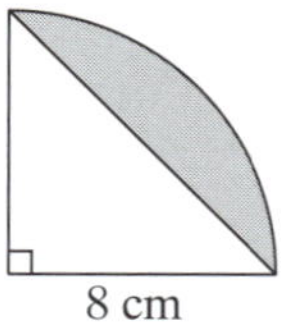

What percentage of the quadrant is shaded? (3 marks)

3 Heron's rule can be used to find the area of a triangle with sides a, b and c.

It states that $A = \sqrt{s(s-a)(s-b)(s-c)}$ where $s = \dfrac{a+b+c}{2}$.

a Use Heron's rule to find the exact area of an equilateral triangle with sides of 6 cm. (1 mark)

b Find the area of a regular hexagon with sides 6 cm long. Give the answer to the nearest square centimetre. (1 mark)

4 This shape is made up of 1 large semicircle with diameter 14 cm, and 2 smaller semicircles of equal size.

Find the area. (3 marks)

5 The perimeter of this figure (not drawn to scale) is 37 cm.

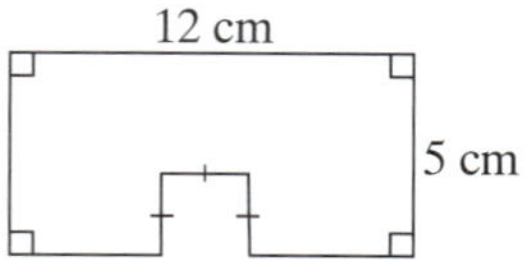

Find its area. (2 marks)

6 A window is in the shape of a square with a semicircle on top.

Find the area of the window. (3 marks)

7 The area of an annulus is 134.3 cm^2. If the radius of the outer circle is 7 cm, find the radius of the inner circle. (2 marks)

8 This shape is made up of a sector and semicircle. The radius of the semicircle is 6 cm.

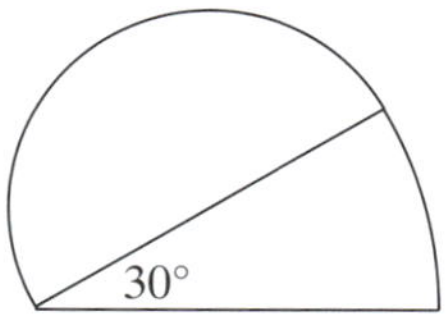

Find the area of the shape. (3 marks)

9 Find the area of this shape.

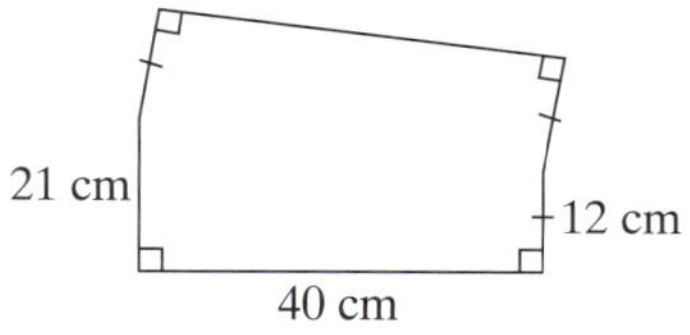

(2 marks)

10 A house is 14 m long and 7.2 m wide. It is to have a verandah, 2.4 m wide, on all sides. The floorboards are 4.6 m long and 96 mm wide. How many of these boards will be needed for the verandah? (3 marks)

Your Feedback

$\dfrac{\square}{25} \times 100\% = \square\ \%$

PAGES 159–161

PAGE 198

SURFACE AREA OF PRISMS

Right Prisms

STUDY NOTES

1 The **surface area** of a solid is the **sum of the areas of all its faces**.

2 Surface area is measured in **square units** (cm^2, m^2, km^2 and so on).

3 A **prism** is a three-dimensional figure with constant cross-section. (If you cut a prism parallel to the base, the slice will be exactly the same as the base.) The type of prism is determined by the base. A rectangular prism has a rectangular base (or cross-section). A triangular prism has a triangular base and so on.

4 The **net of a prism** is the two-dimensional shape that can be folded to form the prism.

5 A **cube** has six congruent faces so the surface area of a cube is $A = 6s^2$ where s is the length of a side.

6 The **surface area of a rectangular prism** of length l, breadth b and height h is given by $A = 2(lb + lh + bh)$. It doesn't matter which dimension is taken to be the length or breadth or height.

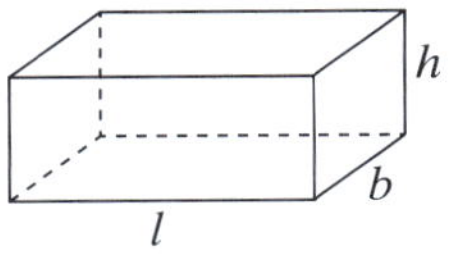

7 The **surface area of other prisms** is found by adding the area of all the faces. For any prism at least two faces will have the same area.

8 For any **right prism**, there are two faces that are the same (the base or cross-section), and all the other faces will be rectangles. The total surface area of these rectangles can be found by multiplying the **perimeter** of the base by the height of the prism.

9 When finding the **surface area of a triangular prism**, we will often need to use **Pythagoras' Theorem** to find the length of one of the sides, or the height of the triangular base. Triangular prisms have two (matching) triangular faces and three rectangular faces. For example, in this diagram two rectangular faces are 17 m by 9 m and one is 16 m by 9 m. To find the area of each triangular face we need the perpendicular height of the triangle. Because the triangle is isosceles, the perpendicular height will be found down the centre of the triangle.

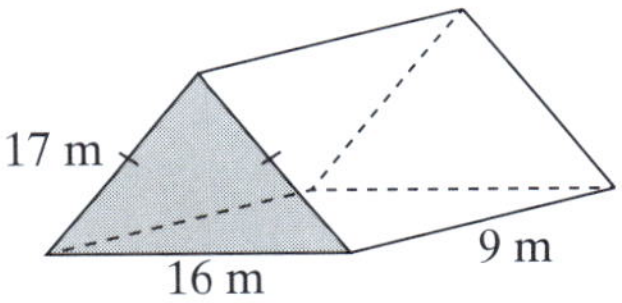

By Pythagoras' Theorem: $h^2 = 17^2 - 8^2$

$= 225$

$h = 15$

The perpendicular height is 15 m.

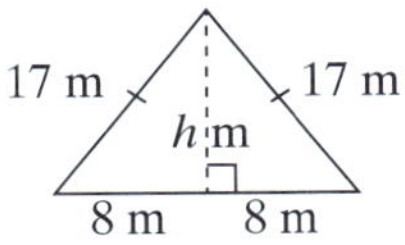

10 When **answering problems related to surface area**, always consider the practical aspect of the question. If you are finding the area of a room that needs to be painted, for example, you are not likely to need the area of the floor. Read the question carefully. Don't just apply the formula for the surface area of a rectangular prism.

Checklist

Can you:

1 *Understand what is meant by surface area?* ☐
2 *Know and understand the units associated with surface area?* ☐
3 *Identify prisms from their nets and draw the net of rectangular and triangular prisms?* ☐
4 *Find the surface area of a cube, rectangular prism or triangular prism?* ☐
5 *Find the surface area of other prisms?* ☐
6 *Solve problems involving surface area?* ☐

SURFACE AREA OF PRISMS

Right Prisms

SKILLS CHECK

1 How many faces has a:

a cube? **b** rectangular prism? **c** triangular prism?

2 **a** Find the area of the shaded face of the rectangular prism.

i

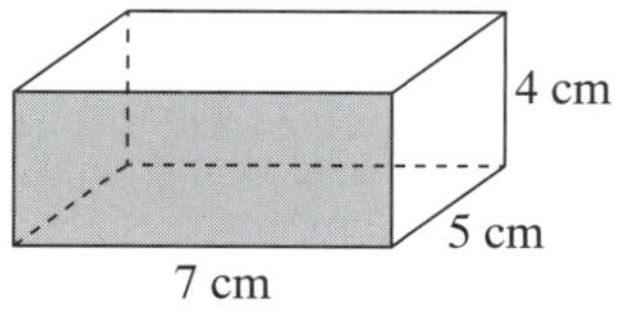

ii

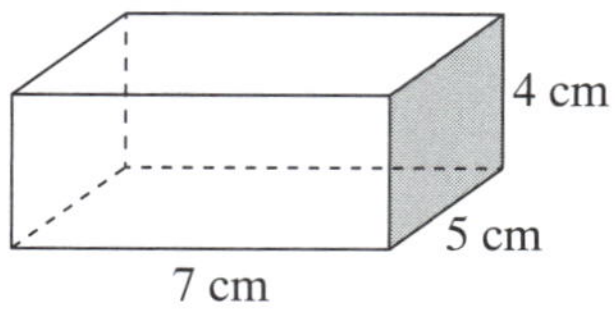

iii

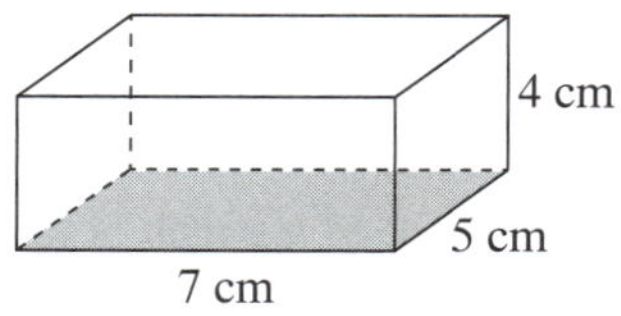

b Find the surface area of a rectangular prism which is 7 cm long, 5 cm wide and 4 cm high.

3 Of what type of prism is each net?

a

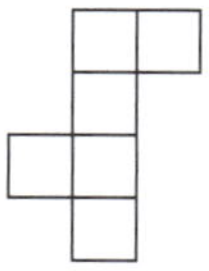

b

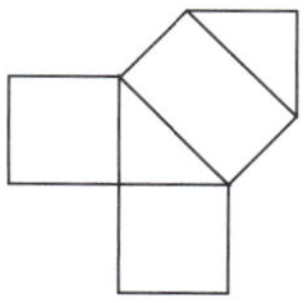

c

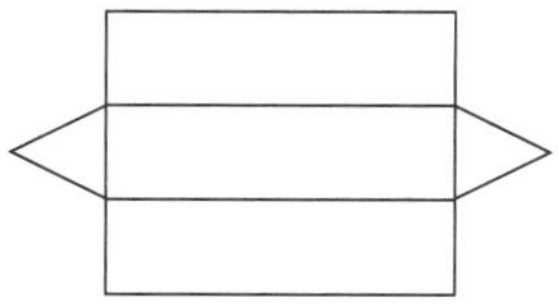

4 This is the net of a rectangular prism. What is the surface area of the prism?

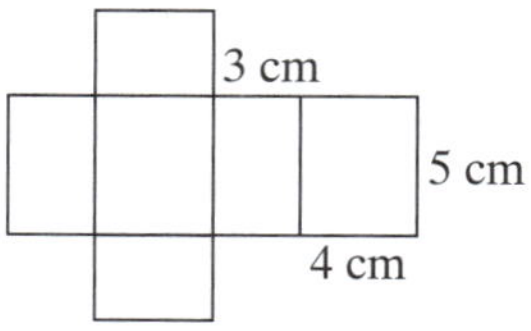

5 Find the surface area of a cube of side length 3 m.

6 Find the surface area of each prism:

a

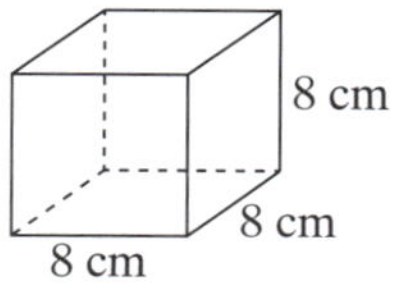

b

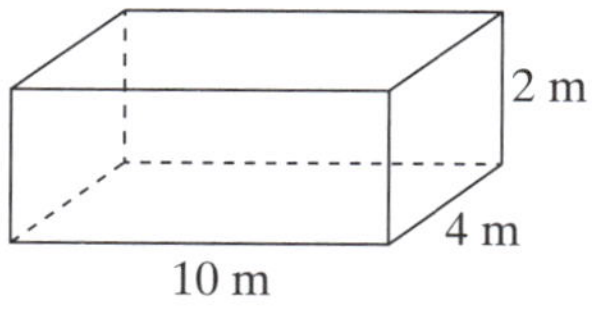

c

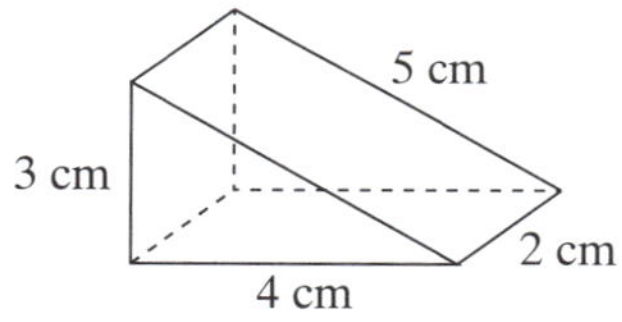

7 Use Pythagoras' Theorem to find the value of x.

a

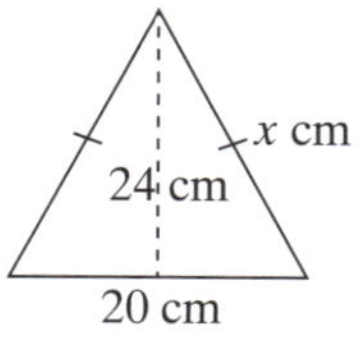

b

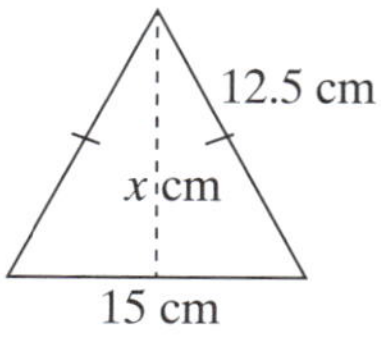

8 The diagram shows a hexagonal prism. The cross-section is a regular hexagon of side length 5 cm.

a How many hexagonal faces does the prism have?

b How many rectangular faces does the prism have?

c What is the area of each of the rectangular faces of the prism?

d What is the total surface area of the prism?

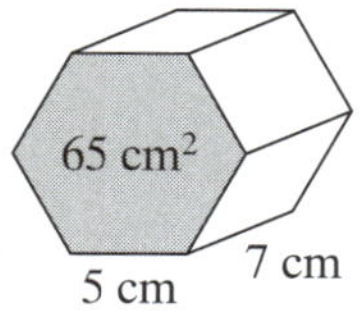

PAGES 161–162

Answers **1 a** 6 **b** 6 **c** 5 **2 a i** 28 cm² **ii** 20 cm² **iii** 35 cm² **b** 166 cm² **3 a** cube **b** triangular prism **c** triangular prism **4** 94 cm² **5** 54 m² **6 a** 384 cm² **b** 136 m² **c** 36 cm² **7 a** 26 **b** 10 **8 a** 2 **b** 6 **c** 35 cm² **d** 340 cm²

SURFACE AREA OF PRISMS

Right Prisms

25 MINUTES

INTERMEDIATE TEST

Part A Multiple Choice

1 A rectangular prism is 7.5 m long, 4.5 m wide and 1.5 m high. What is its total surface area?
A 50.625 m^2 **B** 54 m^2 **C** 101.25 m^2 **D** 103.5 m^2 (1 mark)

2 Which has the greatest surface area? (1 mark)

A

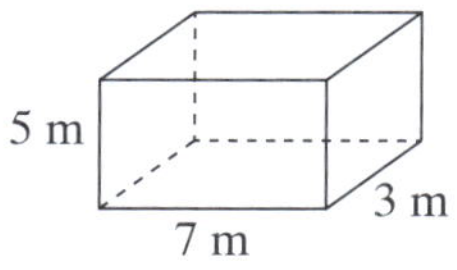

B

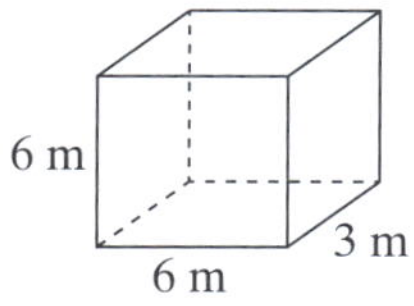

C

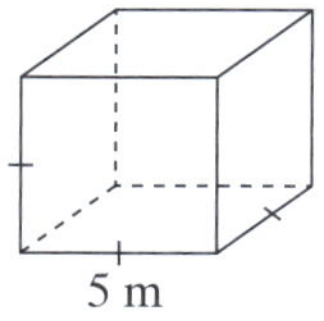

D

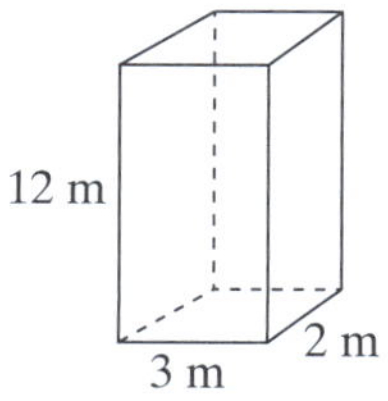

3 A wedge is in the shape of a triangular prism. What is the surface area of the wedge?
A 150 cm^2 **B** 180 cm^2
C 185 cm^2 **D** 276 cm^2 (1 mark)

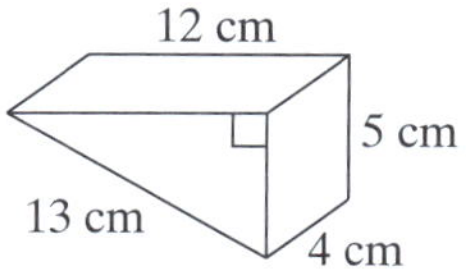

Part B Short Answer

4 A cube has a surface area of 384 cm^2. What length are the sides of the cube? *Hint 1* (2 marks)

5 **a** Find the area of the shaded face of the triangular prism. (1 mark)
b Find the length of the equal sides of the triangle. *Hint 2* (2 marks)
c Find the surface area of the triangular prism. (3 marks)

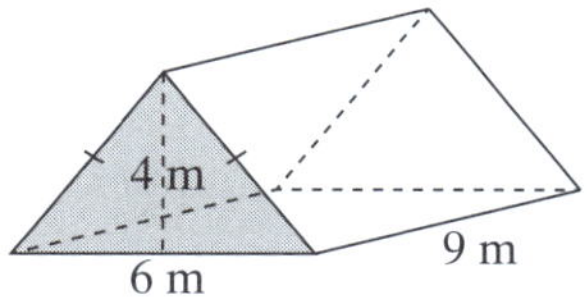

6 The four walls and ceiling of a room of a house are to be painted. The room is 5 m long and 3 m wide. The ceiling is at a height of 2.4 m above the floor.
a Find the total surface area of the four walls and ceiling. *Hint 3* (2 marks)
b One wall of the room has a window which is 1.5 m long and 1 m high and there is one door which is 85 cm wide and 2 m high. If neither the door nor window are to be painted, find the area that needs to be painted. *Hint 4* (3 marks)
c If the walls and ceiling are to receive two coats of paint and if one litre of paint covers an area of 15 m^2, find the amount of paint needed to paint the room. (2 marks)

7 The diagram shows the shell of a swimming pool.
a Find the area of the shaded face. *Hint 5* (2 marks)
b Find the total area of the four walls of the pool. (2 marks)

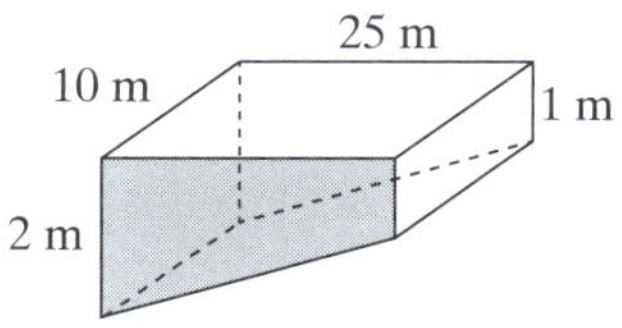

8 The diagram shows a hexagonal prism.
a What is the area of the hexagonal face? *Hint 6* (1 mark)
b What is the surface area of the prism? *Hint 7* (2 marks)

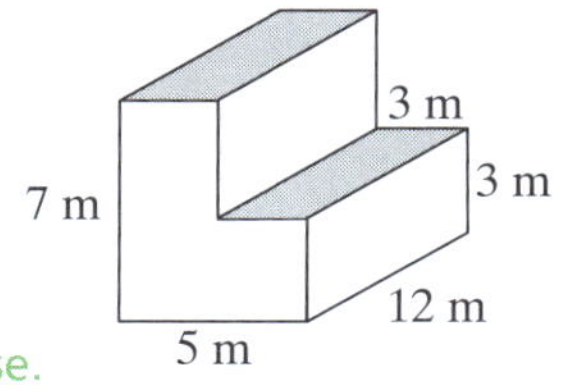

Hint 1: Form an equation and solve.
Hint 2: Use Pythagoras' Theorem. Don't forget to halve the base.
Hint 3: The floor is not counted.
Hint 4: The width of the door is given in centimetres where all other measurements are in metres.
Hint 5: It is a trapezium.
Hint 6: Divide it into two rectangles.
Hint 7: First find the perimeter of the hexagon.

Your Feedback

$\frac{\square}{25} \times 100\% = \square\%$

PAGES 162–163

PAGE 198

SURFACE AREA OF PRISMS

Right Prisms

25 MINUTES

ADVANCED TEST

1 **a** Find the surface area of a cube of side length 5 cm. (1 mark)
b How much longer would each side need to be if the surface area is to double? (1 mark)

2 Find the surface area, in square metres, of each prism.

a

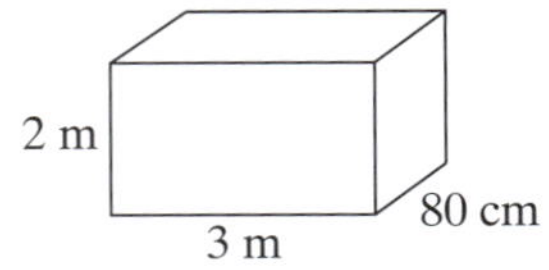

b

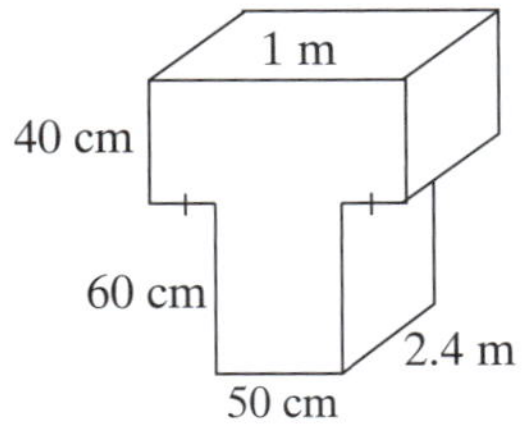

c

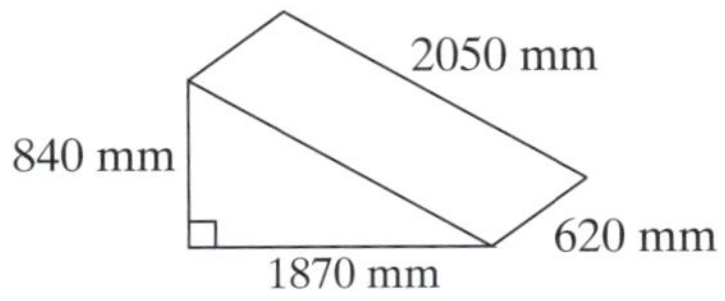

(2 marks each)

3 **a** Find the surface area of a rectangular prism with dimensions 5 cm, 4 cm and 3 cm. (1 mark)
b If each of the dimensions doubles, what will the surface area be? (1 mark)
c What happens to the surface area of any rectangular prism if each dimension doubles? (1 mark)
d Find the surface area of the rectangular prism that has each of its dimensions 3 times larger than those in part **a**. (1 mark)
e What happens to the surface area of a rectangular prism if each dimension is multiplied by 3? (1 mark)
f Explain what will happen to the surface area of a rectangular prism if each dimension is multiplied by the same number. (1 mark)
g Is the result in part **e** true for all prisms? (1 mark)

4 The area of the shaded face of this prism is 4489 cm^2.
What is the surface area of the prism? (2 marks)

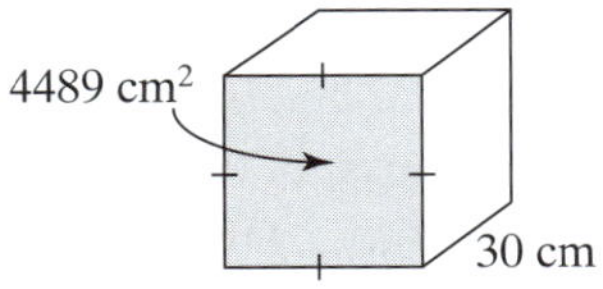

5 **a** Find the area of this pentagon. (2 marks)
b Find the length of the equal sides. (1 mark)
c If the pentagon is the base of a pentagonal prism of height 5.4 m, find the total surface area of the prism. (2 marks)

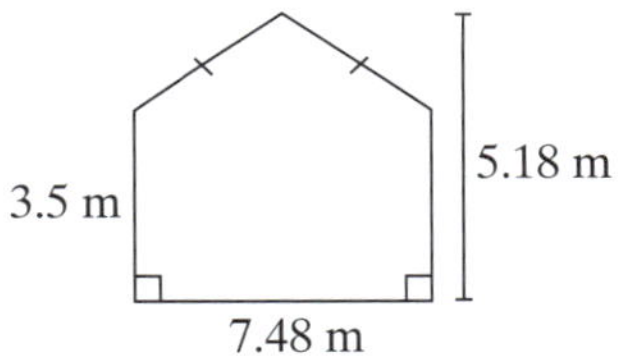

6 A rectangular prism had a hole placed right through it as shown in the diagram.
Find the total surface area of this prism (including the surfaces of the hole). (3 marks)

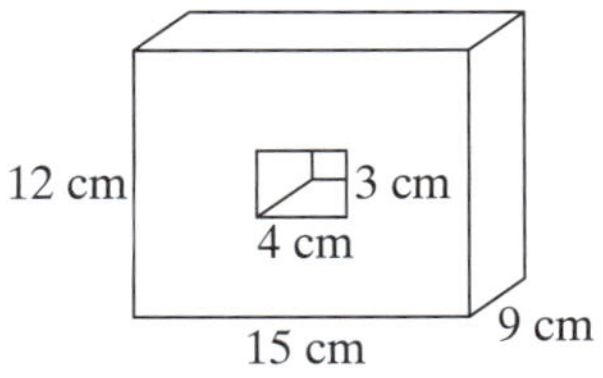

Your Feedback

$\dfrac{\square}{25} \times 100\% = \square\ \%$

PAGES 163–164

PAGE 198

VOLUME OF PRISMS

Right Prisms

STUDY NOTES

1 **Volume** is the measurement of the amount of space inside an object.

2 Volume is measured in **cubic units** (m^3, cm^3, mm^3, km^3 and so on).

3 A **prism** is a three-dimensional figure with constant cross-section.

4 The **volume of any prism** can be found by multiplying the area of the base (cross-section) by the perpendicular height. $V = Ah$

5 The **height** of a prism is the dimension at right angles to the cross-section. It is not necessarily the distance above the ground or the vertical measurement on a diagram.

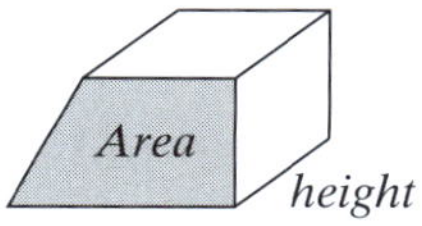

6 The **base of a rectangular prism** is a rectangle. The area of the rectangle is found by multiplying the length (l) by the breadth (b) so the volume of the prism is $V = lbh$.

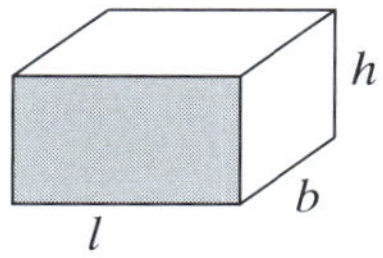

7 The **volume of a cube** can be found by cubing the length of the side. $V = s^3$

8 The **volume of a triangular prism** is $V = \frac{1}{2}bhH$. (There are two different heights: the perpendicular height of the triangular base and the height of the prism.)

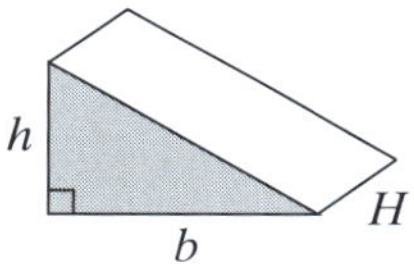

9 If a prism has a **cross-section that is a composite figure** then the cross-section can be divided into triangles or special quadrilaterals so that its area can be found. The **volume** is then found by multiplying that area by the height of the prism.

10 The **volume of composite prisms** can also be found by adding the volumes of each comprising prism.

11 The **capacity** is the amount that an object can hold. It is measured in litres (L, mL, kL and so on).

12 There is a direct relationship between **volume and capacity**. A space with volume 1 m^3 will hold 1000 L. 1 cm^3 holds 1 mL.

Checklist

Can you: ✓

1 *Understand what is meant by volume?* ☐
2 *Know and understand the units associated with volume?* ☐
3 *Recognise and name prisms?* ☐
4 *Know which dimension is the height?* ☐
5 *Find the volume of a prism?* ☐
6 *Find volumes of composite prisms?* ☐
7 *Understand what is meant by capacity?* ☐
8 *Find the capacity of a prism?* ☐
9 *Solve problems involving volume and capacity?* ☐

VOLUME OF PRISMS

Right Prisms

SKILLS CHECK

1 Which unit (cm^3 or m^3) would be best to measure the volume of a:

a shoe box? **b** shipping container? **c** room? **d** book?

2 How many:

a cm^3 in 1 m^3? **b** mm^3 in 1 cm^3?

3 Find the area:

a

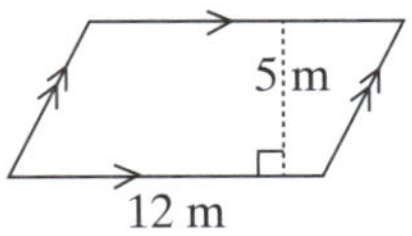

b

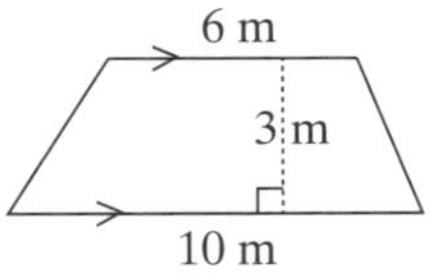

c

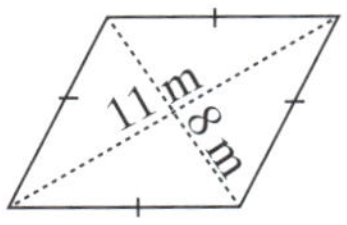

4 Find the volume of each prism, given the area A of the shaded face and the perpendicular height h:

a

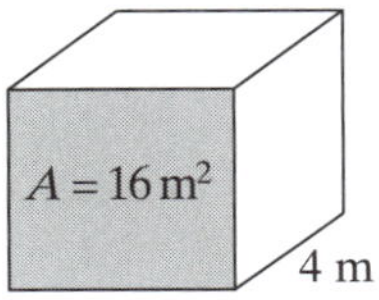

b

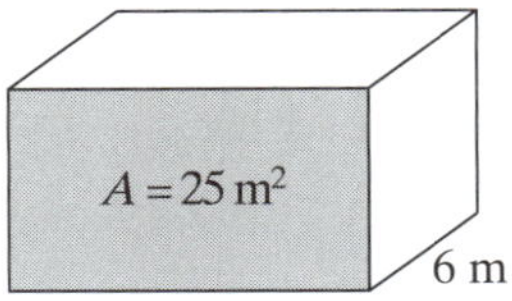

c

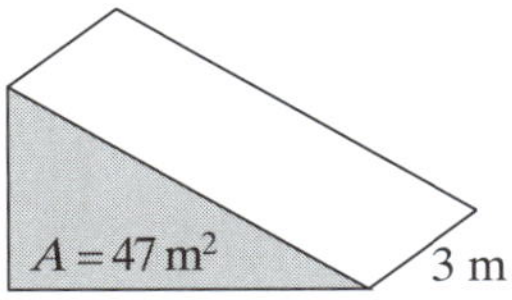

d

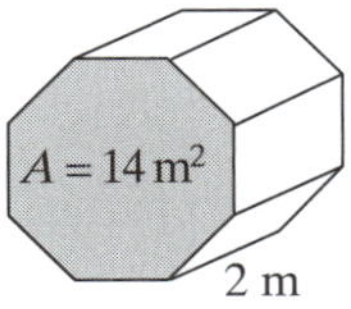

e

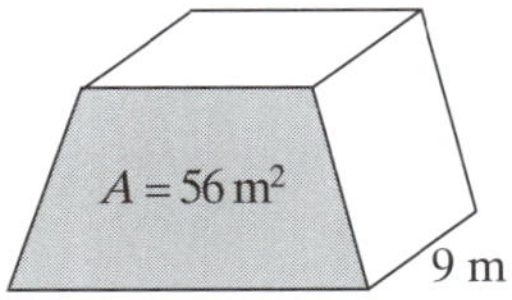

f

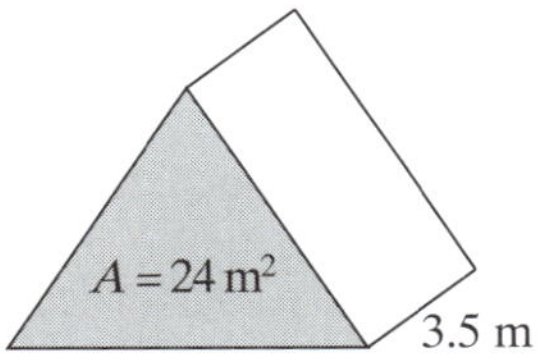

5 Find the volume of this prism made from cubic centimetres:

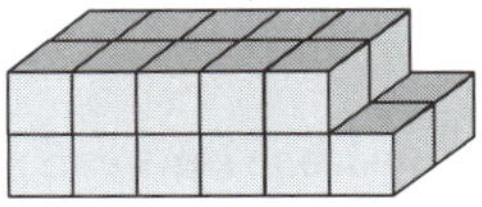

6 Find the volume of each prism:

a

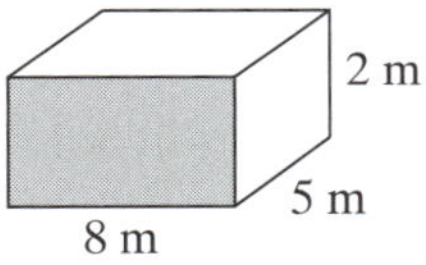

b

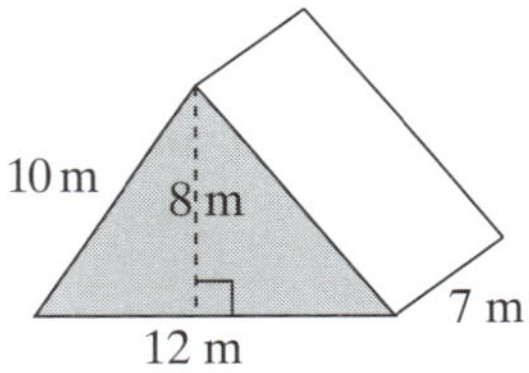

c

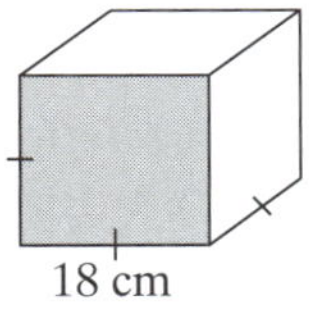

d

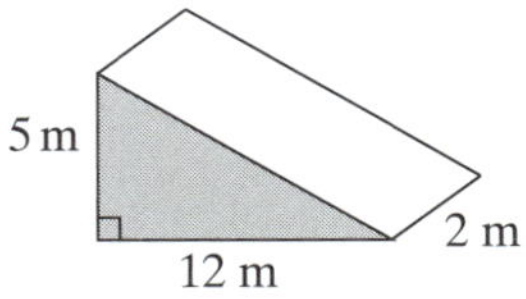

e

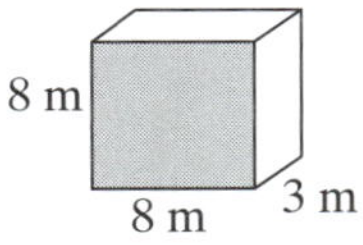

f

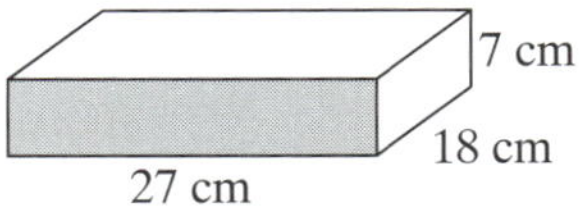

7 Given that 1 m^3 = 1000 L and 1 cm^3 = 1 mL, find:

a the capacity in litres of a volume of 5.6 m^3

b the capacity in millilitres of a volume of 15 cm^3

PAGE 164

Answers **1 a** cm^3 **b** m^3 **c** m^3 **d** cm^3 **2 a** 1 000 000 **b** 1000 **3 a** 60 m^2 **b** 24 m^2 **c** 44 m^2 **4 a** 64 m^3 **b** 150 m^3 **c** 141 m^3 **d** 28 m^3 **e** 504 m^3 **f** 84 m^3 **5** 22 cm^3 **6 a** 80 m^3 **b** 336 m^3 **c** 5832 cm^3 **d** 60 m^3 **e** 192 m^3 **f** 3402 cm^3 **7 a** 5600 **b** 15

VOLUME OF PRISMS
Right Prisms

25 MINUTES

INTERMEDIATE TEST

Part A Multiple Choice

1 If A is the area of the shaded face, in which diagram is the volume found by multiplying A by h?

A
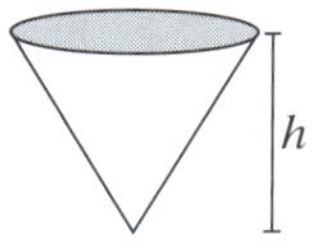

B
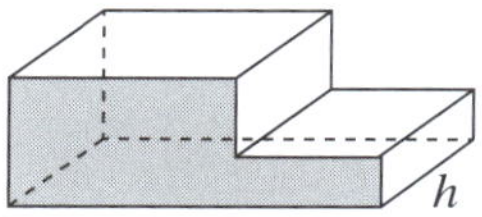

C
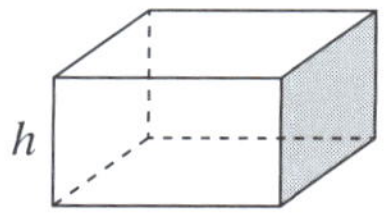

D
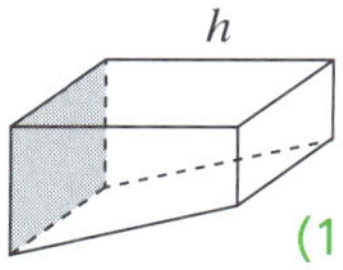

(1 mark)

2 A rectangular garden bed is 4.2 m long and 2.6 m wide. Topsoil is going to be added to the bed to a depth of 30 cm. Which is closest to the amount or topsoil needed? *Hint 1*

A 3.3 m^3 **B** 16.4 m^3
C 32.8 m^3 **D** 163.8 m^3 (1 mark)

3 What is the volume of this triangular prism in cm^3? *Hint 2*

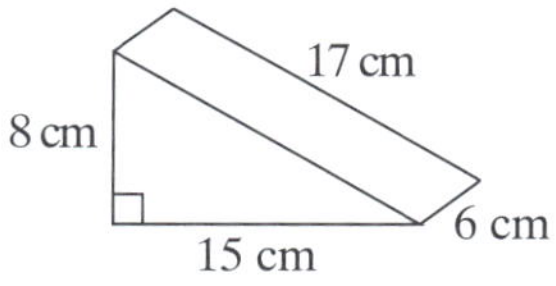

A 360
B 408
C 765
D 1020 (1 mark)

4 A drink carton has a rectangular base 10 cm long and 10 cm wide. The carton can be filled to a height of 20 cm. Given that 1 cm^3 = 1 mL, how much can the carton hold?

A 600 mL **B** 1 L
C 2 L **D** 3 L (1 mark)

Part B Short Answer

5 The volume of a hexagonal prism is 90 cm^3. If the height of the prism is 15 cm, find the area of the base. *Hint 3* (2 marks)

6 When full, a swimming pool is 2 m deep at one end and 1.2 m deep at the other. The pool is 15 m long and 5.5 m wide.

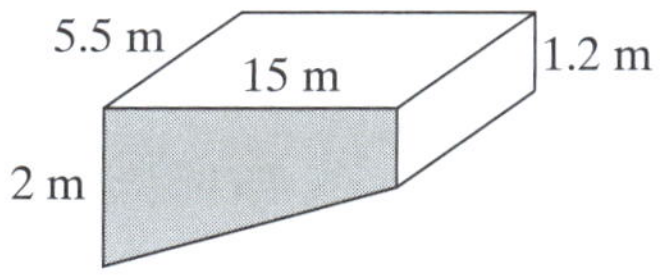

a Find the volume of the pool. *Hint 4* (3 marks)

b Find the amount of water needed to fill the pool. *Hint 5* (1 mark)

7 This set of steps is to be made from concrete.

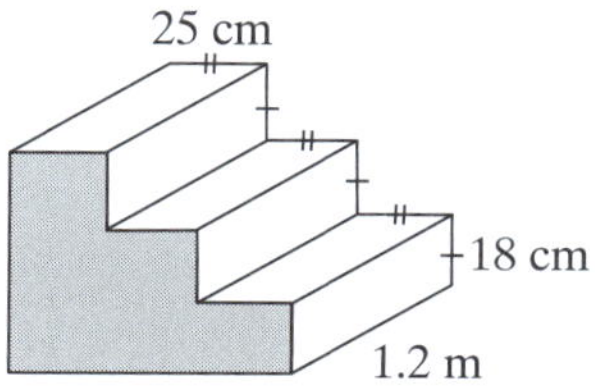

a What is the area of the shaded face in m^2? *Hint 6* (2 marks)

b How many cubic metres of concrete will be needed for the steps? Give the answer to one decimal place. (2 marks)

8

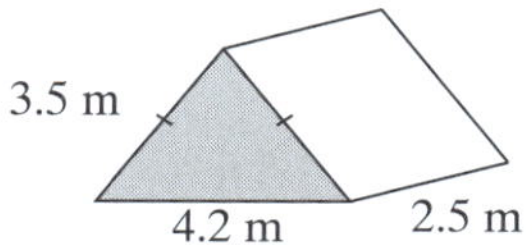

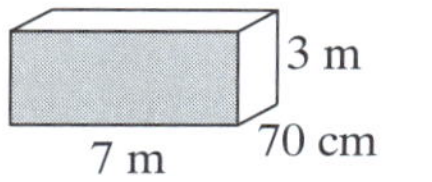

a Find the area of the shaded face of this triangular prism. *Hint 7* (2 marks)

b Find the volume of the triangular prism. (1 mark)

c Show that the rectangular prism has the same volume. (1 mark)

d Which prism has the greater surface area? Justify your answer. (4 marks)

9 Find the volume of this prism. *Hint 8*

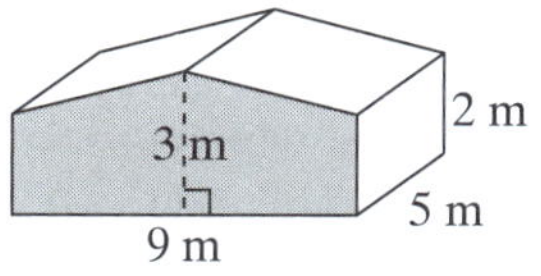

(3 marks)

Hint 1: All measurements need to be in the same units.
Hint 2: Not all of the given measurements are needed.
Hint 3: It doesn't matter what type of prism it is.
Hint 4: The cross-section is a trapezium.
Hint 5: 1 m^3 = 1000 L
Hint 6: Change cm to m before multiplying.
Hint 7: Divide the triangle in half and use Pythagoras' Theorem to find the height.
Hint 8: Divide the cross-section into a rectangle and triangle to find its area.

Your Feedback

$\frac{\square}{25} \times 100\% = \square\%$

PAGES 165–166

PAGE 198

VOLUME OF PRISMS

Right Prisms

25 MINUTES

ADVANCED TEST

1 Find the volume, in cubic metres, of each prism:

a

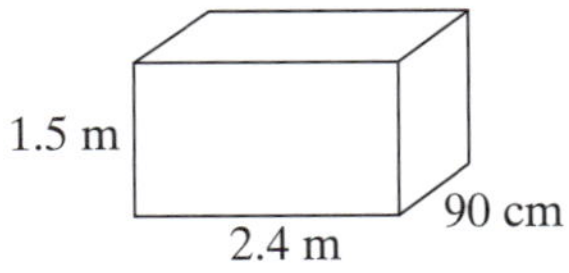

b

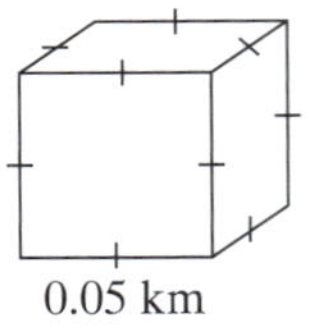

c

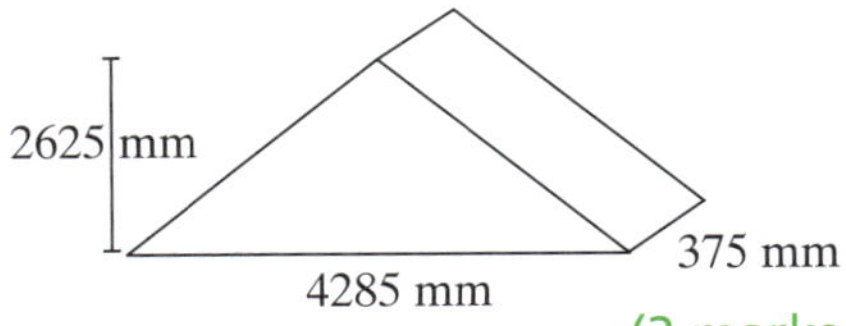

(2 marks each)

2 Find the volume of each of these prisms:

a

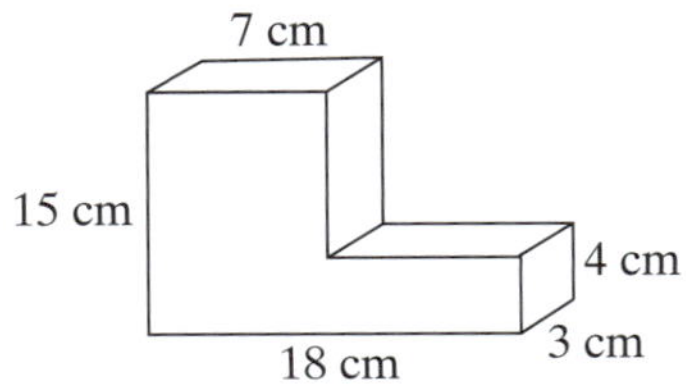

b

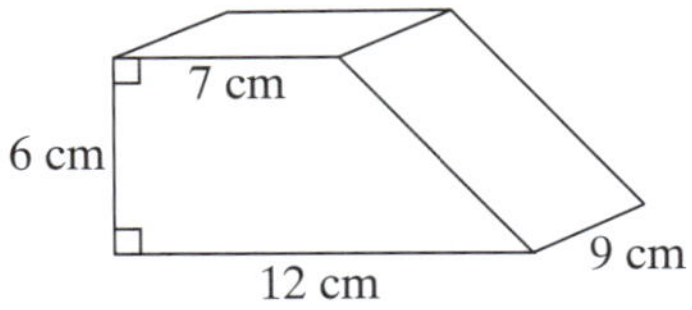

c

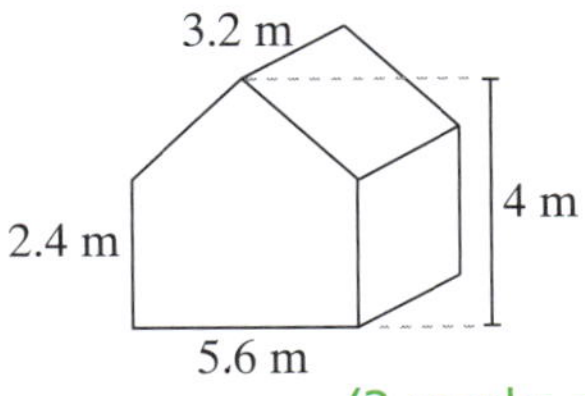

(2 marks each)

3 A cube has surface area of 433.5 cm^2. What is its volume? (2 marks)

4 A rectangular fish tank is 2.5 m long and 60 cm wide. When filled to the recommended height it holds 600 litres of water.

a What is the depth of water in the tank when it is filled to the recommended height? (2 marks)

b If the fish tank is actually 55 cm high, how many more litres of water could it hold? (2 marks)

5 A statue is going to sit on a concrete base. The diagrams show the front and top view of that base.

How much concrete will be needed for the base?

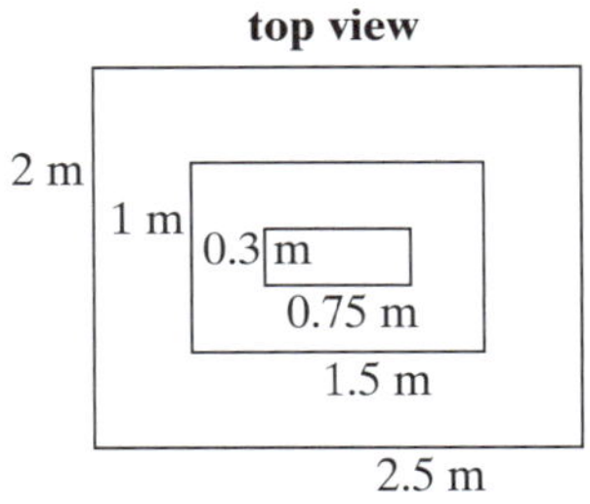

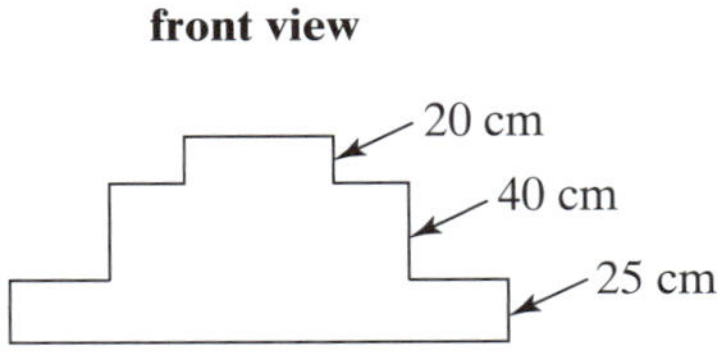

(3 marks)

6 A trapezium, three sides of which are 8 cm long, can be divided into a rectangle and two isosceles triangles as shown in the diagram.

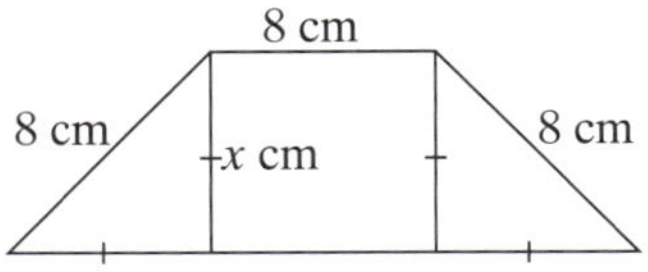

a What is the value of x? Give the answer to three decimal places. (1 mark)

b Find the area of the trapezium to two decimal places. (Use the rounded value of x found in part **a**.) (1 mark)

c Find the area of a regular octagon of side 8 cm. (1 mark)

d Find the volume of an octagonal prism whose base is a regular octagon of side 8 cm and which is 8 cm high. (1 mark)

Your Feedback

$\frac{\square}{25} \times 100\% = \square\%$

PAGE 166
PAGE 198

SURFACE AREA AND VOLUME OF CYLINDERS

Cylinders

STUDY NOTES

1 The **radius** of a circle is a line from the centre to the outside boundary.

2 The **diameter** of a circle is a line right across a circle through the centre. The length of the diameter is twice the length of the radius.

3 The **circumference** is the perimeter of a **circle**. The circumference is given by $C = \pi d$ where d is the diameter, or $C = 2\pi r$ where r is the radius.

4 The **area of a circle** is given by $A = \pi r^2$, where r is the radius.

5 A **semicircle** is half a circle. The perimeter of a semicircle includes half the circumference of the circle plus the diameter. A **quadrant** is one-quarter of a circle. The perimeter of a quadrant is one quarter of the circumference of the circle plus two radii. (Radii is the plural of radius.)

6 An **arc** of a circle is part of the circumference. The length of an arc can be found by finding a fraction of the length of the circumference (the actual fraction depending on the angle at the centre of the circle).

7 The **volume of a cylinder** is given by $V = \pi r^2 h$. The base of a cylinder is a circle so the area of the base is πr^2. When finding the volume the area of the base is multiplied by the height of the cylinder.

8 The **surface area of a closed cylinder** of radius r and height h is given by $A = 2\pi r^2 + 2\pi rh$. The **curved surface area** is $2\pi rh$. If you open out the curved surface of a cylinder you have a rectangle. The length of the rectangle is the circumference of the circle: $2\pi r$. The breadth of the rectangle is the height of the cylinder h so the area of the rectangle is $2\pi r \times h$. The area of each circle (top and bottom) is πr^2.

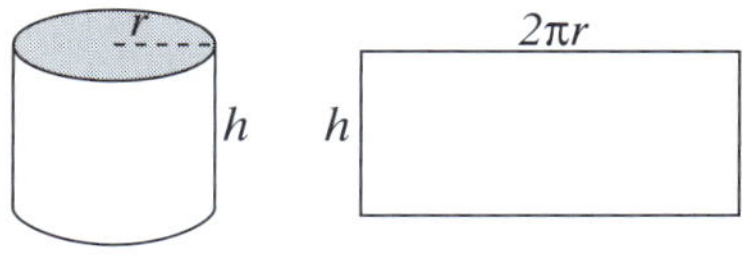

9 A **closed cylinder** means both the top and bottom circles are included in the surface area.

10 Always take care when finding the **surface area of a cylinder** to determine whether it is a closed cylinder, a cylinder **open** at the top ($A = \pi r^2 + 2\pi rh$) or a completely open cylinder.

11 The **exterior surface area of a cylinder** is the outside surface area. The **interior** surface area is the inside surface area and is sometimes required for an open cylinder such as a pipe. The interior surface area will be slightly less than the exterior surface area because the radius of the exterior is greater by the thickness of the pipe.

Checklist

Can you:

1 *Understand the terms 'radius', 'diameter', 'circumference', 'area', 'volume' and 'cylinder'?* ☐
2 *Know and understand the units associated with volume?* ☐
3 *Know and understand the units associated with surface area?* ☐
4 *Know which dimension is the height of a cylinder?* ☐
5 *Find the volume of a cylinder?* ☐
6 *Solve problems involving the volumes of cylinders?* ☐
7 *Understand the difference between open and closed cylinders?* ☐
8 *Find the surface area of both open and closed cylinders?* ☐
9 *Understand the difference between interior and exterior surface area?* ☐
10 *Solve problems involving the surface area of cylinders?* ☐

SURFACE AREA AND VOLUME OF CYLINDERS

Cylinders

SKILLS CHECK

1 The length of the radius of a circle is 7.6 cm. How long is the diameter?

2 The length of the diameter of a circle is 856 mm. How long is the radius?

3 Find the circumference, correct to one decimal place, of a circle with:
a radius 8 cm　　**b** diameter 1.5 m

4 Find the area, correct to one decimal place, of a circle with:
a radius 9 m　　**b** diameter 3.6 m

5 For this circle, find correct to one decimal place, the:
a circumference　　**b** area

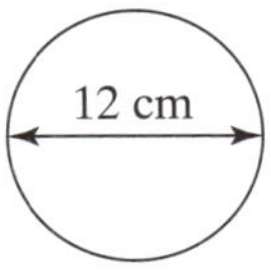

6 Find the area (correct to one decimal place) of:

a
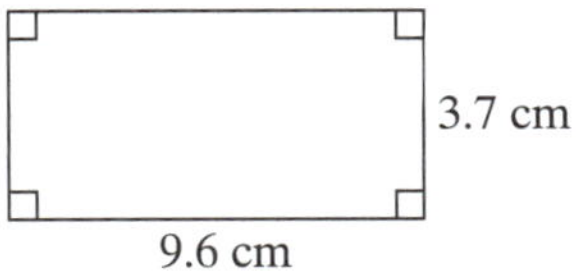

b
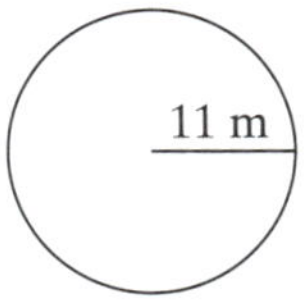

c
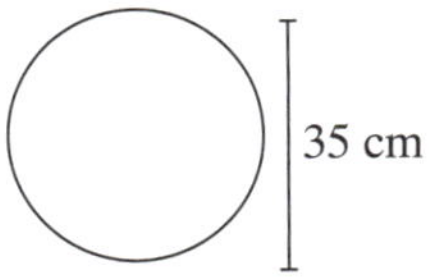

d
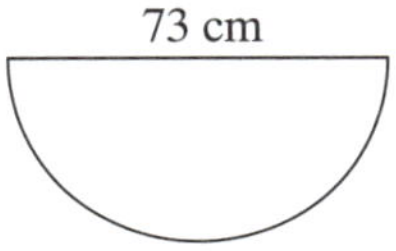

e
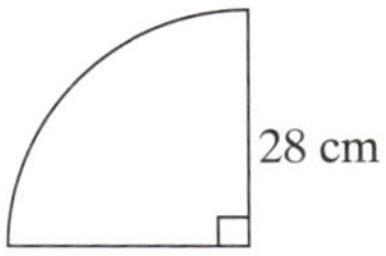

f
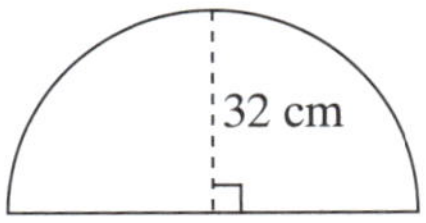

7 Find the perimeter, correct to one decimal place:

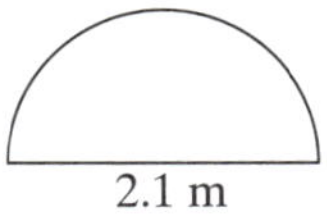

8 Find the volume of this cylinder:

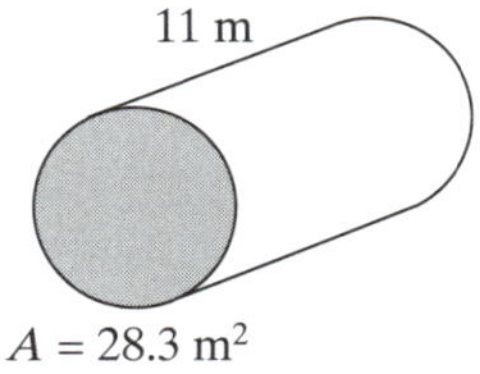

9 Find the volume of a cylinder with radius 9 m and height 2 m.

10 For a cylinder with radius 4 cm and height 9 cm, find (to one decimal place):
a the curved surface area
b the area of the circular base
c the surface area if the cylinder is open at the top
d the total surface area if the cylinder is closed

PAGES 166–167

Answers **1** 15.2 cm **2** 428 mm **3 a** 50.3 cm **b** 4.7 m **4 a** 254.5 m^2 **b** 10.2 m^2 **5 a** 37.7 cm **b** 113.1 cm^2 **6 a** 35.5 cm^2 **b** 380.1 m^2 **c** 962.1 cm^2 **d** 2092.7 cm^2 **e** 615.8 cm^2 **f** 1608.5 cm^2 **7** 5.4 m **8** 311.3 m^3 **9** 508.9 m^3 (1 d.p.) **10 a** 226.2 cm^2 **b** 50.3 cm^2 **c** 276.5 cm^2 **d** 326.7 cm^2

SURFACE AREA AND VOLUME OF CYLINDERS
Cylinders

20 MINUTES

INTERMEDIATE TEST

Part A Multiple Choice

1 Which cylinder has the greatest volume?

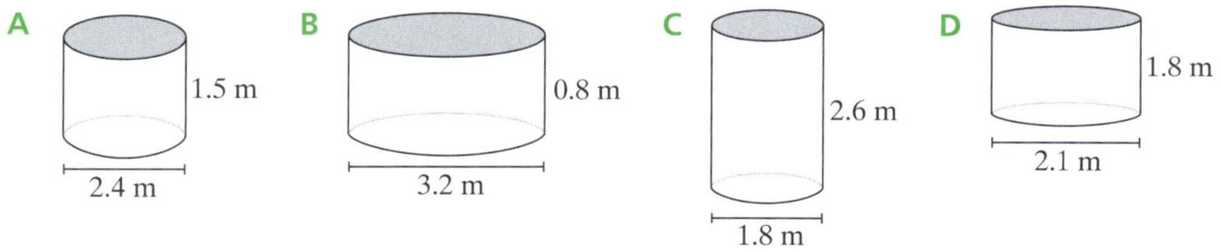

(1 mark)

2 An open pipe is 12 m long and has radius 40 cm. What is its exterior surface area, correct to two decimal places? *Hint 1*

A 30.16 m^2 **B** 6.03 m^2 **C** 7.04 m^2 **D** 3015.93 m^2 (1 mark)

3 Which closed cylinder has the largest surface area?

A diameter 10 m, height 28 m **B** diameter 12 m, height 22 m
C diameter 14 m, height 18 m **D** diameter 16 m, height 13 m (1 mark)

Part B Short Answer

4 A large cylindrical drum has diameter 60 cm and height 0.9 m.

a Find the volume of the drum. *Hint 2* (2 marks)

b Approximately how many drums of this size would be needed to hold 1 m^3 of sand? (1 mark)

5 A fish pond is going to be made from iron. It will be cylindrical in shape with a base and sides, but open at the top. Find the amount of iron that will be used if the pond has radius 1.2 m and height 0.2 m. *Hint 3* (2 marks)

6 A rainwater tank has a capacity of 22 700 litres.

a Given that 1 m^3 = 1000 L, what is the volume of the tank? (1 mark)

b The radius of the tank is 1.7 m. What is the height of the tank? *Hint 4* (3 marks)

7 **a** Find the surface area of a closed cylinder with radius 5 m and height 18 m. (2 marks)

b A community hall has a large rectangular room with a semi-cylindrical roof. Find the area of the four walls and roof. *Hint 5* (3 marks)

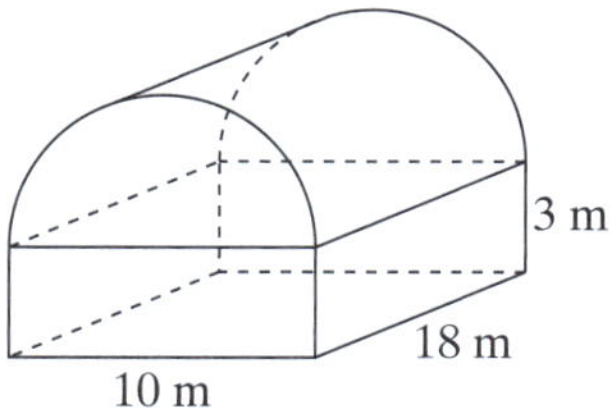

8 Find the volume of this breadbin.
Give the answer to the nearest cubic centimetre. *Hint 6* (3 marks)

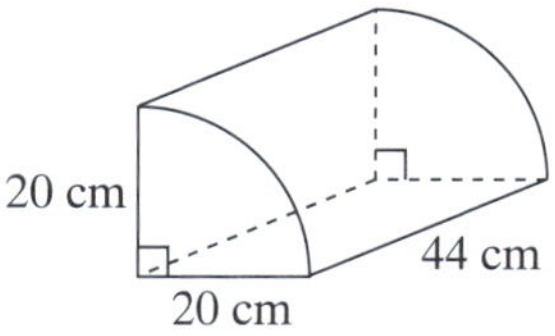

Hint 1: Change the radius to metres.
Hint 2: The diameter is given in centimetres while the height is given in metres. What is the radius in metres?
Hint 3: The surface area includes the curved surface area and one circle.
Hint 4: Substitute the known values into the formula for the volume and solve an equation.
Hint 5: You should be able to use the answer to part **a**.
Hint 6: The breadbin is one-quarter of a cylinder.

Your Feedback

$\frac{\square}{20} \times 100\% = \square\,\%$

PAGES 167–168

PAGE 198

SURFACE AREA AND VOLUME OF CYLINDERS

Cylinders

30 MINUTES

ADVANCED TEST

1 The curved surface of a cylinder of height 8 cm was calculated to be 226.2 cm^2.

- **a** What is the volume of the cylinder? (2 marks)
- **b** What is the total surface area of the closed cylinder? (1 mark)

2 A cylindrical rainwater tank has diameter 2.6 m and height 2 m.

- **a** What is its capacity to the nearest litre? (2 marks)
- **b** The tank is sold as a 10 000 litre tank. What is the difference between the stated capacity and the calculated capacity as a percentage of the calculated capacity? (1 mark)
- **c** Give a reason why the stated capacity would be less than the calculated capacity. (1 mark)

3 A cylinder is open at one end. The height of the cylinder is 825 mm and the volume is 0.13 m^3. Find the surface area of the cylinder. (3 marks)

4 A round bale of silage can be approximated as a cylinder approximately 115 cm high and 110 cm in diameter.

- **a** Find the volume to the nearest hundred cubic centimetres. (1 mark)
- **b** If a bale has a mass of 0.6 t, find its approximate density in g/cm^3. (2 marks)
- **c** Find the surface area of each bale, in square metres to one decimal place. (1 mark)
- **d** Each bale is wrapped in a plastic-type material in such a way as to cover every part of the surface and is to be wrapped in at least five layers of the plastic. The plastic wrap is 750 mm wide and comes in rolls 1500 m long. About how many bales can be wrapped from one packet of plastic wrap? (2 marks)

5 The diagram shows an annular cylinder.

- **a** What is the total of all the surfaces of the cylinder? (3 marks)
- **b** The cylinder is a pipe made of concrete. How much concrete is in the pipe? (2 marks)

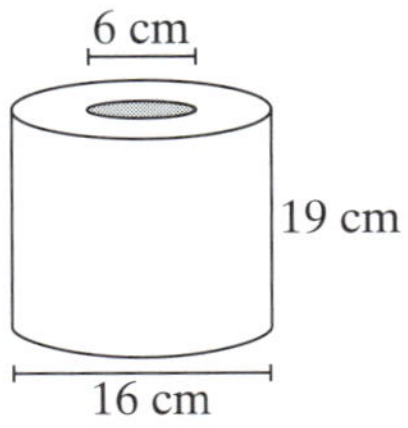

6 Find the surface area of each solid:

a

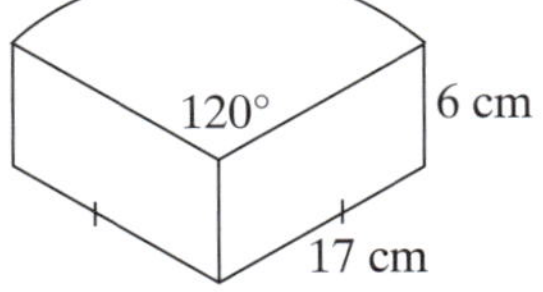

b

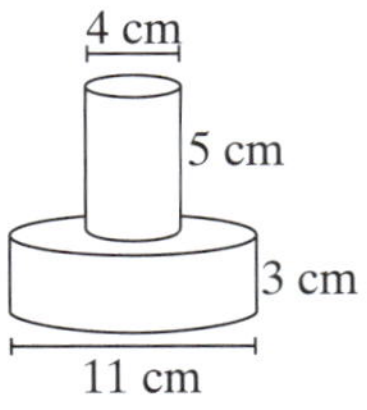

(3 marks each)

7 A 3.2 m long water trough has a semicircular cross-section of diameter 50 cm. Find its capacity in litres. (3 marks)

Your Feedback

$\frac{\square}{30} \times 100\% = \square\,\%$

PAGES 168–169

PAGE 198

ENLARGEMENT AND SIMILARITY

Geometry

STUDY NOTES

1 Under the **enlargement transformation** every point on a figure might move to a new position so that the whole figure is enlarged (or reduced). The figure and its enlargement are similar.

2 **Congruent figures** are exactly the same shape and exactly the same size.

3 **Similar figures** are exactly the same shape but not necessarily the same size. Two figures are similar if an enlargement of one is congruent to the other. The symbol ||| means 'is similar to'.

4 In any pair of similar triangles, the **corresponding angles are equal** and the **corresponding sides are in the same ratio**. Note: we use the term corresponding to mean the matching angles. The angles are in the same position in each figure (when they face the same way). This is also what we mean when we talk about corresponding angles on parallel lines. By talking about corresponding angles in triangles we don't mean that there are any parallel lines.

5 Similar figures are related by a **scale factor**. If two figures are related by a scale factor of 2, then every side of the larger figure is twice as long as the corresponding side of the smaller figure. If the second figure is larger than the first, it is an **enlargement** of the first. If the second figure is smaller than the first it is a **reduction**. If the scale factor is 1 the two figures are the same size and we say they are **congruent**.

6 When naming two similar triangles, the **vertices should be listed in the same order**, matching corresponding angles. If $\Delta ABC \,|||\, \Delta DEF$ then $\angle A = \angle D$, $\angle B = \angle E$ and $\angle C = \angle F$.

7 **If the three sides of one triangle are proportional to the three sides of another triangle** then the two triangles are similar.
$\frac{AB}{DE} = \frac{BC}{EF} = \frac{AC}{DF} = \frac{1}{2}$. The sides must match correctly, the 2 smallest sides, the 2 largest sides and the 2 other sides.

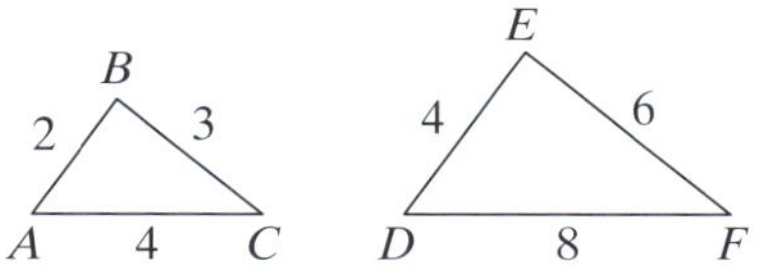

8 **If two sides of one triangle are proportional to two sides of another triangle and the included angles are equal**, then the triangles are similar.
$\angle B = \angle E.\ \frac{AB}{DE} = \frac{BC}{EF} = \frac{2}{3}$

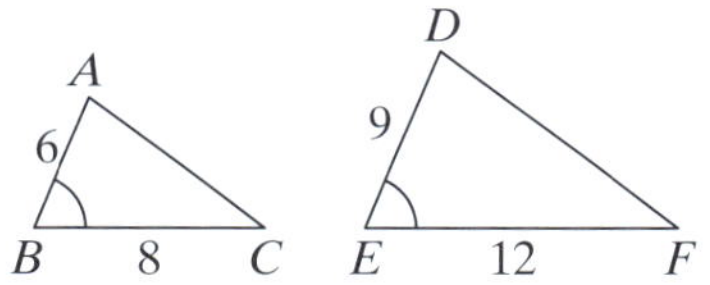

9 **If two angles of one triangle are equal to two angles of another triangle** then the triangles are similar. $\angle A = \angle E,\ \angle B = \angle F$
$\therefore\ \Delta ABC \,|||\, \Delta EFD$
In fact $\angle C = \angle D$ as well. Whenever 2 angles of one triangle equal 2 angles of another triangle the third angles must also be equal (since the angle sum must be 180°).

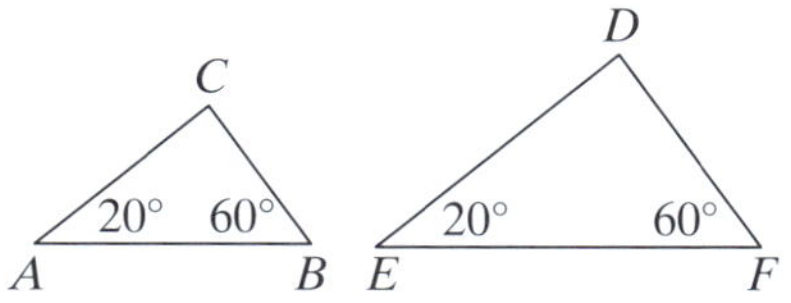

10 **If the hypotenuse and one side of a right-angled triangle are proportional to the hypotenuse and a side of another right-angled triangle** then the two triangles are similar.
$\angle B = \angle E = 90°$ $\quad \frac{AB}{DE} = \frac{AC}{DF} = \frac{4}{3}$

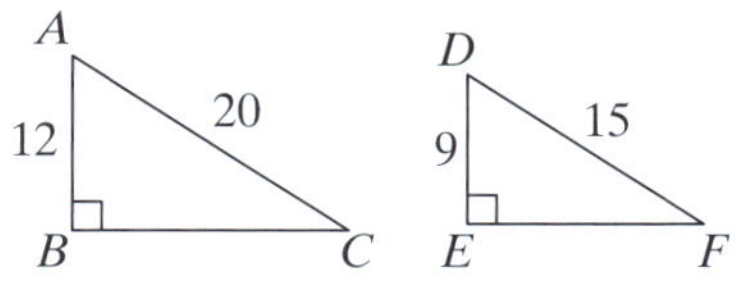

Checklist

Can you:

1 *Understand what is meant by an enlargement and use the enlargement transformation?* ☐
2 *Understand what it means for two figures to be similar?* ☐
3 *Match sides and angles of similar figures and know to name the vertices in matching order?* ☐
4 *Show that two triangles are similar?* ☐

ENLARGEMENT AND SIMILARITY

Geometry

SKILLS CHECK

1 Quadrilateral $ABCD$ has been enlarged by this transformation.

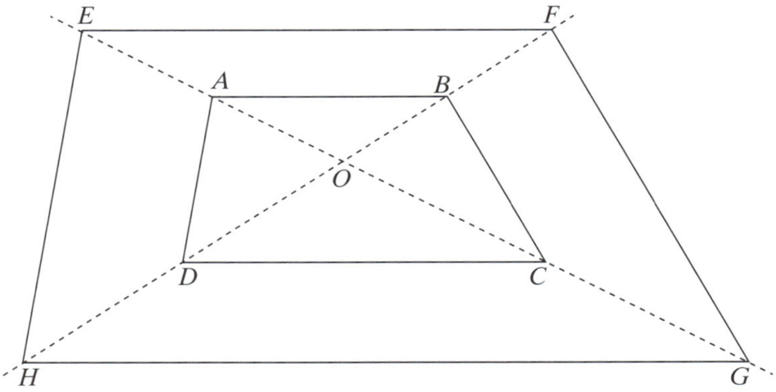

- **a** Find the length, in mm, of:
  i AB **ii** BC **iii** CD **iv** DA
- **b** Find the length, in mm, of:
  i EF **ii** FG **iii** GH **iv** HE
- **c** How many times larger is each side of $EFGH$ than the corresponding side of $ABCD$?
- **d** What is the size, to the nearest degree, of: **i** $\angle DAB$? **ii** $\angle ABC$? **iii** $\angle BCD$? **iv** $\angle CDA$?
- **e** What is the size, to the nearest degree, of: **i** $\angle HEF$? **ii** $\angle EFG$? **iii** $\angle FGH$? **iv** $\angle GHE$?
- **f** What conclusion can be reached about the corresponding angles?

2 Answer true or false:

- **a** Any two squares are similar.
- **b** Any two rectangles are similar.
- **c** Any two circles are similar.
- **d** Any two hexagons are similar.

3 These two triangles are similar.
Which side of triangle DEF corresponds to:

a AB? **b** BC? **c** AC?

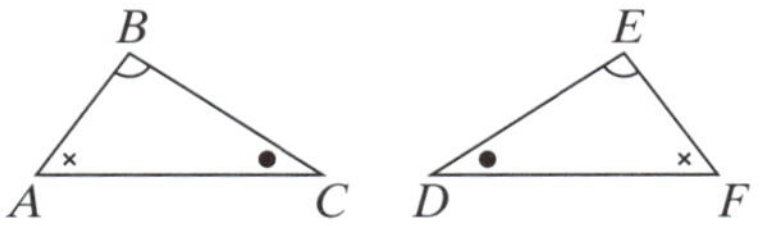

4 Complete (making sure that the vertices appear in the correct order):

a $\Delta ABC \mathbin{|||} \Delta$ ______ **b** $\Delta PQR \mathbin{|||} \Delta$ ______ **c** $\Delta JKL \mathbin{|||} \Delta$ ______

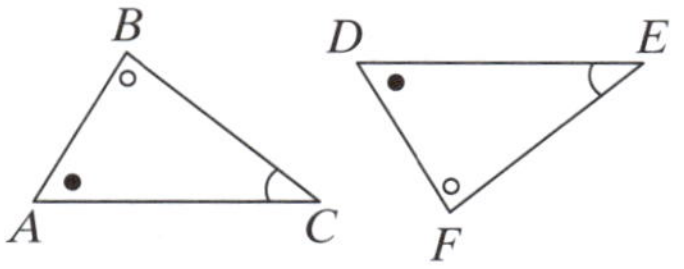

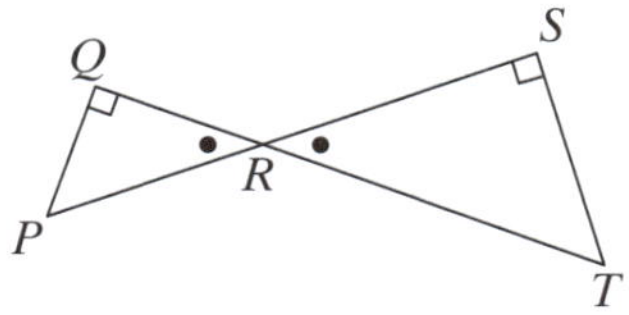

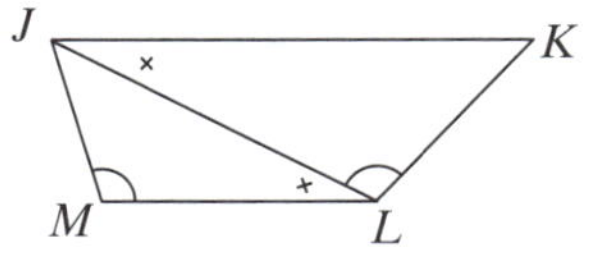

5 The pair of triangles is similar. True or false?

a

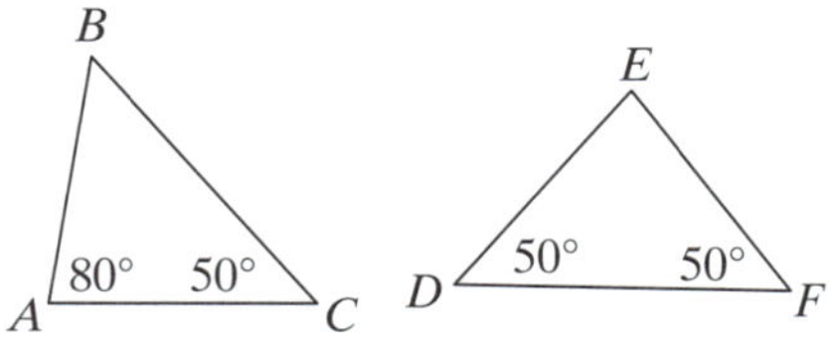

b

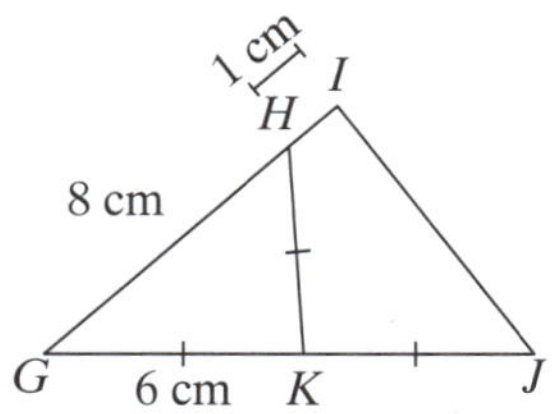

c

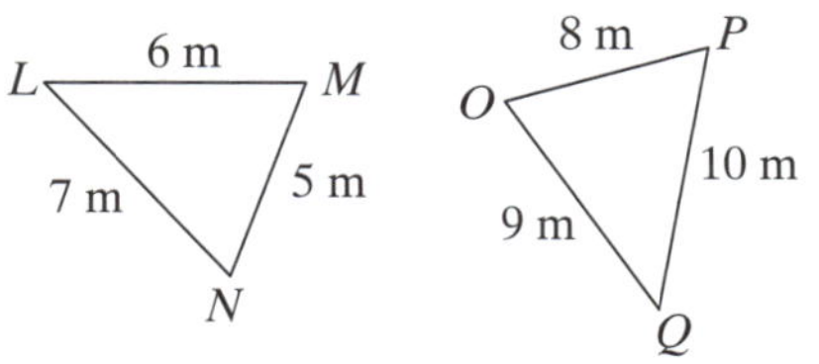

d

PAGES 169–170

Answers **1 a i** 30 mm **ii** 25 mm **iii** 46 mm **iv** 22 mm **b i** 60 mm **ii** 50 mm **iii** 92 mm **iv** 44mm **c** 2 **d i** 100° **ii** 120° **iii** 60° **iv** 80° **e i** 100° **ii** 120° **iii** 60° **iv** 80° **f** equal **2 a** true **b** false **c** true **d** false **3 a** FE **b** ED **c** FD **4 a** DFE **b** TSR **c** LJM **5 a** true **b** true **c** false **d** true

ENLARGEMENT AND SIMILARITY

Geometry

20 MINUTES

INTERMEDIATE TEST

Part A Multiple Choice

1 These two quadrilaterals are similar.

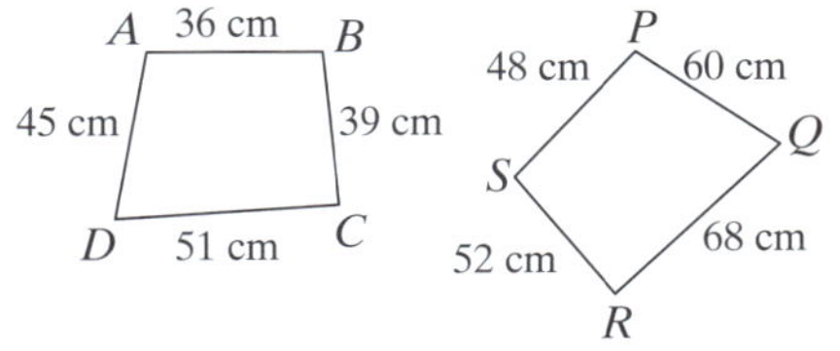

Which angle of $PQRS$ corresponds to $\angle ABC$? Hint 1

A $\angle PQR$ **B** $\angle QRS$
C $\angle PSR$ **D** $\angle QPS$ (1 mark)

2 Which statement is correct?

A Any two right-angled triangles are similar.
B Any two isosceles triangles are similar.
C Any two equilateral triangles are similar.
D No two scalene triangles are similar. (1 mark)

3 $FGEH$ is similar to $ABCD$.

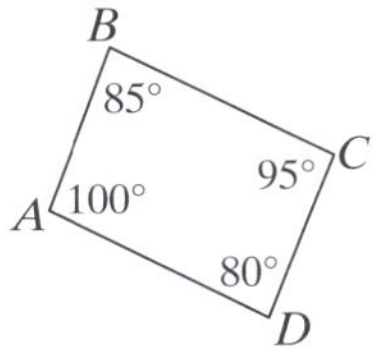

What is the size of $\angle HEG$? Hint 2

A 80° **B** 85°
C 95° **D** 100° (1 mark)

Part B Short Answer

4

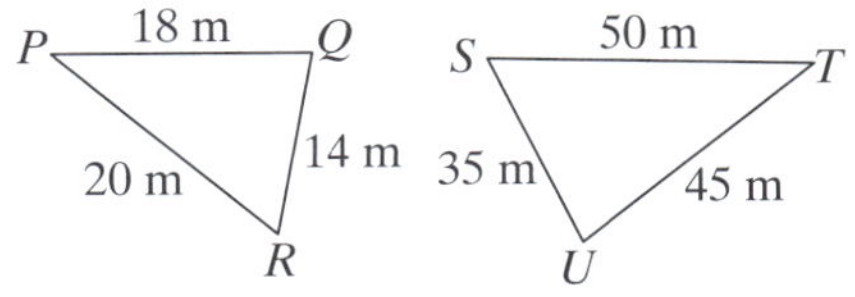

a How many times larger is the largest side of ΔSTU than the largest side of ΔPQR? (1 mark)

b How many times larger is the smallest side of ΔSTU than the smallest side of ΔPQR? (1 mark)

c Jimmy measures the angles of ΔPQR. He finds that $\angle PQR = 76°$, $\angle QRP = 61°$ and $\angle RPQ = 43°$. What is the size of $\angle STU$? Justify your answer. (2 marks)

5

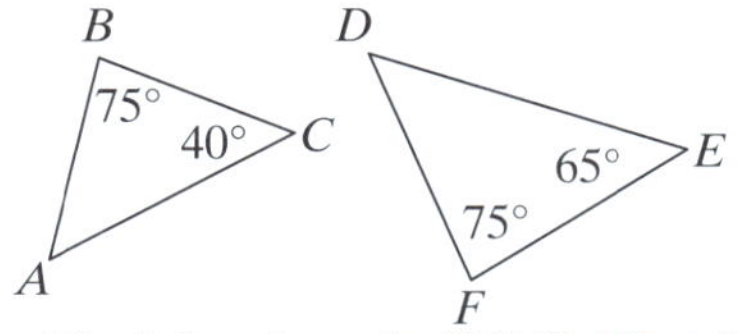

a Find the size of $\angle FDE$. Hint 3 (1 mark)

b Briefly explain why the triangles are similar. (1 mark)

c Which side of ΔDEF corresponds to AC? (1 mark)

6 In the diagram $AE = 6$ cm, $EB = 3$ cm, $AC = 12$ cm and $AD = 8$ cm.

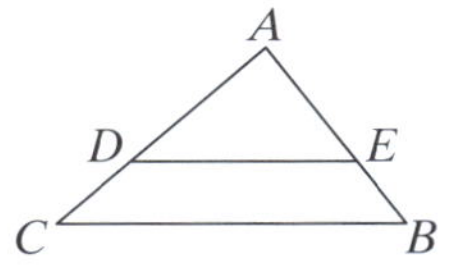

a Show that ΔABC is similar to ΔAED. Hint 4 (3 marks)

b Is DE parallel to CB? Justify your answer. Hint 5 (2 marks)

7 In triangle GHI, $GH = 85$ mm and $GI = 221$ mm. In triangle IJK, $IJ = 156$ mm and $IK = 169$ mm.

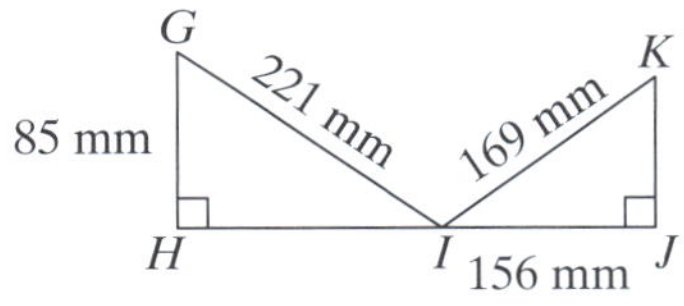

a Find the length of HI. Hint 6 (2 marks)

b Show that the triangles are similar. (3 marks)

Hint 1: Use the measurements to match up the sides.
Hint 2: Which angle of *ABCD* does $\angle HEG$ correspond to? Match the vertices.
Hint 3: The angle sum of a triangle is 180°.
Hint 4: First find the length of *AB*.
Hint 5: Are the corresponding angles equal?
Hint 6: Triangle *GHI* is a right-angled triangle. Use Pythagoras' Theorem.

Your Feedback

$\frac{\square}{20} \times 100\% = \square\%$

PAGES 170–171

PAGE 198

ENLARGEMENT AND SIMILARITY

Geometry

30 MINUTES

ADVANCED TEST

1 Determine whether the given triangles are congruent, similar or not necessarily either congruent or similar. Justify your answer and if they are similar or congruent show this in an expression, matching vertices.

a ΔABC and ΔPQR

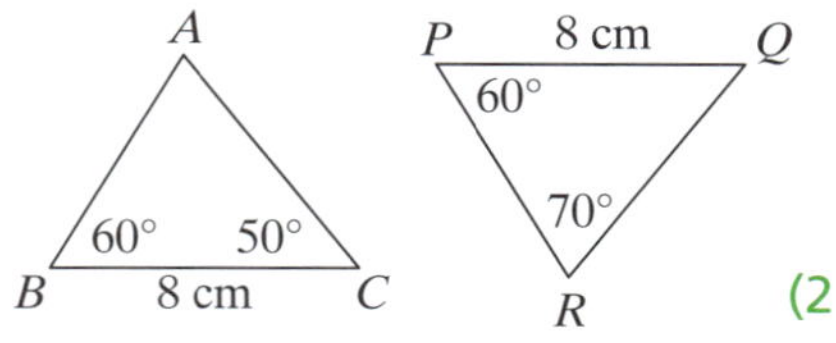

(2 marks)

b ΔJKL and ΔLMN

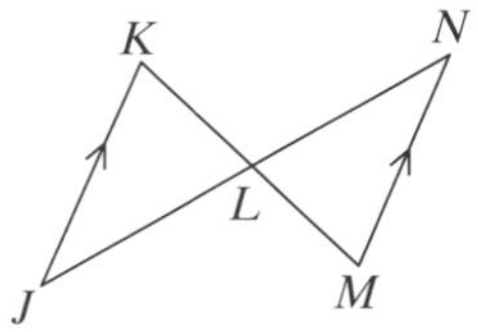

(2 marks)

c ΔUVW and ΔXYZ

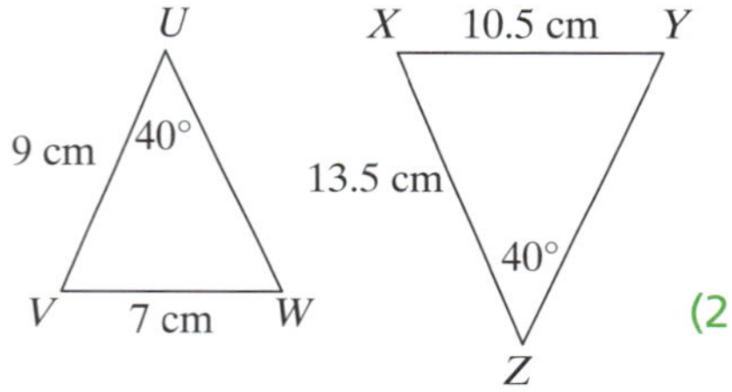

(2 marks)

2 $AB = 8$ cm, $BC = 22.5$ cm, $CD = 18$ cm, $AD = 15$ cm and $BD = 12$ cm.

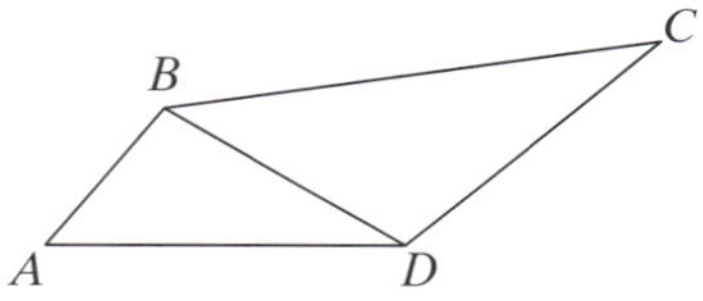

a Show that the two triangles are similar. (2 marks)

b If $\angle ABD = 95°$ and $\angle ABC = 148°$, find the size of $\angle ADC$. (3 marks)

3 **a** Explain why having two angles equal is sufficient to show that two triangles are similar. (1 mark)

b Show that $\Delta ADC \,|||\, \Delta ACB$.

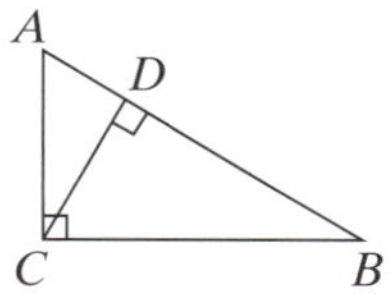

(2 marks)

4

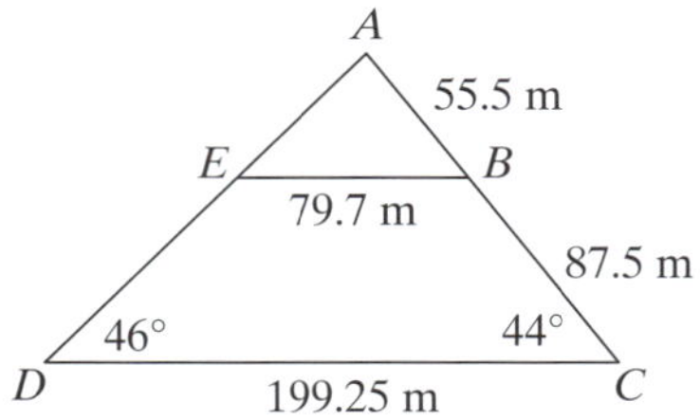

a What type of triangle is ACD? (1 mark)

b Find the length of AD. (1 mark)

c Are the two triangles similar? Justify your answer. (3 marks)

5 $ABCD$ is a trapezium with AB parallel to DC. The diagonals meet at E.

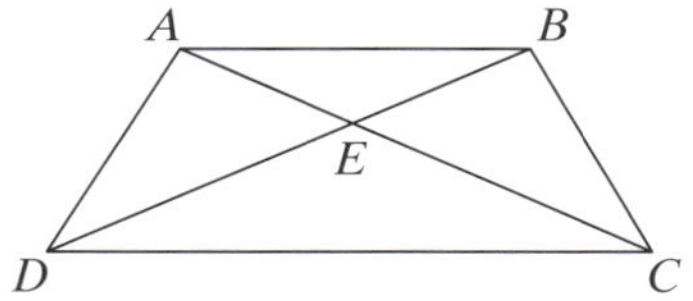

a Show that ΔABE is similar to ΔCDE. (2 marks)

b Justin said: "$\angle AED$ and $\angle BEC$ are vertically opposite. Because ΔABE and ΔCDE are similar we know that the sides are in proportion. So ΔAED must be similar to ΔBEC because the sides are in proportion and the included angles are equal." Briefly explain why Justin is wrong. (2 marks)

6 A square of side 4 cm is enlarged by a factor of 1.5.

a What is the percentage increase in its:

- **i** perimeter? (2 marks)
- **ii** area? (2 marks)

b Will this percentage increase be the same for all squares? Justify your answer. (3 marks)

Your Feedback

$\frac{\square}{30} \times 100\% = \square\,\%$

PAGES 171–172

PAGE 198

RATIO AND SCALE FACTORS

Geometry

STUDY NOTES

1 A **ratio** is a comparison of two (or more) quantities of the same kind. For example, if the numbers of boys and girls in a class is in the ratio 3 to 2 it means that there are 3 boys for every 2 girls.

2 **Order** is very important in a ratio. Whatever is mentioned first is the first part of the ratio and whatever is mentioned second is the second part. For example, if cordial and water are mixed in the ratio 1:5 there is 1 part cordial for every 5 parts water.

3 **Ratios** should be **expressed in simplest form**. So each part should always be divided by any common factors. For example, the ratio 6:8:10 = 3:4:5 after dividing each part by 2.

4 The **ratio of two items** can be **expressed** as a **fraction** and the fraction key on a calculator can be used when simplifying ratios.

5 The **same units** must be used in a ratio. For example, the ratio of 30 cents to 1 dollar is not 30 to 1 but 30 to 100. In simplest form it is 3:10.

6 If items are in **proportion** they are in the same ratio.

7 A **statement of proportion** is an equation involving two equal ratios. For example, 2:3 = 4:6. Such a statement is often given in fractional form. $\frac{2}{3} = \frac{4}{6}$

8 To **find the value of an unknown in a proportion statement** we often **cross-multiply**. That means we multiply the numbers that are diagonally opposite on either side of the equals sign. If $\frac{a}{b} = \frac{c}{d}$ then $ad = bc$.

9 To **solve a problem involving a ratio or proportion**, first express the problem as a proportion statement, cross-multiply to find the value of the unknown and then answer the question.

10 Whenever **two triangles are similar** the sides are in proportion. We can use the proportions to find the size of an **unknown side**.

11 A **scale factor** tells us how many times bigger one object is than another. For example, if an object has been enlarged by a factor of 2 then every side of the new object is twice the length of the corresponding side of the original object.

12 In a **scale drawing**, the drawing and the actual object are similar figures. All the corresponding lengths are in proportion and all the corresponding angles are equal.

13 The **scale on a map** is usually given as a ratio (with the first part of the ratio 1). A scale of 1:100 000 for example means that 1 cm on the map represents a real distance of 100 000 cm.
Now 100 000 cm = 1000 m or 1 km. So on the map 1 cm would represent 1 km.

Checklist

Can you:

1 *Understand what is meant by a ratio?* ☐
2 *Understand that order is very important in a ratio?* ☐
3 *Express ratios in simplest form?* ☐
4 *Solve problems involving ratios?* ☐
5 *Understand what is meant by proportion?* ☐
6 *Cross-multiply?* ☐
7 *Solve problems involving proportions?* ☐
8 *Use proportions to find sides of similar triangles?* ☐
9 *Understand scale factors?* ☐
10 *Construct scale drawings and solve problems involving scales and scale drawings?* ☐

RATIO AND SCALE FACTORS

Geometry

SKILLS CHECK

1 Write each ratio in simplest form:

a 8:10 b 12:6 c 25:35

d 7:14 e 16:20 f 24:8

2 Write as a ratio in simplest form:

a 9 months to 1 year b 25 cm to 1 m c 30 g to 2 kg

d 1 cm to 1 m e 5 cm to 100 m f 1 cm to 1 km

3 Find the value of x if:

a $\frac{x}{8} = \frac{1}{2}$ b $\frac{x}{6} = \frac{2}{3}$ c $\frac{x}{3} = \frac{4}{5}$

d $\frac{3}{4} = \frac{x}{8}$ e $\frac{6}{72} = \frac{1}{x}$ f $\frac{2}{x} = \frac{5}{7}$

4 This quadrilateral is going to be enlarged by a factor of 9.

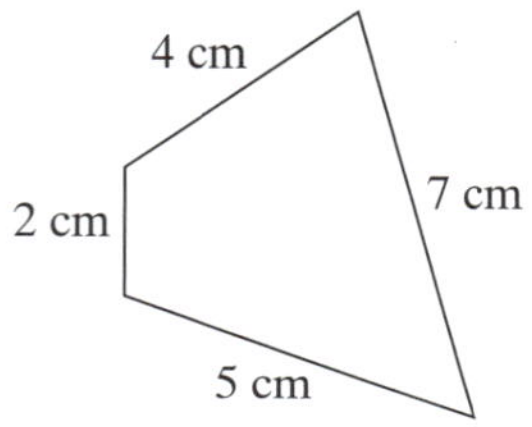

On the enlarged quadrilateral what is the length of:

a the shortest side? b the longest side? c the perimeter?

5 Given that each pair of triangles is similar, find the value of x:

a

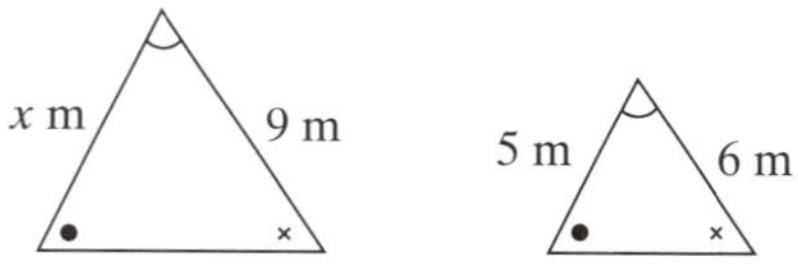

b

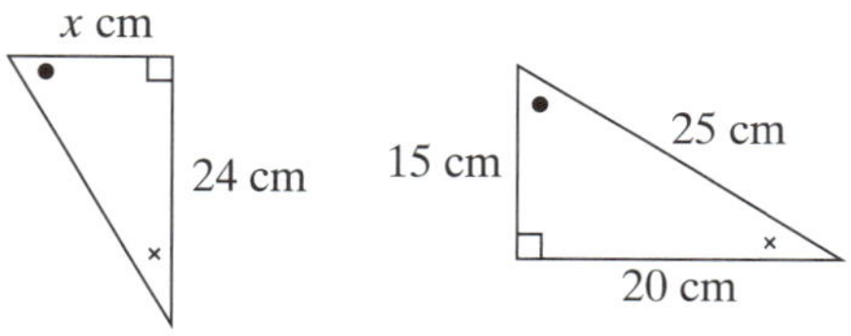

c

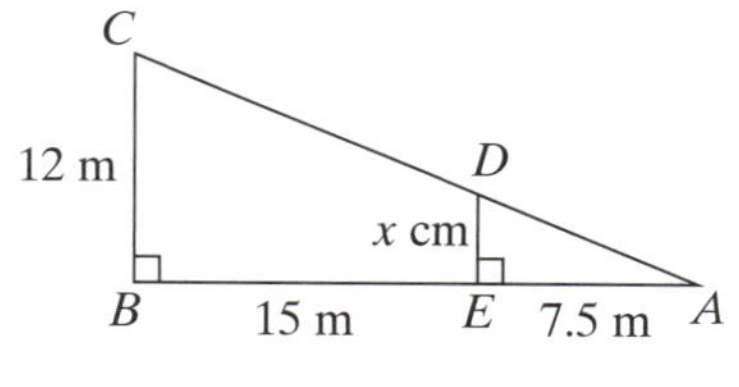

d

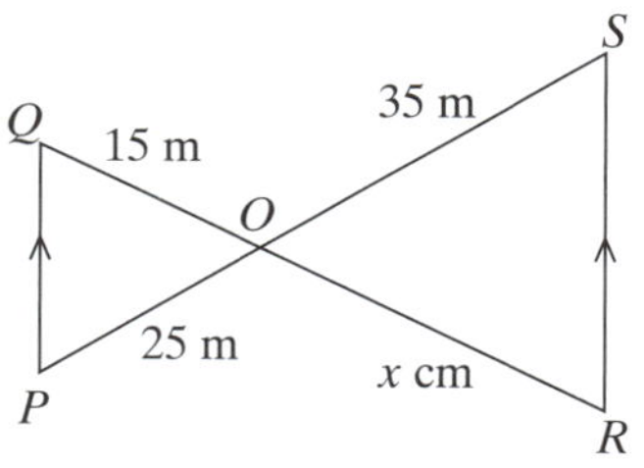

PAGES 172–173

Answers 1 a 4:5 b 2:1 c 5:7 d 1:2 e 4:5 f 3:1 2 a 3:4 b 1:4 c 3:200 d 1:100 e 1:2000 f 1:100000 3 a 4 b 4 c 2.4 d 6 e 12 f 2.8 4 a 18 cm b 63 cm c 162 cm 5 a 7.5 b 18 c 4 d 21

RATIO AND SCALE FACTORS

Geometry

20 MINUTES

INTERMEDIATE TEST

Part A Multiple Choice

1 The sides of a triangle are in the ratio 3:4:5. If the shortest side is 60 mm long, how long is the longest side?

A 70 mm B 80 mm

C 90 mm D 100 mm (1 mark)

2 A photograph is 10 cm long and 6 cm wide. It is enlarged so that it is now 90 cm wide. How long will the enlarged photo be?

A 54 cm B 94 cm

C 150 cm D 180 cm (1 mark)

3 The distance between two towns on a map is 4 cm. The actual distance between the towns is 80 km. What is the scale on the map?

A 1:2000 B 1:20 000

C 1:200 000 D 1:2 000 000 (1 mark)

Part B Short Answer

4 The sides of two rectangles are in proportion. One rectangle is 69 mm long and 39 mm wide. The other rectangle is 91 mm wide. How long is it? (1 mark)

5 A plan (scale drawing) has been made of a paddock. The actual length of the northern boundary is 72 m. This boundary is 144 mm long on the map.

a How many times longer than the drawing is the northern boundary?
Hint 1 (1 mark)

b The western boundary is 45 m long. How long should it be on the plan? (1 mark)

c Another side of the drawing is 170 mm long. How long is the actual boundary? (1 mark)

6 For each pair of similar triangles find the value of x. (All measurements are in cm.)

a

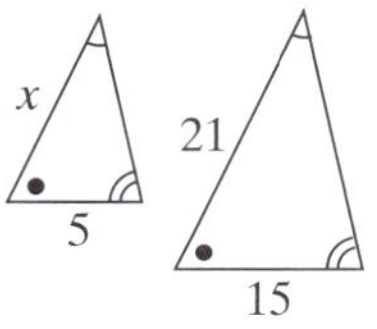

(1 mark)

b

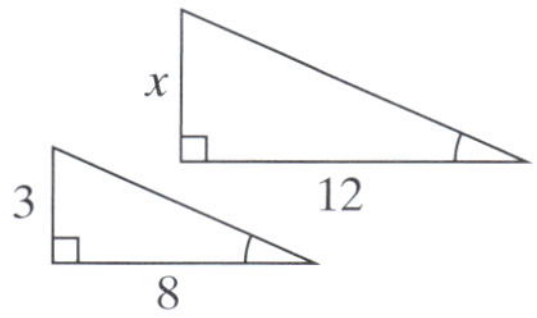

(1 mark)

c

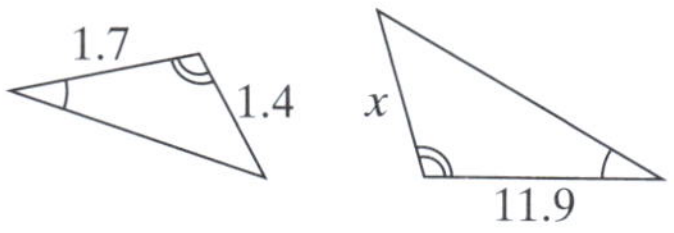

(2 marks)

7 A scale drawing was made of a rectangular piece of land.

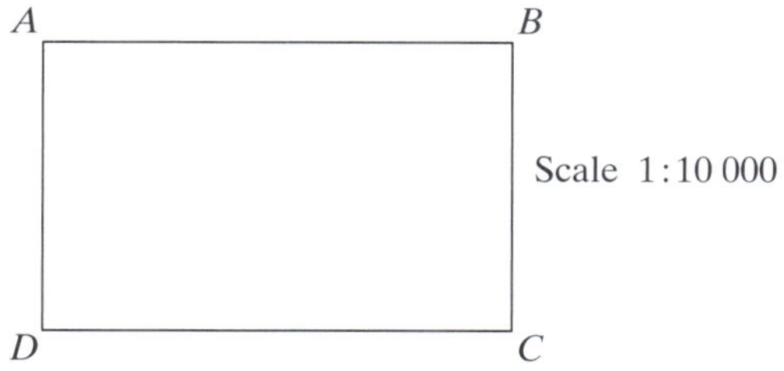

a What actual length, in metres, is represented by 1 cm? (1 mark)

b What is the actual length of side AB?
Hint 2 (1 mark)

c What is the perimeter of the land? (2 marks)

d What is the area in hectares? (2 marks)

8 Triangles ABC and DEC are similar.

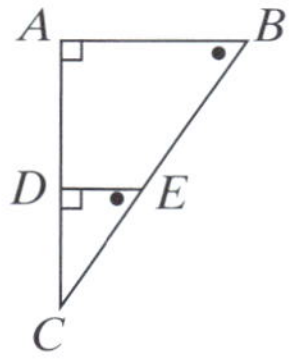

a Which side of triangle ABC corresponds to side EC of triangle DEC? (1 mark)

b If $CD = 8$ cm, $DE = 6$ cm and $AD = 12$ cm, find the length of AB.
Hint 3 (2 marks)

Hint 1: One measurement is in metres and one in millimetres. First convert to the same unit.
Hint 2: You will need to measure the diagram.
Hint 3: Be careful to use the measurement for the whole of side AC.

Your Feedback

$\frac{\square}{20} \times 100\% = \square\%$

PAGES 173–174

PAGE 198

RATIO AND SCALE FACTORS

Geometry

25 MINUTES

ADVANCED TEST

1 The scale on a plan is 1 : 1 000 000. A length on the plan is measured to be 4.75 cm. What is the actual distance in kilometres? (1 mark)

2 This is a scale drawing of a triangular block of land. (Scale 1 : 2000)

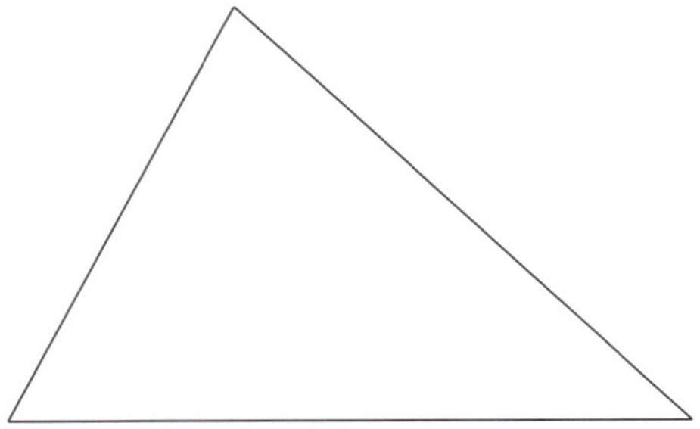

a By measurement and calculation determine the perimeter of the land. (3 marks)

b What is the area of the land? (2 marks)

3 It is known that $\Delta FDE \,|||\, \Delta ABC$.

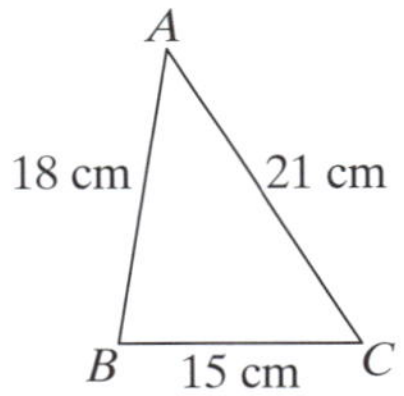

a Which angle of ΔFDE corresponds to $\angle BAC$? (1 mark)

b If side $DE = 48$ cm what is the enlargement factor? (1 mark)

c How long is side DF? (1 mark)

4 A tree casts a shadow of length 28 m. At the same time a 1 m post casts a shadow of length 70 cm.

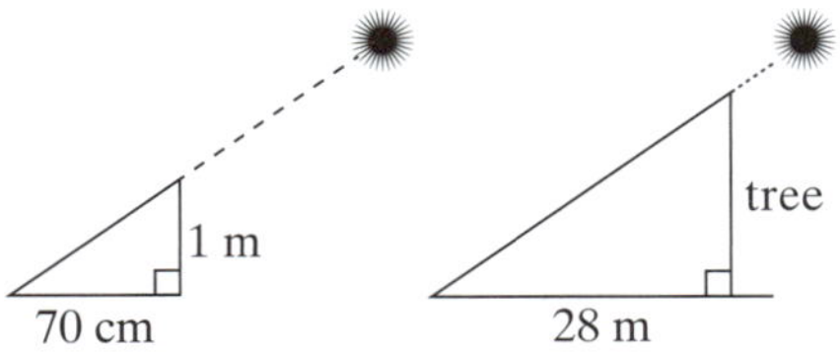

a Briefly explain why the triangles are similar. (1 mark)

b How tall is the tree? (2 marks)

5

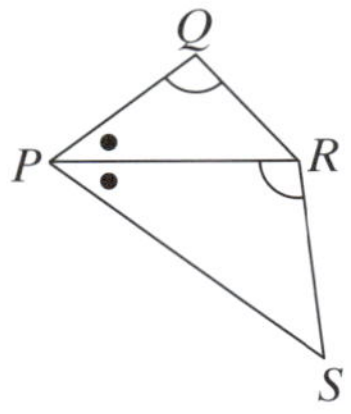

a Show that ΔPQR is similar to ΔPRS. (2 marks)

b If $PQ = 8$ cm and $PS = 18$ cm, find the length of PR. (2 marks)

6 The circumference of a circle of radius 8 cm is approximately 50 cm in length. Without using a calculator, find the approximate radius of a circle with circumference 2.5 m in length. (2 marks)

7 A rectangular photograph is in a frame 12 cm by 10 cm. There is a 1 cm wide border around the photo inside the frame. The photo is removed from the frame and enlarged by a factor of 7.5. What are the dimensions of the new frame needed if it is again to have a 1 cm wide border on all sides? (3 marks)

8 This is a rough plan of a building site, not drawn to scale.

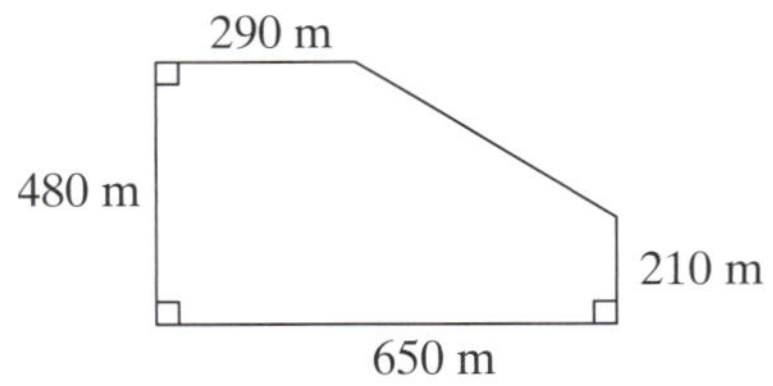

a Make a scale drawing of the site using the scale 1 : 10 000. (2 marks)

b Using the scale drawing, find the actual length of the remaining boundary of the building site. (2 marks)

Your Feedback

$\frac{\square}{25} \times 100\% = \square\ \%$

PAGE 174

PAGE 198

TRIGONOMETRY OF RIGHT-ANGLED TRIANGLES

Trigonometry

STUDY NOTES

1 The **hypotenuse** is the longest side of a right-angled triangle. It is the side opposite the right angle. The **opposite** and **adjacent sides** depend on which angle we are considering. The opposite side is opposite the angle and the adjacent side is next to the angle.

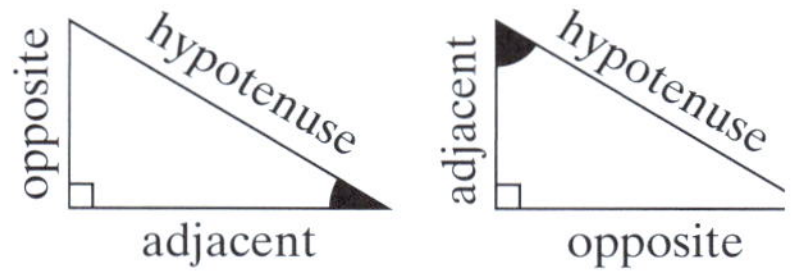

2 Upper-case letters (capitals) are used to name **angles** and lower-case letters to name **sides**. The side a is opposite angle A. Angle B is opposite side b.

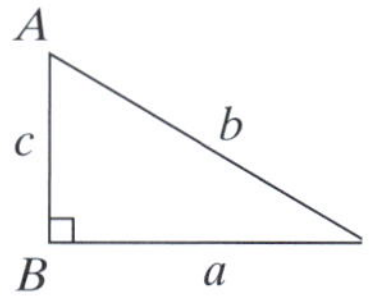

3 The **ratios of the sides** of a right-angled triangle are the same whenever the angles are equal. For example, for every angle of 30°, the opposite divided by the hypotenuse will always be 0.5.

4 **tangent** is $\dfrac{\text{opposite}}{\text{adjacent}}$ **sine** is $\dfrac{\text{opposite}}{\text{hypotenuse}}$ **cosine** is $\dfrac{\text{adjacent}}{\text{hypotenuse}}$

These ratios are abbreviated to tan, sin and cos.

5 We use a **calculator** to find the trigonometric ratio of any angle, or to find the size of the angle for which we know the ratio. Make sure you know how to use your calculator with the trigonometric ratios.

6 There are **60 minutes in a degree**, and **60 seconds in a minute**. Remember to round up to the next degree if there are 30 or more minutes and round up to the next minute if there are 30 or more seconds.

7 To find the **length of a side** of a right-angled triangle using trigonometry, write the ratio and solve a simple equation. For example, to find the length of the side AB we want to find the side opposite the given angle. The side we know is the hypotenuse. So we use the ratio linking opposite and hypotenuse—sine.

$$\begin{aligned}\sin 28^\circ &= \frac{x}{3}\\ x &= 3 \times \sin 28^\circ\\ &= 1.408\,414\,688\ldots\\ &= 1.4 \quad (1 \text{ d.p.})\end{aligned}$$

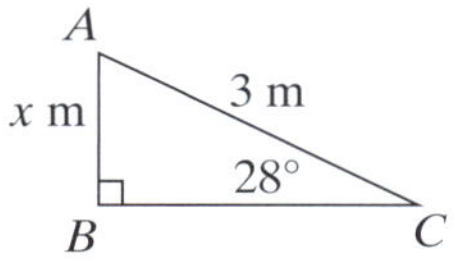

$\therefore AB = 1.4$ m, correct to one decimal place

8 To find the **size of an angle**, write down the ratio and use your calculator. For example, to find the size of angle A as the two sides we know are the hypotenuse and the side adjacent to the required angle we use the ratio linking adjacent and hypotenuse—cosine.

$$\begin{aligned}\cos A &= \frac{7}{12}\\ A &= \cos^{-1}\left(\frac{7}{12}\right)\\ &= 54.314\,665\,29\ldots^\circ\\ &= 54^\circ \quad (\text{nearest degree})\end{aligned}$$

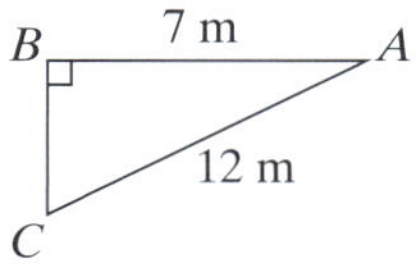

Checklist

Can you:

- [ ] **1** *Identify the hypotenuse, opposite and adjacent sides in a right-angled triangle?*
- [ ] **2** *Name sides and angles?*
- [ ] **3** *Understand that the trigonometric ratios are constant for angles of the same size?*
- [ ] **4** *Know the ratios for tan, sin and cos?*
- [ ] **5** *Use your calculator with the trig. ratios?*
- [ ] **6** *Find sides and angles?*

TRIGONOMETRY OF RIGHT-ANGLED TRIANGLES

Trigonometry

SKILLS CHECK

1 In relation to angle P, name the opposite and adjacent sides and the hypotenuse:

a 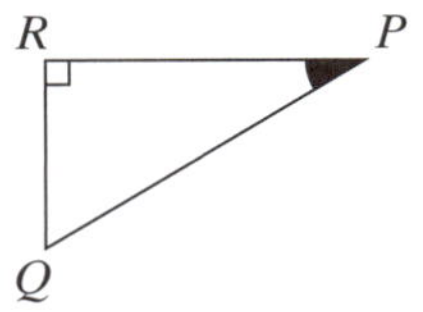

b 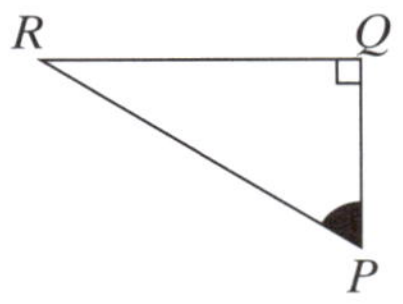

c 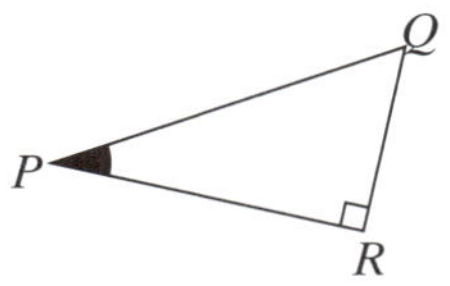

2 Complete:

a The side opposite angle A is side ______

b The angle opposite side c is angle ______

c The angle opposite side b is angle ______

d The side opposite angle Q is side ______

3 Referring to the diagram, write down the correct ratio:

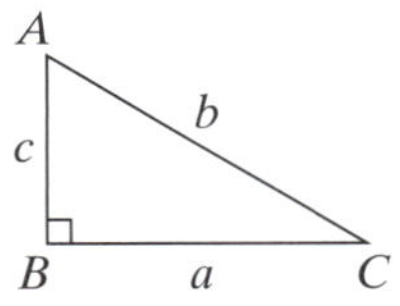

a $\tan A$ b $\sin A$

c $\cos A$ d $\tan C$

e $\sin C$ f $\cos C$

4 Find, correct to three decimal places:

a $\sin 32°$ b $\tan 65°$ c $\cos 49°$ d $\tan 21°$

5 Find the value of $\sin 23°48'$ correct to four decimal places.

6 Find θ, correct to the nearest degree, if:

a $\tan\theta = 0.2375$ b $\sin\theta = 0.7183$ c $\cos\theta = 0.0795$ d $\tan\theta = 2.4$

7 Find the size of angle A to the nearest minute if $\cos A = 0.2785$.

8 Find the value of x, correct to one decimal place, if:

a $\cos 58° = \frac{x}{8}$ b $\tan 23° = \frac{x}{9}$ c $\sin 33° = \frac{3}{x}$

9 Find the size of angle A, correct to the nearest whole degree, if:

a $\tan A = \frac{2}{9}$ b $\cos A = \frac{5}{7}$ c $\sin A = \frac{37}{61}$

PAGES 174–175

Answers **1 a** PQ hypotenuse, PR adjacent, QR opposite **b** PR hypotenuse, PQ adjacent, QR opposite **c** PQ hypotenuse, PR adjacent, QR opposite **2 a** a **b** C **c** B **d** q **3 a** $\frac{a}{c}$ **b** $\frac{a}{b}$ **c** $\frac{c}{b}$ **d** $\frac{c}{a}$ **e** $\frac{c}{b}$ **f** $\frac{a}{b}$ **4 a** 0.530 **b** 2.145 **c** 0.656 **d** 0.384 **5** 0.4035 **6 a** 13° **b** 46° **c** 85° **d** 67° **7** 73°50' **8 a** 4.2 **b** 3.8 **c** 5.5 **9 a** 13° **b** 44° **c** 37°

TRIGONOMETRY OF RIGHT-ANGLED TRIANGLES

Part A Multiple Choice

1 This display was shown on a calculator `65°41′3.34″`. What is this to the nearest degree?

A 34° **B** 41° **C** 65° **D** 66° (1 mark)

2 Which statement is correct?

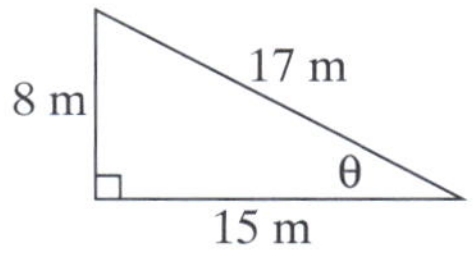

A $\sin\theta = \frac{15}{17}$ **B** $\cos\theta = \frac{17}{15}$

C $\tan\theta = \frac{8}{15}$ **D** None are correct. (1 mark)

3 Which expression will give the correct value of x?

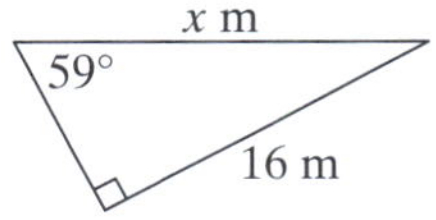

A $\frac{16}{\sin 59°}$ **B** $16\sin 59°$

C $16\tan 59°$ **D** $\frac{16}{\tan 59°}$ (1 mark)

Part B Short Answer

4 Find the length of the side AB.
Give the answer correct to one decimal place. Hint 1 (2 marks)

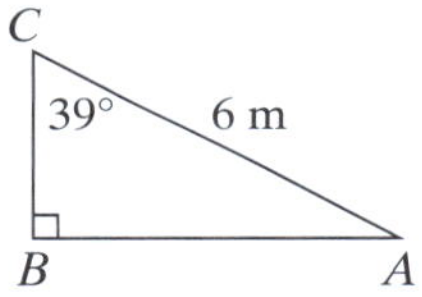

5 Find the size of the angle marked θ to the nearest whole degree. Hint 2 (2 marks)

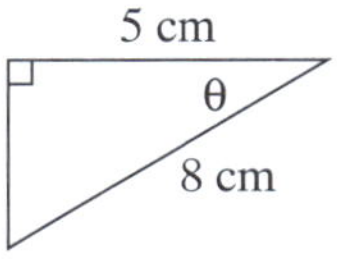

6 Find the length of the hypotenuse. Hint 3 (2 marks)

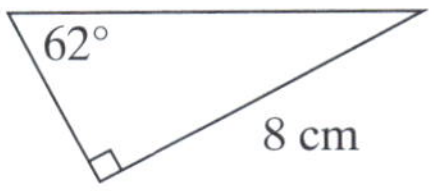

7 Find the value of the unknown. (2 marks each)

a

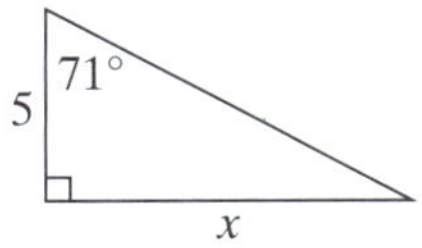

b

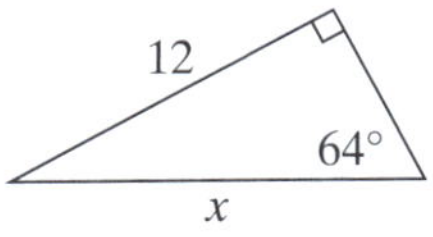

c

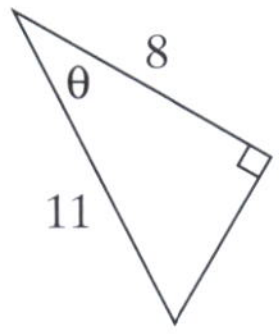

d

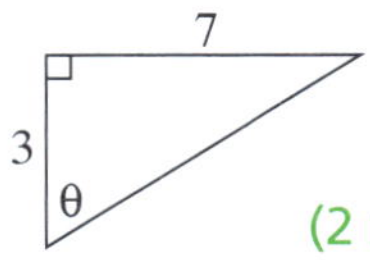

8 To find the width of a dam at its widest part a surveyor places pegs at points A and B either side of the dam. C is a point 30 m from A such that $\angle CAB = 90°$ and $\angle ACB = 52°$. Find the width of the dam to the nearest metre. Hint 4 (3 marks)

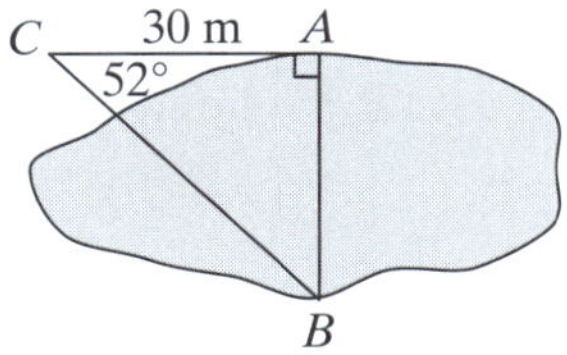

Hint 1: You need to find the side opposite the angle. You know the hypotenuse.
Hint 2: The sides you know are the one adjacent to the angle and the hypotenuse.
Hint 3: The unknown is in the denominator. You will need to divide, not multiply, when solving an equation.
Hint 4: You are looking for the side opposite the angle. You know the length of the adjacent side.

Your Feedback

$\frac{\square}{20} \times 100\% = \square\%$

PAGES 175–176

PAGE 198

TRIGONOMETRY OF RIGHT-ANGLED TRIANGLES

Trigonometry

30 MINUTES

ADVANCED TEST

1 Find the value of x, giving each answer correct to one decimal place.

a
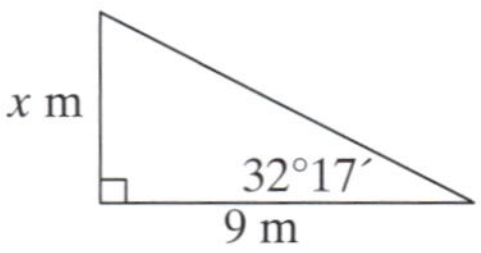

b
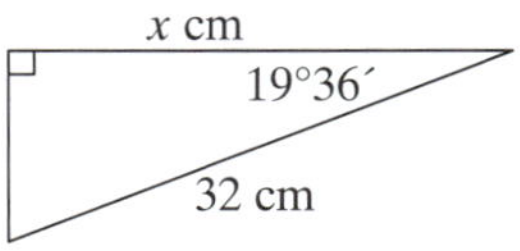

c
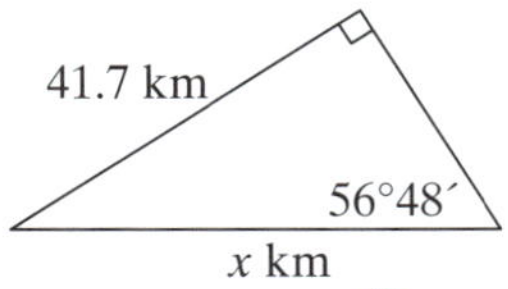

(2 marks each)

2 Find the value of θ, giving each answer correct to the nearest minute.

a
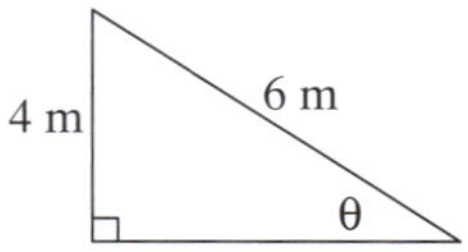

b
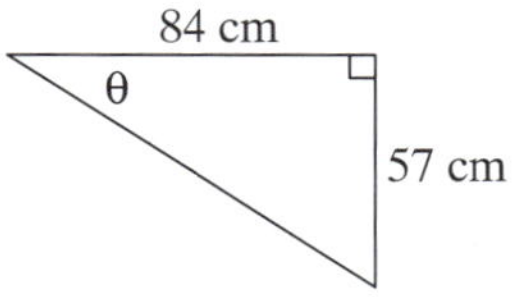

c
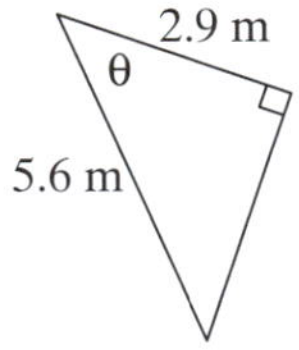

(2 marks each)

3 A rectangular gate has a diagonal brace which makes an angle of 28° with the horizontal. If the gate is 2.5 m long, find the length of the brace. (3 marks)

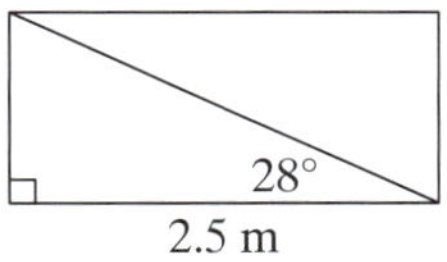

4 A gable roof of a building is 7 m wide and 1.5 m high at its highest point. Find the size of the angle the roof makes with the ceiling. (3 marks)

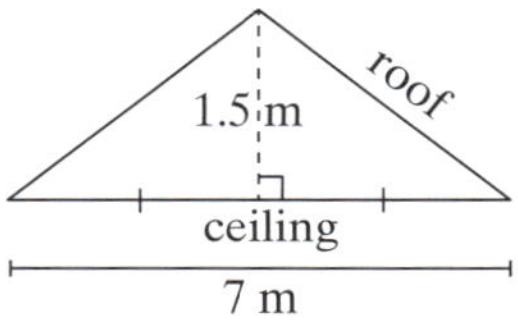

5 a Using ΔBCD, find the length of DC to one decimal place. (2 marks)

b Using ΔABC, find the length of AC to one decimal place. (2 marks)

c Find the length of AD to the nearest metre. (1 mark)

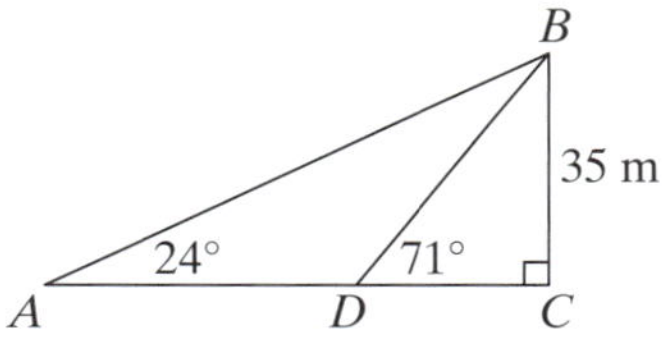

6 a Find the length of BD to one decimal place. (2 marks)

b What is the size of $\angle BAD$ to the nearest degree? (2 marks)

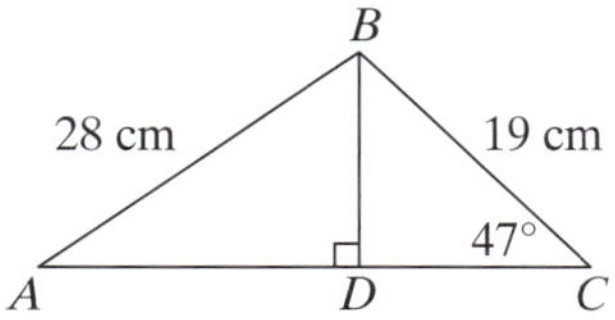

7 a Show that $h = b\sin A$. (1 mark)

b Show that $a\sin B = b\sin A$. (2 marks)

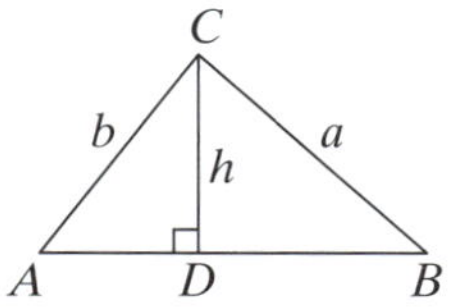

Your Feedback

$\frac{\square}{30} \times 100\% = \square\,\%$

PAGES 176–177

PAGE 198

COLLECTING AND EVALUATING DATA

Single Variable Data Analysis

STUDY NOTES

1 **Data** is information. It can be collected, arranged and analysed. Data can be **displayed** in charts and graphs such as column graphs, sector graphs (pie charts) and picture graphs.

2 **Tally marks** are sometimes used when collecting data. These are arranged in groups of five. 𝍸

3 The **frequency** is the number of times a score occurs.

4 A **survey** can be undertaken to collect opinions, facts or figures.

5 **Data** can be collected by **observation**. For example, observing the traffic to count the number of trucks, cars, buses and motorbikes.

6 **Data** can be collected in a **census**. In a census every member of the population is surveyed. A census gives more accurate information about the population but it is not always a practical way to collect information.

7 **Data** can be collected from a **sample** of the population. This might be a random sample where there is no order or reason for choosing those who take part. Or there could be some system or reasoning behind the choice of participants. It is much easier to collect information from a sample than from a census.

8 **Sample size** is an important consideration when conducting a survey. Generally, the more participants who take part in a sample the more accurate the information will be. If the sample size is too small, then we cannot attach a lot of importance to the result of the survey.

9 A sample should also be **representative**. For example, if trying to determine the favourite state netball team, the survey will be biased if only people in one state are questioned.

10 It is important to try and eliminate **bias** from a survey. If a survey is biased then its results can be misleading. Questions in a survey can be biased if worded in such a way as to produce a desired response. The way in which a sample is chosen can also lead to biased responses.

11 **Questions of ethics** should also be considered.

12 **Numerical data** is data that uses numbers. It involves counting or amounts.

13 **Categorical data** is data that can be placed in categories. For example, categories could be colours, states, countries, days of the week or makes of car.

14 **Published surveys or data** should be carefully considered as to their reliability and validity. When considering the reliability the source of the data is important. Is it a reliable source? (Did it come from a recognised research institute such as a university, or did it come from some anonymous source?) Who paid to have the research done? For validity, consider the size and make-up of the sample. Is the data useful? Are any conclusions drawn from the data valid?

15 **Predictions** can be made from a sample to apply to the whole population. (Sample size should be considered when making such predictions.)

Checklist

Can you: ✓

1 *Understand the terms 'data', 'census', 'sample' and 'survey'?* ☐
2 *Understand what it means to collect data by observation, census or sample?* ☐
3 *Understand the importance of sample size and of a representative sample?* ☐
4 *Recognise bias in a survey?* ☐
5 *Understand the difference between numerical and categorical data?* ☐
6 *Investigate data and critically evaluate published data?* ☐
7 *Make predictions for the whole population based on a sample?* ☐

COLLECTING AND EVALUATING DATA

Single Variable Data Analysis

SKILLS CHECK

1 Determine whether the following surveys are of a sample or census.

a Asking everyone in one class about a question involving the whole school

b Asking everyone in one class about a question involving the whole class

2 This graph was drawn to show the numbers of cars in households.

a How many households had 2 cars?

b How many households had more than 2 cars?

c How many households were surveyed?

d How many cars were there in total?

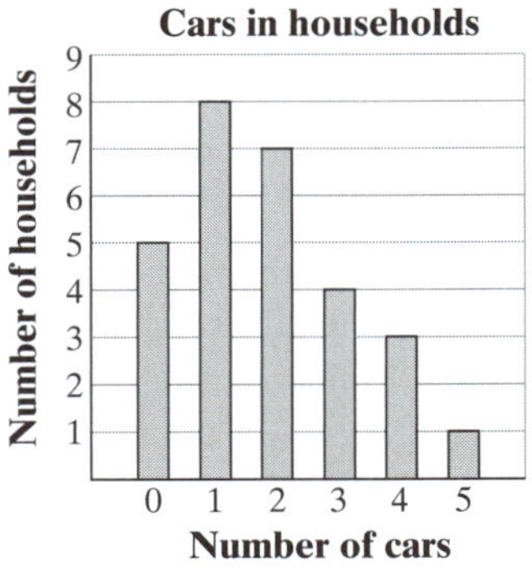

3 This table shows the number of votes received by 5 candidates in a ballot.

Candidates	Number of votes received
Annabel	18
Ethan	24
Henry	31
Laura	37
Pia	26

a How many votes did Pia get?

b Who won the ballot?

c How many more votes did Henry get than Ethan?

d How many people voted?

4 Give a reason why each of these graphs is misleading.

a

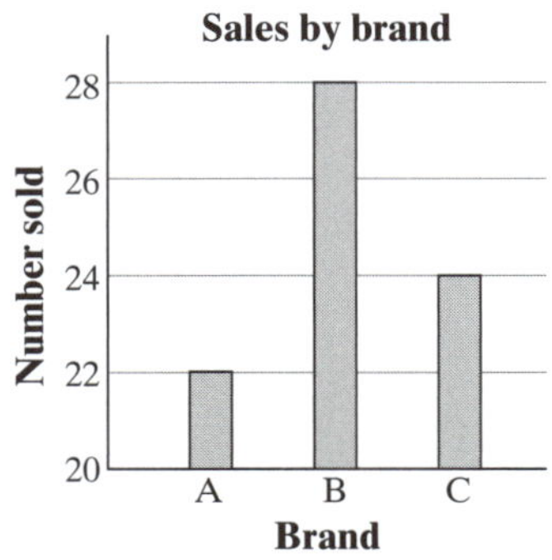

b

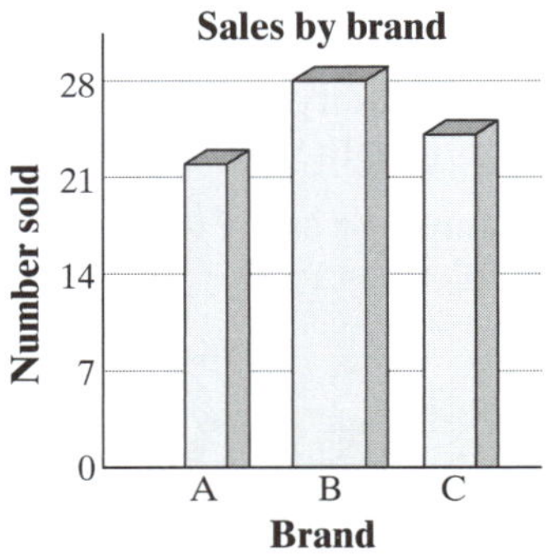

c

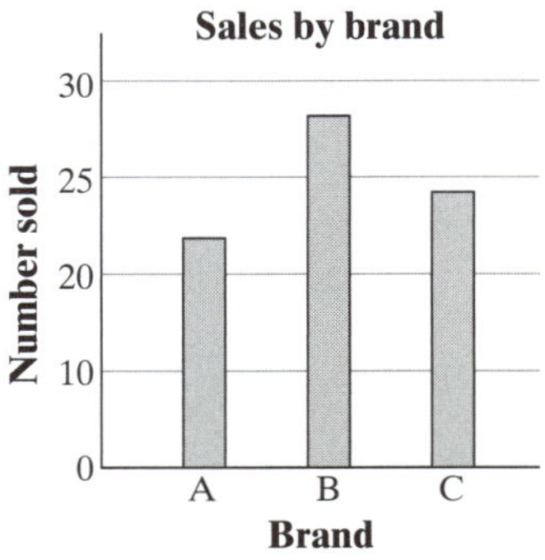

PAGES 177–178

Answers **1 a** sample **b** census **2 a** 2 **b** 8 **c** 28 **d** 51 **3 a** 26 **b** Laura **c** 7 **d** 136 **4 a** The scale on the vertical axis does not start at zero. **b** The use of 3D columns and different widths of the columns. **c** The scale on the vertical axis is not even.

COLLECTING AND EVALUATING DATA
Single Variable Data Analysis

INTERMEDIATE TEST

Part A Multiple Choice

1 Which type of data would be most desirable in a survey?
A unreliable B unbiased C unethical D unrepresentative (1 mark)

2 Which is NOT numerical data?
A populations of countries
B numbers of eggs laid by chooks
C height of children
D positions in a race (1 mark)

3 Which would NOT result in a random sample?
A picking some names from those in a hat
B using a computer to select names from a list
C asking everyone as they come in the door.
D They would all be random samples. (1 mark)

Part B Short Answer

4 This table shows the average rainfall (in mm) for each month at Laurieville.

Summer		Autumn			Winter			Spring			Summer
Jan	**Feb**	**Mar**	**Apr**	**May**	**Jun**	**Jul**	**Aug**	**Sep**	**Oct**	**Nov**	**Dec**
113	204	180	137	111	133	64	62	58	101	147	115

a Which month has the highest average rainfall? (1 mark)
b Which season has the least average rainfall? *Hint 1* (1 mark)
c On average, how much more rain falls in summer than in winter? (1 mark)
d Two other consecutive months have the same total average rainfall as April and May. Which two months are they? *Hint 2* (2 marks)
e What is the average annual rainfall for Laurieville? (1 mark)
f 'Summer is the wettest season, so next year we can expect to get more rain in summer than in any other season.' Do you agree? Justify your answer. (2 marks)
g Ellen lives in Laurieville and is getting married in September next year. She thinks that she is less likely to have a wet day in September than in other months. Give a reason why her theory might be wrong. What other information would you suggest Ellen finds? (2 marks)

5 A survey of students appeared in a school newspaper.

The school uniform is horrible. Should the uniform be changed?	
Response	**Number of students**
Strongly in favour	35
Partly in favour	43
Partly against	20
Strongly against	22
Don't know/don't care	30

a How many students were surveyed? (1 mark)
b How many students were in favour of changing the uniform? (1 mark)
c What percentage of students didn't know (or care)? (1 mark)
d How many more students were strongly in favour than were strongly against? (1 mark)
e 'Based on these results the school uniform should be changed,' Jon said. Give three reasons why Jon could be wrong. *Hint 3* (3 marks)

Hint 1: The seasons are summer, autumn, winter and spring. Count all 3 summer months.
Hint 2: Consecutive means 'following'. In this question it is two months that follow one another.
Hint 3: Consider the sample as well as the figures.

Your Feedback

$\frac{\square}{20} \times 100\% = \square\%$

PAGE 178
PAGE 198

COLLECTING AND EVALUATING DATA

Single Variable Data Analysis

25 MINUTES

ADVANCED TEST

1 A census was taken of every household in a country town to see how many children under 10 lived in the town. The results are shown in the table:

Number of children under 10	0	1	2	3	4	5	6
Number of households	1512	823	764	189	75	21	3

a How many households are in the town? (1 mark)

b How many children under 10 live in the town? (2 marks)

c A neighbouring town is very similar to this one but slightly larger. It has 5300 households. Approximately how many households would you predict have no children under 10? (2 marks)

2 Before buying her car, Molly collected some information about the dimensions of different cars she was considering. The table shows the measurements, all in millimetres.

Make	**Length**	**Width**	**Height**	**Ground clearance**
Ozdrive	4485	1735	1590	200
Zippy	4610	1785	1710	170
Fortuna	4415	1825	1775	205
Prestige	4600	1815	1695	195

a What make of car was both the shortest and widest? (1 mark)

b What was the ground clearance of the longest car? (1 mark)

c How long was the car that had a difference of 120 mm between its width and height? (1 mark)

d What was the biggest difference between the length and width of any one car? (1 mark)

e Molly had believed that the better the ground clearance the greater the overall height of the car. Which make best shows that that is not the case? Justify your answer. (1 mark)

3 A company published this table.

Revenue	**This year $m**	**Last year $m**
Premiums	6672	6423
Dividends	138	112
Interest	425	396
Trusts	59	41
Other	218	178
Total		7150

a What was the total income this year? (1 mark)

b Last year, what percentage of income was interest? (1 mark)

c What percentage of income was interest this year? (1 mark)

d What was the percentage increase in income from interest from last year to this? (1 mark)

4 Five girls nominated for the position of club captain. The graph shows the result of the election.

Votes

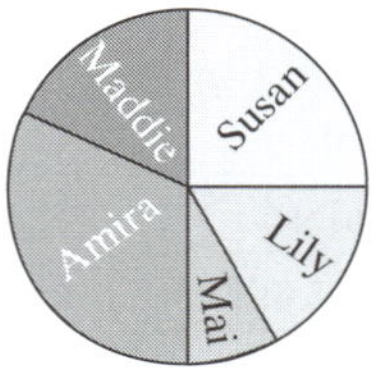

a Who won? (1 mark)

b What fraction of votes were for Mai? (1 mark)

c Lily received 12 votes. How many votes did Susan receive? (3 marks)

d How many more votes did Maddie get than Lily? (2 marks)

5 'Now for the result of last night's survey. A staggering 93% of people wanted to see the new policy changed.'

a Give three reasons why no notice can be taken of this survey. (3 marks)

b Comment on the wording of part **a** of this question. (1 mark)

Your Feedback

$\frac{\square}{25} \times 100\% = \square\%$

PAGES 178–179

PAGE 198

STEM-AND-LEAF PLOTS, HISTOGRAMS AND DESCRIBING DATA

Single Variable Data Analysis

STUDY NOTES

1 **Data** is often sets of scores. The **frequency** is the number of times a score occurs. Data can be displayed or represented in plots and graphs.

2 A **histogram** is like a column graph but without gaps between columns. In a frequency histogram the scores are shown on the horizontal axis (they go across the bottom) and the frequency is shown on the vertical axis.

3 A **dot plot** is another way to represent data.

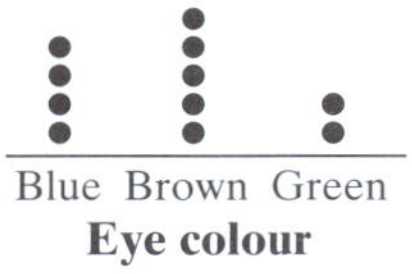

4 A **stem-and-leaf plot** is a way of showing all the scores (usually, but not necessarily, in order). The stem is the first part of the score and the leaf is the second part of the score. For example, this stem-and-leaf plot shows scores between 73 and 98. There are 13 scores altogether and two scores of 91.

Stem	Leaf
7	3 7 8 9
8	0 2 4 5 6
9	1 1 6 8

Key: 7|3 = 73

5 The **mode** of a set of data is the score with the highest **frequency**. It is the score that happens the most. In the stem-and-leaf plot shown, the mode is 91.

6 A **back-to-back stem-and-leaf plot** is a way of showing two sets of scores at the same time. There is a common stem. For example, this back-to-back stem-and-leaf plot shows two groups of scores. The stems are 1, 2, 3 and 4. Group A has scores between 13 and 47 and group B has scores between 18 and 46.

Group A		Group B
6 3	1	8
9 7 4 0	2	1 5 8
8 6 6 3 1	3	2 4 4 9
7 5 2	4	0 0 1 2 6

7 A set of data is **bimodal** if there are two modes. In the back-to-back stem-and-leaf plot shown, group A has a single mode of 36 but the data for group B is bimodal. The two modes are 34 and 40.

8 A data display can show symmetry. If the two sides of the display appear to match we say it is **symmetric**.

9 A data display is said to be **skewed** if it is not symmetric and more scores appear on one side than on the other. It is positively skewed if there are more lower scores and negatively skewed if there are more higher scores.

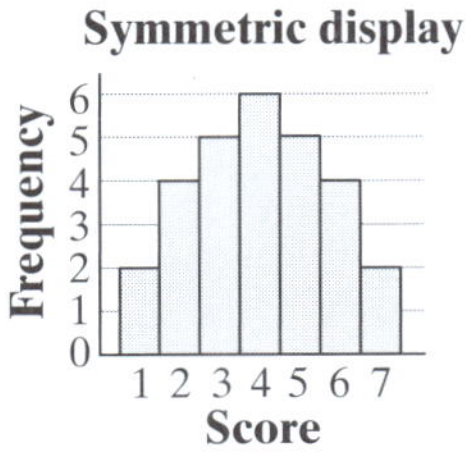

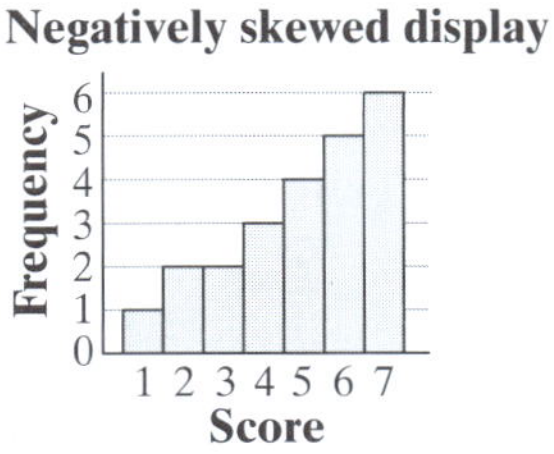

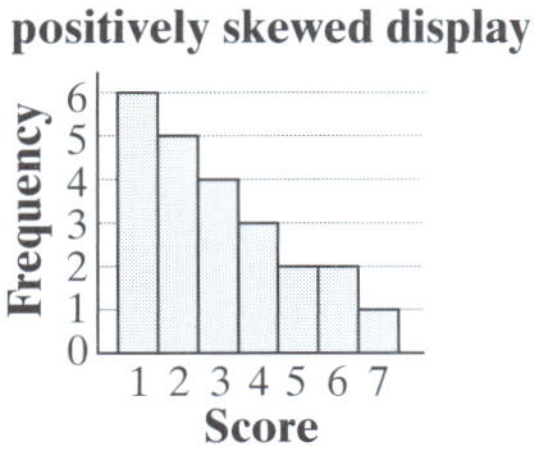

10 The **shape** of a graph or plot can be used when **describing data**, using terms such as 'symmetric', 'skewed' and 'bimodal'.

Checklist

Can you:

1 *Understand the terms 'data', 'frequency' and 'mode'?* ☐
2 *Construct histograms?* ☐
3 *Construct stem-and-leaf plots?* ☐
4 *Understand the terms 'symmetric', 'skewed' and 'bimodal'?* ☐
5 *Understand the difference between the terms 'positively skewed' and 'negatively skewed'?* ☐
6 *Describe data?* ☐

STEM-AND-LEAF PLOTS, HISTOGRAMS AND DESCRIBING DATA

Single Variable Data Analysis

SKILLS CHECK

1 Write down the mode(s) for each set of scores:

a 4, 7, 8, 8, 9, 12

b 2, 6, 3, 8, 4, 7, 2, 5, 3, 9, 6, 1, 4, 6

c 1, 1, 1, 2, 2, 3, 4, 4, 4, 5, 5, 6

d 23, 37, 48, 39, 37, 23, 26, 48, 37, 23

2 The scores achieved by every student in a class test are shown in this stem-and-leaf plot.

a How many students are in the class?

b What was the lowest score?

c What was the highest score?

d What was the mode?

e How many students scored more than 75?

5	3 8
6	0 4 8
7	1 2 3 3 5 7 9
8	0 1 4 6 9
9	2

3 The histogram shows the results in a quiz.

a How many scored 5?

b What was the lowest score?

c What was the highest score?

d Five people achieved one of the scores. What score was that?

e What was the mode?

f How many people took part in the quiz?

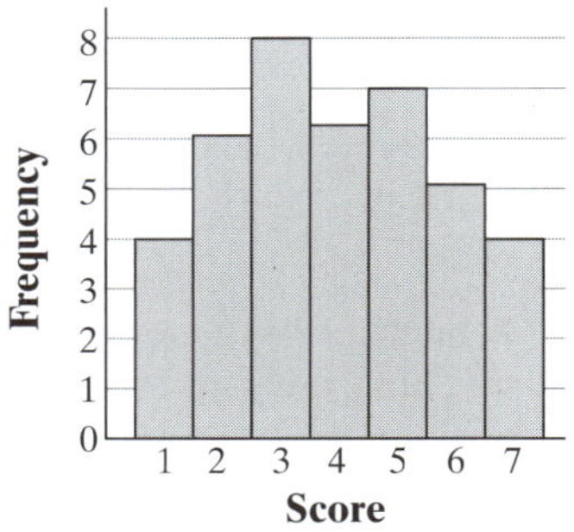

4 This back-to-back stem-and-leaf plot was drawn up to show the heights (in centimetres) of students in a class.

a How many boys are in the class?

b How tall is the tallest boy?

c What is the difference in height between the shortest girl and the shortest boy?

d What height is the mode for girls?

e What height is the mode for boys?

f How many boys are taller than the tallest girl?

Girls		Boys
7 4	13	
9 6 5 2 0	14	
8 7 3 3 1	15	3 4 5 6 8 9
6 4 2	16	1 2 4 5 5 7 9
	17	0 3 6
	18	1

5 Determine whether the data is symmetrical or skewed and, if skewed, whether positively or negatively skewed:

a

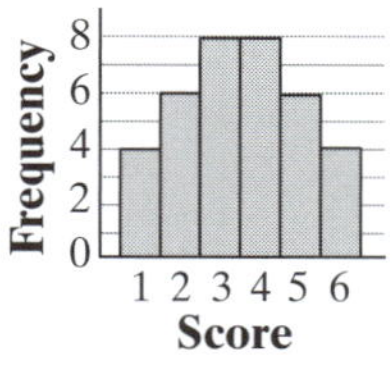

b

1	2 4 6 6 7 9
2	1 3 3 5 5 8 9
3	0 2 6 7
4	3 4 8
5	1 2

c

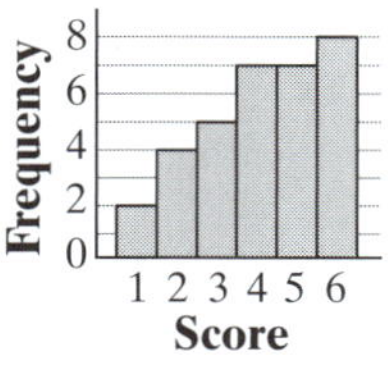

d

1	4 7 8
2	2 3 6 7 9
3	0 1 4 8 8 9
4	0 2 2 2 7
5	3 5 6

e

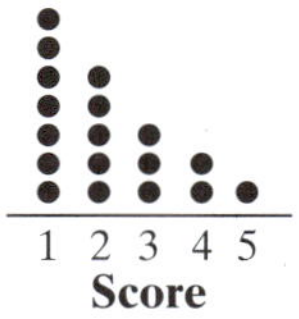

6 For each data set determine whether there is a single mode or whether it is bimodal:

a

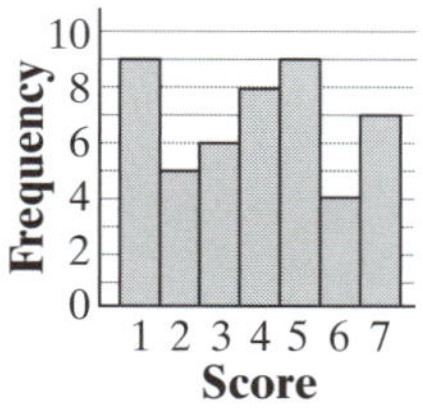

b

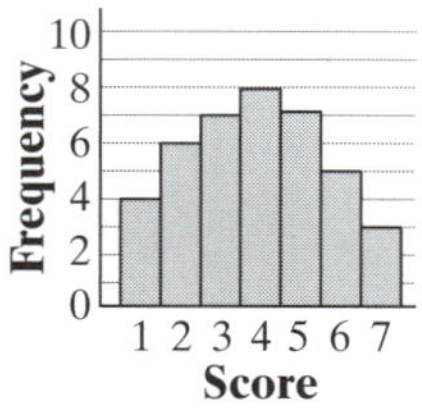

c

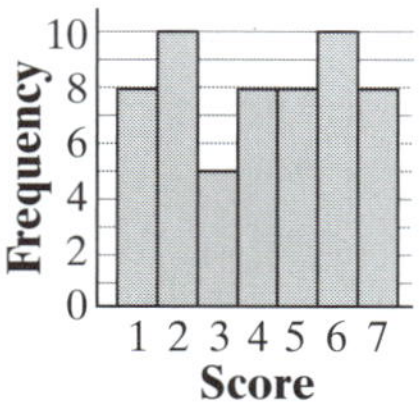

d

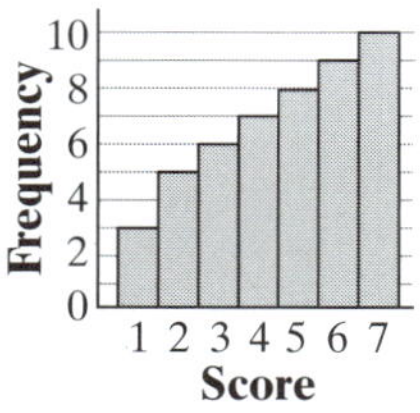

PAGES 179–180

Answers **1 a** 8 **b** 6 **c** 1 and 4 **d** 23 and 37 **2 a** 18 **b** 53 **c** 92 **d** 73 **e** 8 **3 a** 7 **b** 1 **c** 7 **d** 6 **e** 3 **f** 40 **4 a** 17 **b** 181 cm **c** 19 cm **d** 153 cm **e** 165 cm **f** 6 **5 a** symmetric **b** positively skewed **c** negatively skewed **d** symmetric **e** positively skewed **6 a** bimodal **b** single mode **c** bimodal **d** single mode

STEM-AND-LEAF PLOTS, HISTOGRAMS AND DESCRIBING DATA

Single Variable Data Analysis

20 MINUTES

INTERMEDIATE TEST

Part A Multiple Choice

1 Which histogram is NOT symmetric?

A
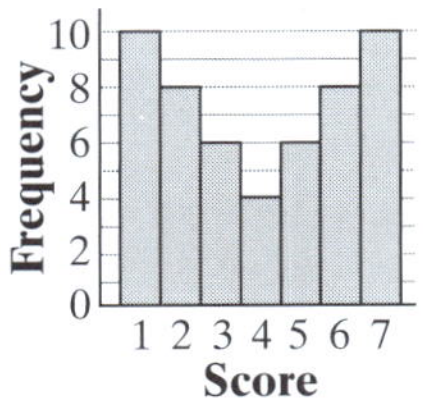

B
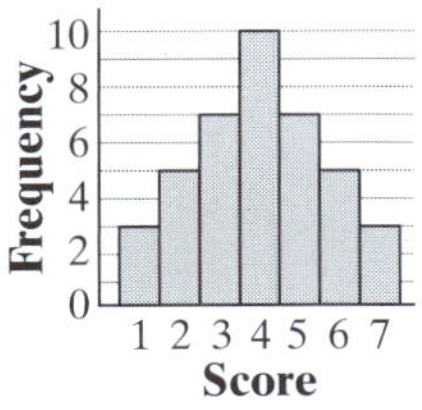

C
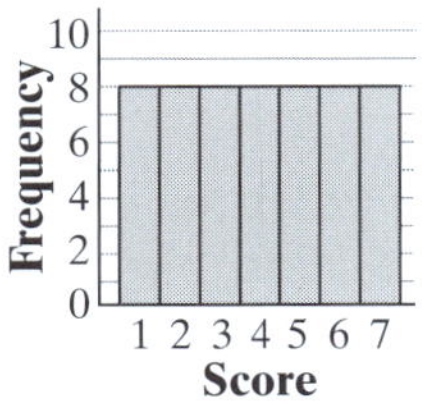

D
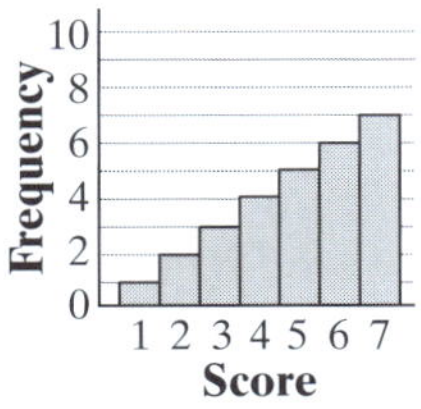

(1 mark)

2 What is the mode of these scores?

A 4
B 48
C 78
D 84

5	4
6	3 7
7	0 2 6 8 9
8	1 4 4
9	2 5

(1 mark)

3 Which shows a histogram that is both skewed and bimodal?

A
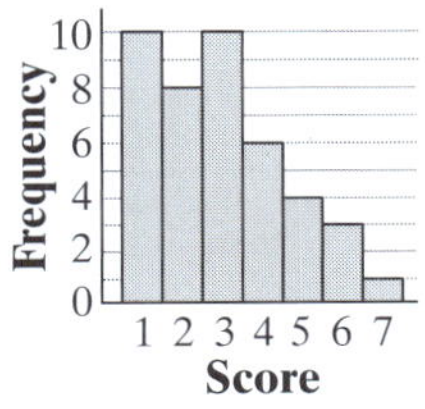

B
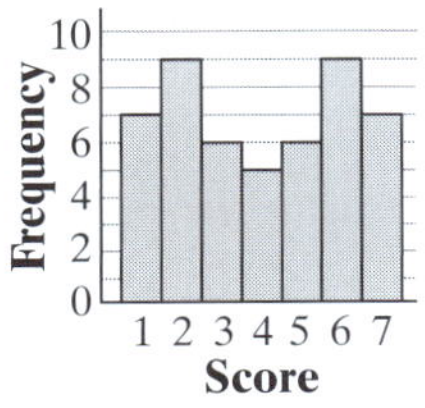

C
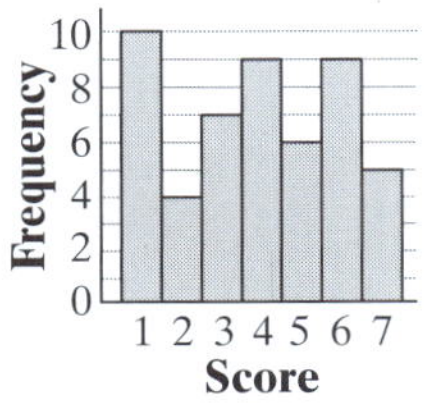

D
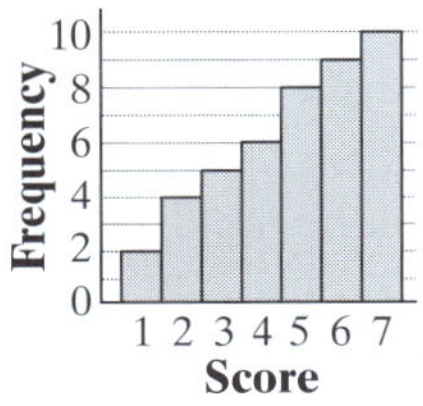

(1 mark)

Part B Short Answer

4 The marks scored by students in an exam were:
Class X: 70, 74, 82, 65, 67, 94, 72, 66, 64, 77, 81, 79, 63, 65, 86, 68, 74
Class Y: 68, 73, 81, 75, 90, 69, 84, 85, 73, 78, 82, 97, 71, 76, 64, 93, 80, 87

a Construct a back-to-back stem-and-leaf plot showing these results. (3 marks)
b Briefly describe the data using the shape of the plot. *Hint 1* (2 marks)

5 A die was rolled. The results are shown in the table.

Score	1	2	3	4	5	6
Frequency	6	5	8	6	8	7

a How many times was the die rolled? *Hint 2* (1 mark)
b Construct a frequency histogram from the data. (3 marks)
c Draw a frequency polygon on the same diagram. (1 mark)
d Briefly describe the shape of the histogram. (1 mark)
e Give a reason why the histogram is better than the table. (1 mark)
f Give a reason why the table is better than the histogram. (1 mark)

6 These are the scores achieved by the members of a class in a spelling test out of 10:
8, 6, 9, 6, 7, 8, 10, 9, 9, 7, 6, 5, 10, 7, 4, 9, 8, 5, 9, 8, 6, 9, 7, 8, 8, 10

a Show these scores on a dot plot. (2 marks)
b Describe the data. (2 marks)

Your Feedback

$\frac{\square}{20} \times 100\% = \square\%$

PAGES 180–181

PAGE 198

Hint 1: Is there symmetry? Are they skewed?
Hint 2: What is the total frequency?

STEM-AND-LEAF PLOTS, HISTOGRAMS AND DESCRIBING DATA

Single Variable Data Analysis

30 MINUTES

ADVANCED TEST

1 This ordered stem-and-leaf plot shows the amount of milk (in L) produced by a sample of cows in a day.

0	8
1	2 3 5 8 8 9
2	0 1 2 ■ 5 6 7
3	1 2 3 4 5 5 6
4	0 1 4 7
5	2

a The amount that Sweetpea produced is covered by a black mark on the display. How many litres might she have produced? (2 marks)

b What percentage of the cows produced more than 40 litres? (1 mark)

c Is the data bimodal? Justify your answer. (1 mark)

d Briefly describe the shape of the display. (1 mark)

e Give two reasons why a stem-and-leaf plot is an appropriate way to display the data. (2 marks)

2 Alice kept a record of the number of minutes it took her to drive to and from work for some weeks. She drew graphs of the results.

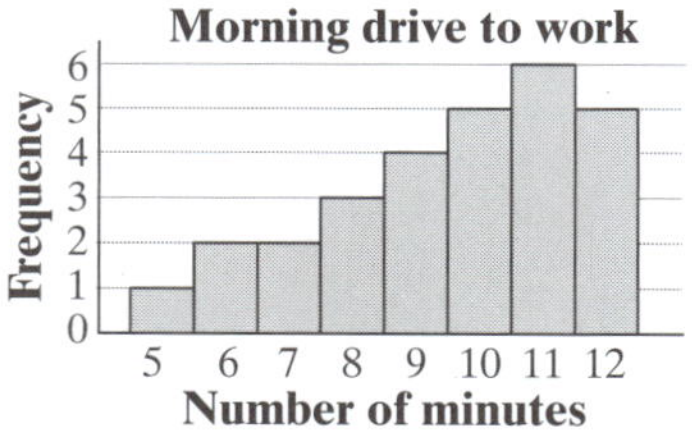

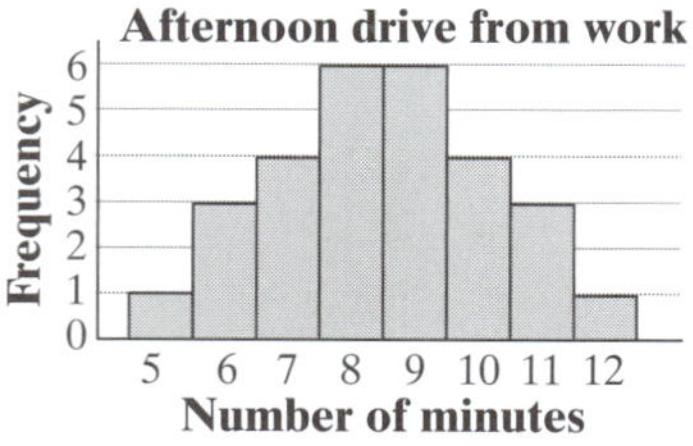

a How many journeys to work did Alice make? (1 mark)

b How many minutes longer is the most common time to travel to work than the most common time to travel home from work? (1 mark)

c One graph is symmetric and one is skewed. Which graph is which? Explain any significance of the shape of the graphs. (2 marks)

3 These are the scores achieved in a class test, marked out of 20.

14, 19, 15, 14, 18, 14, 15, 13, 17, 15, 20, 16, 14, 13, 16, 17, 14, 18, 15, 13, 19, 16, 14, 14, 15, 17, 15, 16

a Show the data in a frequency distribution table. (2 marks)

b Draw a histogram. (2 marks)

c Describe the shape of the display and give a reason for this shape. (2 marks)

d Briefly explain the effect on the display if every score decreased by two marks. (1 mark)

4 These dot plots show the number of goals four players scored in each game last season, e.g. Carlo scored 1 goal in each of 4 games.

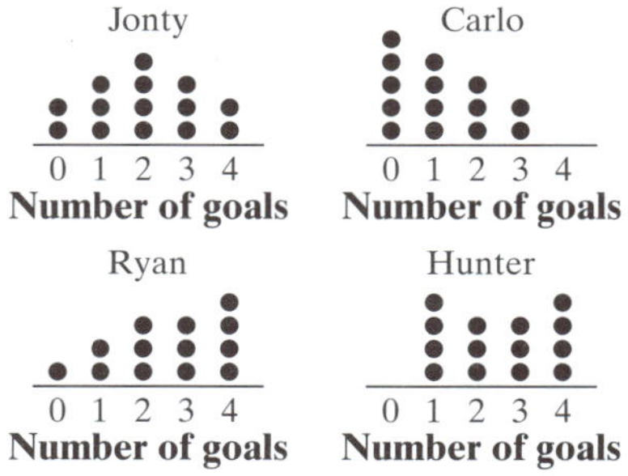

a One player missed one game through injury. Who was that player? (1 mark)

b How many goals did Jonty score altogether? (1 mark)

c Which person has a display that is both symmetric and bimodal? (1 mark)

d Which person's display is positively skewed? (1 mark)

e Who is the weakest player? Justify your answer. (1 mark)

5 The heights (in cm) of some students were recorded.

Boys: 165, 156, 176, 179, 168, 146, 155, 174, 172, 161, 181, 153, 177, 168, 144, 175, 164, 182, 173, 157, 185

Girls: 153, 145, 176, 140, 137, 163, 166, 148, 151, 135, 159, 155, 142, 163, 174, 160, 157, 132, 168, 159, 142

a Draw an ordered back-to-back stem-and-leaf plot of the data. (3 marks)

b What is the modal height of all students? (1 mark)

c What is the difference in height between the tallest and shortest student? (1 mark)

d Comment on any similarities and differences. (2 marks)

Your Feedback

$\frac{\square}{30} \times 100\% = \square\%$

PAGES 181–182

PAGE 198

COMPARING DATA DISPLAYS AND USING MEAN, MEDIAN AND RANGE

Single Variable Data Analysis

STUDY NOTES

1 **Statistical data** can be collected, arranged and analysed.

2 The **mean** is commonly known as the average. To find the mean, add all the scores together and divide by the number of scores. The symbol for mean is $\bar{x}$. The total of all the scores can be found by multiplying each score (x) by its frequency (f). The number of scores is the total frequency.

So $\bar{x} = \dfrac{\Sigma fx}{\Sigma f}$ (Σ means 'sum').

3 **Calculators** can be used in statistics mode to find the mean. Because calculators work in different ways in statistics mode, make sure you know how to use **your** calculator.

4 The **mode** is the score with the highest frequency. It is the score that happens the most.

5 The **median** is the middle score when the scores are arranged in order. If there is an odd number of scores, there is one middle score. When there is an even number of scores, the median is the average of the two middle scores.

6 The **range** is the difference between the highest and lowest score. It is the highest score minus the lowest score.

7 **Mean and median** are measures of **location**. If we know either the mean or median (or both) we have a good idea of where the scores are positioned. For example, if we know that the mean of a set of exam marks is 30% then we know that the marks are not particularly good!

8 The **range** is a measure of **spread**. From the range we can tell whether the scores are all reasonably close together or whether they are spread out. This can help us to determine whether the results are consistent or not. For example, the scores in a test where the range is 20 marks are much more consistent than the scores in a test where the range is 70 marks.

9 In a **stem-and-leaf plot** the highest and lowest scores can easily be seen, so the range can be calculated. The mode can be found by observation and the median can be found by counting the scores and determining the middle score, or by counting from both ends of the plot. Don't forget to write the whole score, not just the leaf part.

10 Two different sets of data can be **compared** and **interpreted**. By comparing the mean of each set, or the median of each set, we have information about the location of both data sets. The range of both data sets will tell us which set is the more consistent and allow us to determine whether there is much difference between them. This information can be combined with information gained from the shape of a display to give us a picture and allow us to make comparisons. For example, in this back-to-back stem-and-leaf plot the range for class C is 33 and the range for class M is 38 so class C is a little more consistent than class M. The median for the two classes is the same so we can conclude that there is not a lot of difference between the two classes.

Class C		Class M
8 6 3	11	0 4 7 7
9 7 5 4	12	1 6 8 8 8
8 7 6 3 2	13	2 3 5 5 7 9
6 4 3 1	14	0 2 2 4 8

Checklist

Can you:

1 *Find the mean, mode, median and range of a set of scores?* ☐

2 *Use a calculator to find the mean of a set of scores?* ☐

3 *Understand that the mean and median are measures of location?* ☐

4 *Understand that the range is a measure of spread?* ☐

5 *Find the mean, median, mode or range from a stem-and-leaf plot?* ☐

6 *Compare different sets of data?* ☐

COMPARING DATA DISPLAYS AND USING MEAN, MEDIAN AND RANGE

Single Variable Data Analysis

SKILLS CHECK

1 Find the mean for each set of scores:

a 5, 7, 8, 16

b 5, 5, 6, 6, 6, 7, 7, 7, 7, 7, 7, 8, 8, 9, 10

2 Find the range for each set of scores:

a 3, 17, 24, 26, 32

b 15, 8, 13, 6, 14, 16, 13, 9, 18, 17, 12, 11

3 Find the median for each set of scores:

a 2, 5, 7, 9, 16

b 15, 23, 31, 27, 19, 34, 20

c 3, 4, 8, 10

d 85, 97, 76, 79, 94, 86

4 For the scores 1, 2, 2, 5, 6, 7, 7, 7, 8 find the:

a mean b mode c median d range

5 For the scores 58, 63, 62, 37, 58, 67 find the:

a mode b median c mean d range

6 Consider the scores in the frequency table.

a What is the mode?

b What is the range?

c What is the mean?

d What is the median?

Score	Frequency
1	6
2	8
3	8
4	11
5	9
6	3
Total	45

7 The scores achieved in a class test are shown in this histogram.

a What is the range?

b What is the mode?

c What is the mean?

d What is the median?

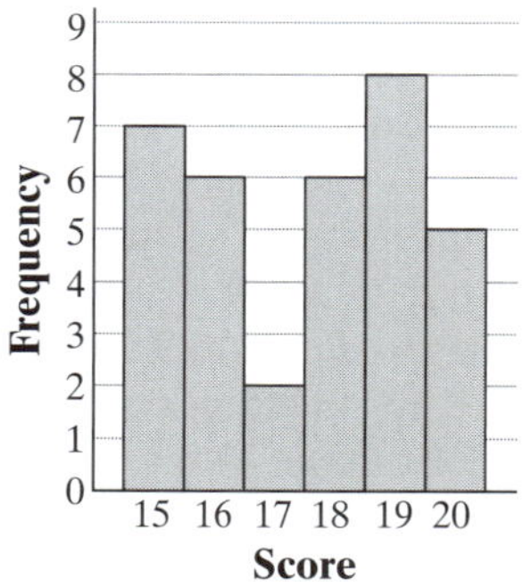

8 Consider the scores in this stem-and-leaf plot:

a What is the mode?

b What is the median?

c What is the range?

d What is the mean?

5	2 3 7
6	0 4 4 5 6
7	1 3 5 7 7 8 9
8	0 0 0 2 3 6 8
9	1 4 5 5 7

PAGE 182

Answers **1** a 9 b 7 **2** a 29 b 12 **3** a 7 b 23 c 6 d 85.5 **4** a 5 b 7 c 6 d 7 **5** a 58 b 60 c 57.5 d 30 **6** a 4 b 5 c 3.4 d 4 **7** a 5 b 19 c 17.5 d 18 **8** a 80 b 78 c 45 d 76.4 (1 d.p.)

COMPARING DATA DISPLAYS AND USING MEAN, MEDIAN AND RANGE

Single Variable Data Analysis

25 MINUTES

INTERMEDIATE TEST

Part A Multiple Choice

1 For the scores 20, 13, 17, 15, 15, 18, 12, 16, 15 which is the largest?
A mean **B** median **C** mode **D** range (1 mark)

2 For the scores 7, 8, 10, 5, 3, 9, 7, 4, 7, 6 which is NOT equal to 7?
A mode **B** range **C** mean **D** median (1 mark)

3 The mean of a set of 5 scores is 8. Another score of 17 is added to the set.
What is the new mean? *Hint 1*
A 8.5 **B** 9 **C** 9.5 **D** 11.4 (1 mark)

4 One set of scores must be more consistent than another if the:
A mean is lower **B** range is lower **C** mean is higher **D** range is higher (1 mark)

Part B Short Answer

5 These are the goals scored per game by two teams in all the games they played last year.

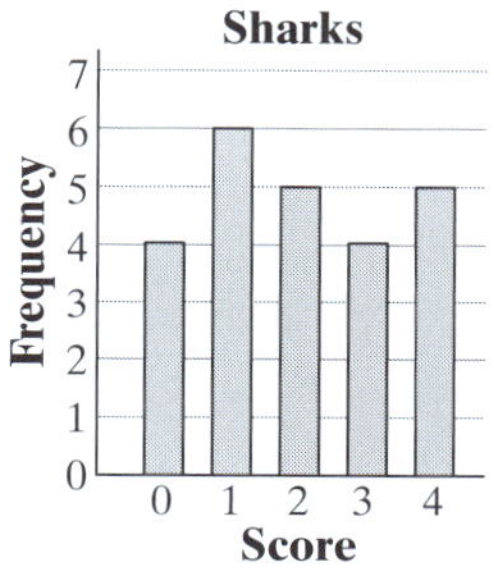

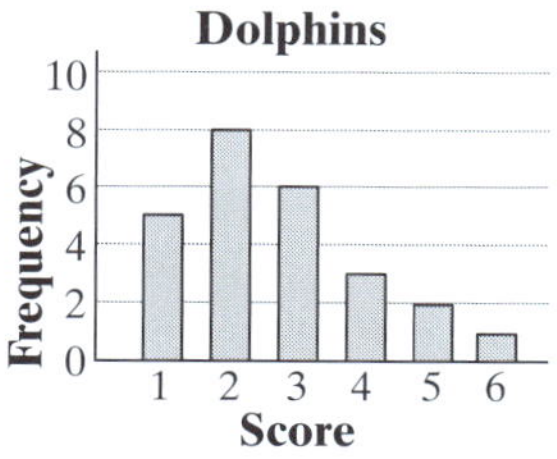

- **a** What is the range for the Sharks? (1 mark)
- **b** What is the range for the Dolphins? (1 mark)
- **c** What is the mode for the Sharks? (1 mark)
- **d** What is the mode for the Dolphins? (1 mark)
- **e** What is the median for the Sharks? *Hint 2* (1 mark)
- **f** What is the median for the Dolphins? (1 mark)
- **g** What is the mean for the Sharks? (1 mark)
- **h** What is the mean for the Dolphins? (1 mark)
- **i** Briefly comment on any comparisons that can be made. (1 mark)

6 Two classes sat for the same exam. The results are shown below:
Class 9P: 70, 85, 76, 52, 74, 68, 81, 62, 66, 75, 59, 68, 71, 78, 82, 63
Class 9S: 77, 55, 61, 73, 64, 67, 72, 59, 84, 60, 79, 75, 80, 73, 65

- **a** Which class has the greater mean? (2 marks)
- **b** Which class has the greater median? (2 marks)
- **c** The range of which class is the greatest, and by how much? (2 marks)
- **d** Daniel is in 9P. He thinks his class did better than 9S. Do you agree? Justify your answer. *Hint 3* (2 marks)

7 The back-to-back stem-and-leaf plots show the scores of two separate classes on the same test.

- **a** Find the median for each set of scores. (2 marks)
- **b** Briefly comment on any similarities or differences with the two sets of data. *Hint 4* (2 marks)

Class T		Class K
7 6	5	6 6 9
6 5 3	6	0 1 1 4 7 7
9 8 4 2 1	7	2 3 5 6 6 8 8 9
8 8 6 5 5 3 1 0	8	1 4 5 6 7 9
4 4 3 2 1 0 0	9	3 4

Hint 1: Both the sum of the scores and number of scores increase.
Hint 2: How many games did the team play? Which game will be the middle one? How many goals were scored in that game?
Hint 3: Consider both the mean and the median.
Hint 4: Do the sets have the same range? Which class performed better?

Your Feedback

$\frac{\square}{25} \times 100\% = \square\%$

PAGES 182–183

PAGE 198

COMPARING DATA DISPLAYS AND USING MEAN, MEDIAN AND RANGE

Single Variable Data Analysis

30 MINUTES

ADVANCED TEST

1 The back-to-back stem-and-leaf plot shows the ages of men and women admitted to the local hospital with heart problems last month.

Males		Females
7 3	4	5
8 6 4 0	5	3 9
9 7 7 7 5 3 2 2 1	6	4 6 8 8 9
8 7 7 6 5 4 4 2 1 1 0	7	0 2 5 7 8 9
7 7 5 4 3 2 0	8	1 3 3 6 8
6	9	0 4 7

a Compare the range of ages for males and for females. (2 marks)

b Discuss the difference in the median age for males and females. (2 marks)

c Discus the difference in the mean age for males and for females. (2 marks)

d Compare the modal age for males and females. (2 marks)

e Briefly comment on the display and the measures that have been found. (1 mark)

f The male patient aged 96 had his age incorrectly recorded. It should have been 76. What difference does this make to the mean, mode, median and range and the whole picture? (3 marks)

2 All of the students in a class did two different quizzes. These dot plots were drawn to show the marks achieved in each quiz.

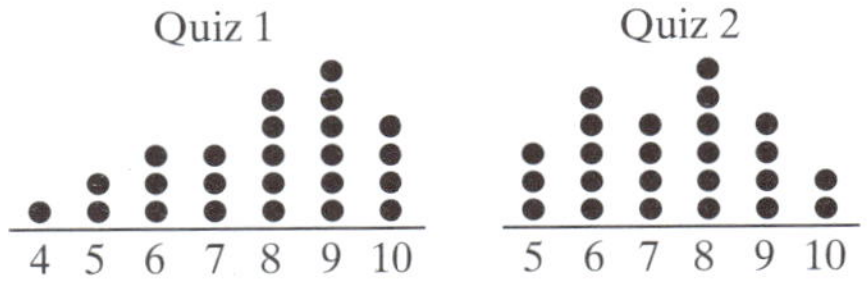

a How many students are in the class? (1 mark)

b Jesse said that the scores were more consistent in quiz 2. Do you agree? Justify your answer. (2 marks)

c On which quiz do you think the students performed better? Justify your answer with reference to the means and medians. (3 marks)

3 A spelling test was given to all the students in a class. The marks were out of 10. Because the students didn't do well, the test was given again the next week. The results are shown here.

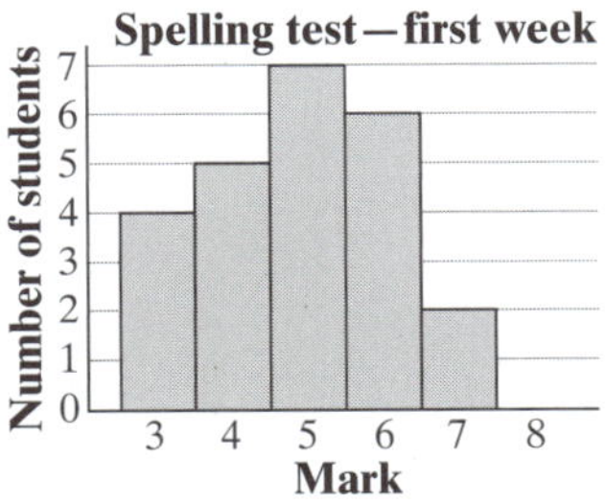

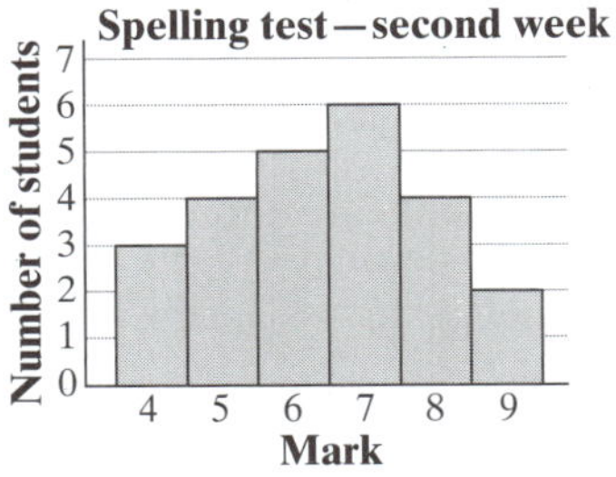

a How many students are in the class? (1 mark)

b By how much did the mean increase in the second test? (2 marks)

c By how much did the median increase in the second test? (2 marks)

d What happened to the range over the two tests? (1 mark)

e Briefly comment on the students' performance in the second test compared to the first. (2 marks)

4 Ernest recorded the number of work-related messages he received on the train on his way to and from work last month.

Number of messages	Frequency in morning	Frequency in afternoon
0	8	5
1	6	7
2	1	4
3	2	3
4	3	1

Compare and contrast with reference to the mean, mode, median and range. (4 marks)

Your Feedback

$\frac{\square}{30} \times 100\% = \square\%$

PAGES 183–184

PAGE 198

RELATIVE FREQUENCIES AND PROBABILITIES

Probability

STUDY NOTES

1 **Outcomes** of an event are the possible results. For example, if a coin is tossed there are two possible outcomes, a head or a tail.

2 Probability often deals with **simple experiments** or **trials**.

3 **Random events** occur completely by chance. There is no system, method or pattern if an event occurs **at random**.

4 The **frequency** of an event is the number of times that event occurs. The **relative frequency** is given by $\frac{\text{frequency of the event}}{\text{total frequency}}$. For example, if a spinner is spun 30 times and 2 appears 7 times, then the relative frequency of 2 is $\frac{7}{30}$.

5 **Relative frequency** can be used to **predict future experimental outcomes**. For example, if the same spinner was spun 120 times we would expect 2 to appear 28 times ($\frac{7}{30} \times 120$). Using a large number of trials gives a more accurate relative frequency, which in turn is more efficient at future predictions.

6 $P(A)$ is the **probability of an event** A. $P(A) = \frac{\text{number of outcomes that give } A}{\text{total number of possible outcomes}}$

For example, when a die is tossed there are 6 possible outcomes (1, 2, 3, 4, 5 or 6). There is one chance in 6 of tossing a 2. We say the probability of tossing a 2 is $\frac{1}{6}$.

7 If an event, A, is **impossible** then $P(A) = 0$. If the event is **certain**, then $P(A) = 1$. Otherwise the probability of an event is always between 0 and 1. $\mathbf{0 \le P(A) \le 1}$

8 **Lists and tables** can be used to keep track of the sample space. For example, a table is the best way to consider the results when two dice are thrown together. Use a system when creating a list. That way you are less likely to leave things out.

9 A **Venn diagram** is another tool used in probability. Intersecting circles are used to keep track of the number of outcomes favourable to certain events. For example, this Venn diagram has been drawn to show the numbers of students studying either history or geography. A total of 39 (18 + 5 + 14 + 2) students were surveyed. 2 studied neither history nor geography. (They are outside the circles.) 5 studied both subjects. (They lie inside both circles.) 18 studied only history but 23 (18 + 5) studied history. 14 studied only geography but 19 (14 + 5) studied geography altogether.

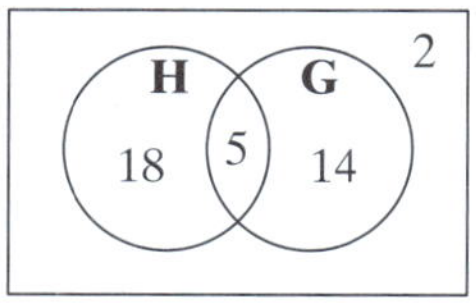

Checklist

Can you:

1 *Understand the terms 'outcomes', 'event', 'at random', 'probability' and 'relative frequency'?* ☐
2 *Find the relative frequency from the results of experiments?* ☐
3 *Find the probability of simple events?* ☐
4 *Understand that probability will never be less than 0 or greater than 1?* ☐
5 *Understand what it means if the probability of an event is 0 or 1?* ☐
6 *Create tables or lists?* ☐
7 *Understand and use Venn diagrams?* ☐

RELATIVE FREQUENCIES AND PROBABILITIES

Probability

SKILLS CHECK

1 When a coin was tossed 20 times, a head appeared 12 times. What is the relative frequency of:

a heads? **b** tails?

2 A fair die is tossed once. What is the probability that the number on the uppermost face is:

a 4? **b** 1? **c** 6? **d** 7?

3 A number is chosen at random from the set {1, 2, 3, 4, 5, 6, 7, 8, 9}. What is the probability that the number is:

a 7? **b** even? **c** odd?
d less than 10? **e** 12? **f** greater than 3?

4 In an experiment a die was tossed 50 times and the number of times each number appeared was recorded. The results are shown in the table below.

Number	1	2	3	4	5	6
Frequency	8	7	8	10	9	8

What is the relative frequency of:

a 2? **b** 5? **c** 6?

5 A bag holds 5 black, 4 yellow and 3 green counters. If one counter is selected at random, what is the probability that it is:

a black? **b** yellow? **c** red? **d** not black?

6 A number of people were surveyed as to whether they played sport, went to a party or went to a movie on a particular weekend. The results are shown in the Venn diagram. How many people:

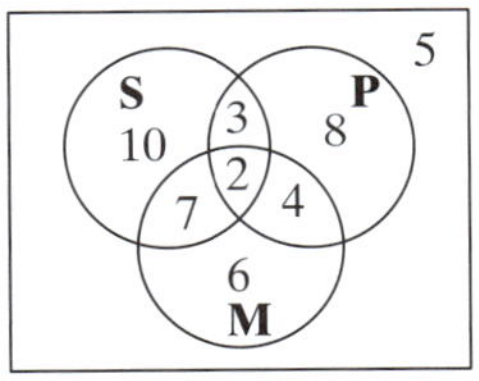

a were surveyed?
b did not participate in any of these activities over the weekend?
c did all three?
d went to both a movie and a party but didn't play sport?
e played sport?

7 The table shows the results when a number of people were surveyed about their weight.

	Overweight	Not overweight	Total
Men	73	56	129
Women	42	60	102
Total	115	116	231

a How many people were surveyed?
b What fraction of men were overweight?
c What fraction of women were not overweight?
d What fraction of overweight people were men?

PAGES 184–185

Answers **1 a** $\frac{3}{5}$ **b** $\frac{2}{5}$ **2 a** $\frac{1}{6}$ **b** $\frac{1}{6}$ **c** $\frac{1}{6}$ **d** 0 **3 a** $\frac{1}{9}$ **b** $\frac{4}{9}$ **c** $\frac{5}{9}$ **d** 1 **e** 0 **f** $\frac{2}{3}$ **4 a** $\frac{7}{50}$ **b** $\frac{9}{50}$ **c** $\frac{4}{25}$ **5 a** $\frac{5}{12}$ **b** $\frac{1}{3}$ **c** 0 **d** $\frac{7}{12}$
6 a 45 **b** 5 **c** 2 **d** 4 **e** 22 **7 a** 231 **b** $\frac{73}{129}$ **c** $\frac{10}{17}$ **d** $\frac{73}{115}$

RELATIVE FREQUENCIES AND PROBABILITIES

Probability

INTERMEDIATE TEST

Part A Multiple Choice

1 A die is thrown once. What is the probability that the number on the uppermost face is 5?

A $\frac{1}{5}$ **B** $\frac{1}{6}$ **C** $\frac{5}{6}$ **D** none of these (1 mark)

2 A bag holds 5 green and 3 yellow marbles. If a marble is chosen at random, what is the probability that it is yellow?

A $\frac{3}{5}$ **B** $\frac{3}{8}$ **C** $\frac{5}{8}$ **D** none of these (1 mark)

3 As a car approached each intersection, a passenger recorded the colour of the traffic lights. She found the lights were red 37 times, green 28 times and amber 9 times. What is the relative frequency of amber lights?

A $\frac{9}{28}$ **B** $\frac{9}{37}$ **C** $\frac{9}{65}$ **D** $\frac{9}{74}$ (1 mark)

Part B Short Answer

4 A table has been drawn to show the responses when a group of men and women were asked which brand of coffee (X or Y) and which brand of chocolate (A or B) they preferred.

	X	Y	Total
A	32	●	63
B	19	56	75
Total	51	87	

a The number of people who preferred brand A chocolate and brand Y coffee can't be read. What should it be? *Hint 1* (1 mark)

b What is the probability that a person:

i who preferred brand A chocolate also preferred brand X coffee? (1 mark)

ii who preferred brand Y coffee also preferred brand B chocolate? (1 mark)

iii chosen at random from the group preferred brand B chocolate and brand X coffee? (2 marks)

5 What is the probability of getting a blue peg when choosing a peg at random from a bag holding 2 blue, 1 green and 3 red pegs? (2 marks)

6 A cardboard spinner is pentagonal in shape. As shown in the diagram it has two red spaces, two grey spaces and one blue space.

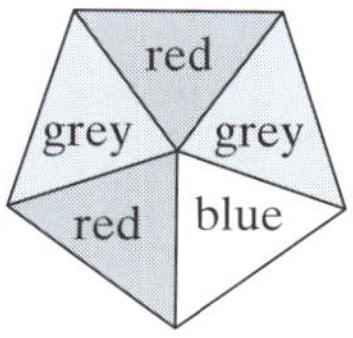

a If the spinner has an equal chance of landing on any of the spaces, what is the probability that it will land on grey? (1 mark)

b The spinner is spun 100 times. It lands on red 37 times and on grey 41 times. What is the relative frequency of blue? *Hint 2* (2 marks)

c The spinner was spun 200 times. Using the relative frequency in part **b**, find the number of times it should land on red? Justify your answer. (2 marks)

7 A survey about pets was taken of 60 households. 38 had at least one dog, 32 had a cat and 14 had both. What is the probability that a household selected at random from this group had neither a cat nor a dog? *Hint 3* (2 marks)

8 A drawing pin was dropped 100 times and a record was made according to whether it landed point down or point up. The results are shown in the table.

Point down	Point up
45	55

a What is the relative frequency, as a percentage, of the pin landing point down? (1 mark)

b If the pin was dropped 120 times, approximately how many times would it land point up? *Hint 4* (2 marks)

Hint 1: Use either row A total or column Y total. Both give the same answer.
Hint 2: First find how many times the spinner must have landed on blue.
Hint 3: Use a Venn diagram.
Hint 4: Use the relative frequency of the pin landing face up.

Your Feedback

$\frac{\square}{20} \times 100\% = \square\%$

PAGES 185–186

PAGE 198

RELATIVE FREQUENCIES AND PROBABILITIES

Probability

30 MINUTES

ADVANCED TEST

1 Ellen rolled a die 60 times. The number of times each score was obtained is shown in the table.

Number	1	2	3	4	5	6
Frequency	8	11	10	9	9	13

a What is the relative frequency, as a fraction in simplest form, of a score:

- **i** of 2? (1 mark)
- **ii** of 5? (1 mark)
- **iii** that is even? (1 mark)
- **iv** greater than 2? (1 mark)
- **v** less than 3? (1 mark)

b How many times would you expect each score to occur? (1 mark)

c Which scores occurred more often than you would expect? (1 mark)

d A second trial is to be conducted, rolling the die 240 times. Ellen commented that she would expect to get a score of one 32 times. Do you agree? Justify your answer. (2 marks)

2 A coin is tossed 24 times. The relative frequency of heads is $\frac{3}{8}$.

a How many times did tails appear? (1 mark)

b In a second trial the same coin was tossed 40 times. The number of tails was the same as in the first trial. What is the relative frequency of heads for the second trial? (1 mark)

c Jonah commented that the overall relative frequency of heads must be $\frac{1}{2}$. Do you agree? Justify your answer. (2 marks)

3 Elsie made this spinner.

a What is the theoretical probability of spinning 3? (1 mark)

b In three different trials, each involving 36 spins, the number 3 was spun 8, 6 and 7 times respectively. Is this what you would expect? Comment. (2 marks)

c In another trial 2 occurred 8 times. The relative frequency of 2 and 3 respectively was $\frac{1}{3}$ and $\frac{1}{4}$. How many times did 1 occur? (2 marks)

4 **a** A bag holds marbles in three different colours. 8 marbles are green, 6 are blue and the rest are red. The probability of randomly choosing a blue marble from the bag is $\frac{2}{9}$. What is the probability of choosing a red marble? (2 marks)

b Another bag also has marbles in the same three colours. The probability of choosing a blue marble is $\frac{1}{4}$ and the probability of choosing a red marble is $\frac{2}{5}$. What is the probability of choosing a green marble? (2 marks)

5 A Venn diagram has been drawn to show the number of students from Year 9 who represented a school in athletics, swimming or cross-country. What is the probability that a randomly chosen student from Year 9 represented the school in:

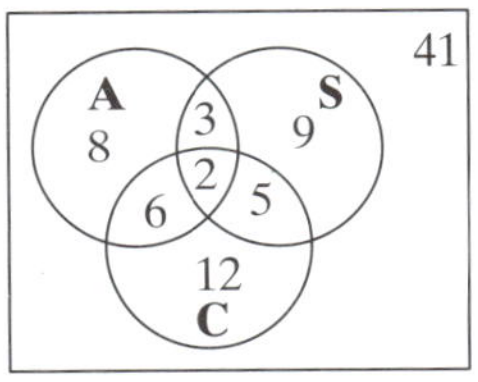

- **a** swimming? (1 mark)
- **b** all three sports? (1 mark)
- **c** only cross-country? (1 mark)
- **d** athletics and cross-country but not swimming? (1 mark)
- **e** exactly two of the sports? (1 mark)
- **f** swimming or athletics but not both? (1 mark)

6 A group of 50 students was asked if they had seen two movies. 19 of the students had seen *Rust* and 23 had seen *Flashy*, while 14 had not seen either movie. If a student is chosen at random from the group, what is the probability that he or she had seen both? (2 marks)

Your Feedback

$\frac{\square}{30} \times 100\% = \square\%$

PAGES 186–187

PAGE 198

TWO-STEP CHANCE EXPERIMENTS

Probability

STUDY NOTES

1 In probability we consider the likelihood of certain events. The set of all possible outcomes is called the **sample space**. For example, if the colour of a card chosen at random from a pack of playing cards is noted, the sample space is red and black.

2 $P(A)$ is the **probability of an event A**. It is the **chance** that the event A occurs.

$$P(A) = \frac{\text{number of outcomes that give } A}{\text{number of outcomes in the sample space}}$$

3 The **probability of an event** is usually expressed as a **fraction**, but it could also be given as a decimal or percentage. If the probability of an event is **0**, the event is **impossible**. If the probability of an event is **1**, the event is **certain**. Otherwise the probability of an event will be **between 0 and 1**. The **sum** of the probabilities of **all possible outcomes** must always be **1**.

4 If two events are such that if one doesn't happen then the other must, the events are **complementary.** The probability of an event is one minus the probability of the complementary event. For example, if the probability that it will rain tomorrow is $\frac{1}{3}$, the probability that it will not rain tomorrow is $1 - \frac{1}{3} = \frac{2}{3}$.

5 When considering **two-step events** we need to know if the likelihood of one part occurring is influenced by the likelihood of the other. It is important to know when balls are chosen from a bag, for example, whether it is **with or without replacement**; that is, whether or not the first ball is replaced before the second ball is drawn. The reason why this is important is because the probability for the second ball will be different if the first ball is not replaced.

6 A **tree diagram** is simply a tool used to help determine the **sample space** and find **probabilities**. For example, if drawing a tree diagram to show the possible outcomes if a coin is tossed twice, the first coin could either be a head or a tail. If the first coin was a head, the second coin could either be a head or a tail and if the first coin was a tail, the second coin can either be a head or tail. So, there are four different possible outcomes.

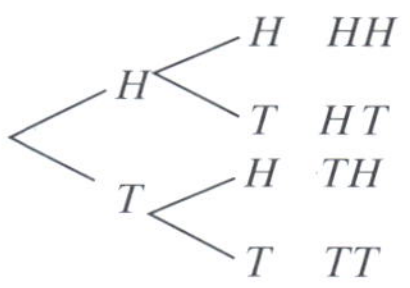

The probability of two heads is $\frac{1}{4}$ and the probability of a head and a tail (in any order) is $\frac{2}{4} = \frac{1}{2}$.

7 In **compound events** if we can find the sample space, we can then determine the **favourable outcomes** and then the probability. For example, if tossing two dice and determining the probability of a sum less than 4, by drawing a table we can find there are 36 possible outcomes of which 3 are favourable (1 and 1, 1 and 2, 2 and 1). So the probability of getting a sum less than 4 is $\frac{3}{36}$ or $\frac{1}{12}$.

Checklist

Can you:

1 *Understand what is meant by probability and sample space?* ☐
2 *Find the sample space and probabilities of simple events?* ☐
3 *Understand what is meant by complementary events?* ☐
4 *Understand the need to check whether certain events occur with or without replacement?* ☐
5 *Draw tree diagrams?* ☐
6 *Find probabilities of compound events?* ☐

TWO-STEP CHANCE EXPERIMENTS

Probability

SKILLS CHECK

1 List all the possible outcomes if:

a a coin is tossed once

b a whole number, greater than 5 and less than 10, is chosen at random

2 List the sample space if:

a a die is thrown once

b a letter is chosen at random from the word 'STRANGE'

3 The probability that a letter chosen at random from the alphabet is 'k' is $\frac{1}{26}$. What is the probability that the letter is not 'k'?

4 The table shows the possible outcomes if two dice are tossed together and the numbers on the uppermost faces added. What is the probability that the total is:

a 8?

b less than 4?

c greater than 6?

d even?

+	1	2	3	4	5	6
1	2	3	4	5	6	7
2	3	4	5	6	7	8
3	4	5	6	7	8	9
4	5	6	7	8	9	10
5	6	7	8	9	10	11
6	7	8	9	10	11	12

5 A box holds 7 blue and 3 red pens. Dan takes out a pen and finds that it is blue. He then takes a second pen at random. What is the probability that the second pen is blue if:

a the first pen is replaced before the second pen is selected?

b the first pen is not replaced before the second pen is selected?

6 A tree diagram has been drawn to show the possible results when two coins are tossed. What is the probability that:

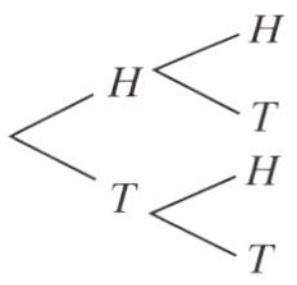

a both coins show heads?

b both coins show tails?

c one coin shows a head and one shows a tail?

d at least one coin shows a head?

7 A bag holds 5 yellow and 4 white golf tees. One tee is chosen at random from the bag.

a What is the probability that it is white?

b A second tee is chosen at random. What is the probability that it is also white (the first tee not having been replaced)?

c If the first tee was replaced, what is the probability that the second tee chosen is white?

8 There are 3 marbles in a box; one is red, one blue and one green. Mia takes a marble from the box, notes the colour and then replaces it. She then again takes a marble from the box.

a Draw a tree diagram to show the possible results.

b What is the probability that one marble is red and one is blue?

9 There are 3 marbles in a box; one is red, one blue and one green. Pia takes a marble from the box and does not replace it. She then again takes a marble from the box.

a Draw a tree diagram to show the possible results.

b What is the probability that one marble is red and one is blue?

PAGES 187–188

Answers **1 a** H, T **b** 6, 7, 8, 9 **2 a** 1, 2, 3, 4, 5, 6 **b** A, E, G, N, R, S, T **3** $\frac{25}{26}$ **4 a** $\frac{5}{36}$ **b** $\frac{1}{12}$ **c** $\frac{7}{12}$ **d** $\frac{1}{2}$ **5 a** $\frac{7}{10}$ **b** $\frac{2}{3}$ **6 a** $\frac{1}{4}$ **b** $\frac{1}{4}$ **c** $\frac{1}{2}$ **d** $\frac{3}{4}$ **7 a** $\frac{4}{9}$ **b** $\frac{3}{8}$ **c** $\frac{4}{9}$ **8 a** see page 188 **b** $\frac{2}{9}$ **9 a** see page 188 **b** $\frac{1}{3}$

TWO-STEP CHANCE EXPERIMENTS

Probability

20 MINUTES

INTERMEDIATE TEST

Part A Multiple Choice

1 Which could NOT be the probability of an event? *Hint 1*

A 0 **B** $\frac{2}{5}$ **C** $\frac{5}{2}$ **D** 1 (1 mark)

2 There are four blue, three red and two green pegs in a bag. Two pegs are taken from the bag, one after the other, without replacement.
The first peg is red. What is the probability that the second peg is also red?

A $\frac{1}{3}$ **B** $\frac{1}{4}$ **C** $\frac{2}{9}$ **D** $\frac{3}{8}$ (1 mark)

3 Two coins are tossed together. What is the probability that both show tails?

A $\frac{1}{4}$ **B** $\frac{1}{3}$ **C** $\frac{1}{2}$ **D** none of these (1 mark)

4 There are three ribbons in a box. Two are pink and one is blue. One ribbon is taken from the box, its colour is noted, and then it is replaced. A second ribbon is then taken. What is the probability that both ribbons are blue?

A 0 **B** $\frac{1}{9}$ **C** $\frac{1}{3}$ **D** none of these (1 mark)

Part B Short Answer

5 There are four calves in a yard. Two calves are black, one is red and one is white. A buyer is choosing two of the calves. Each of the calves is equally likely to be chosen.

a Draw a tree diagram to show the possible pairings of calves. *Hint 2* (2 marks)

b What is the probability that both chosen calves are black? (1 mark)

c What is the probability that at least one of the chosen calves is red? (1 mark)

d The buyer makes her first choice and chooses one of the black calves. What is the probability that the other calf will also be black? *Hint 3* (1 mark)

6 Paris and Dane play a game. They throw two dice together and subtract the lower value from the higher to form the score. (The score is zero if the numbers on the dice are the same.) Paris wins the game if the score is 1 or 3. Dane wins if the score is 2, 4 or 5. Who has the greater chance of winning? Justify your answer. *Hint 4* (3 marks)

7 There are three pens in a case: one black, one blue and one red. Without looking, Julie takes two pens from the case.

a What is the probability that the pens are black and blue? (2 marks)

b James knows that the first pen Julie took was black. What is the probability that the pens are black and blue? (1 mark)

8 There are five cards in a hat. The letters A, B, C, D and E are written on the cards, one letter on each card. One card is taken, without looking, from the hat. The first card is not replaced and a second card is then drawn.

a What is the probability that the first card shows B? (1 mark)

b What is the probability that the second card shows C? (1 mark)

c Make a list of all the possible outcomes for the two cards. (1 mark)

d What is the probability that the cards are D and E, in any order? (1 mark)

e If the first card was replaced before the second card was drawn, what is the probability that the cards are D and E (in any order)? *Hint 5* (1 mark)

Hint 1: Between what two values must all probabilities lie?

Hint 2: There is no replacement.

Hint 3: The first calf is black. Only look at that part of the tree diagram.

Hint 4: Draw a table to show possible outcomes.

Hint 5: Add more outcomes to your list.

Your Feedback

$\frac{\square}{20} \times 100\% = \square\%$

PAGE 188

PAGE 198

TWO-STEP CHANCE EXPERIMENTS

Probability

30 MINUTES

ADVANCED TEST

1 One red, one green and two blue pencils are on a desk.

- **a** Draw a tree diagram and show the sample space if two pencils are taken (without replacement). (2 marks)
- **b** What is the probability that:
  - **i** both pencils are blue? (1 mark)
  - **ii** both pencils are the same colour? (1 mark)
  - **iii** the first pencil is red? (1 mark)
  - **iv** exactly one pencil is blue? (1 mark)
  - **v** neither pencil is blue? (1 mark)
  - **vi** neither is red? (1 mark)
  - **viii** one is green and one is blue? (1 mark)
- **c** If the first pencil is replaced before the second is taken will the probability of getting 2 blue pencils be greater or less than before? Justify your answer. (2 marks)

2 There are six mugs on a table, four are white and two are blue. Jack takes one of the mugs and then Cath takes one of the remaining ones.

- **a** What is the probability that Cath's mug is blue? (1 mark)
- **b** Jack's mug was white. What is the probability that Cath's mug is blue? (1 mark)
- **c** Briefly explain the significance of knowing the colour of Jack's mug for the probability of Cath's mug. (1 mark)

3 This spinner is spun twice.

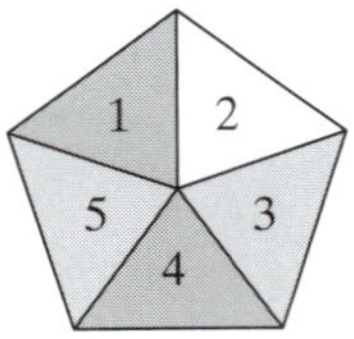

What is the probability that:

- **a** the same number will be spun each time? (1 mark)
- **b** the two numbers add to 6? (1 mark)
- **c** the first number is greater than the second number? (1 mark)

4 A bucket holds 5 green, 3 red and 2 yellow balls. Two balls are taken from the bucket to be used in a game. What is the probability that:

- **a** both balls are red? (1 mark)
- **b** at least one ball is green? (2 marks)
- **c** exactly one ball is green? (1 mark)

5 There are 10 songs on a playlist. Choosing 'play' a song is randomly selected from the playlist and played. Two songs are played, one after the other.

- **a** What is the probability that the same song is played twice? (1 mark)
- **b** How many songs would need to be on the playlist so that the probability of the same song being repeated is less than 5%? (1 mark)

6 Taseefa has these ten cards.

1	2	3	4	5	6	7	8	9	10

- **a** What is the probability that the sum of the numbers on two cards, chosen randomly one after the other, is 7 if the cards are chosen:
  - **i** with replacement? (1 mark)
  - **ii** without replacement? (1 mark)
- **b** What is the probability that a randomly chosen card is both odd and prime? (1 mark)
- **c** Taseefa makes up a game where two cards are chosen one after the other with replacement. She wins if both cards are not both odd and prime. If Taseefa continued to play the game and bet money on the result would she make a profit or lose money? Justify your answer with appropriate calculations. (2 marks)
- **d** If, in Taseefa's game, the first card is not replaced will it make a difference to whether she would make a profit or loss? Justify your answer. (2 marks)

Your Feedback

$\frac{\square}{30} \times 100\% = \square\%$

PAGES 189–190

PAGE 198

TIPS FOR THE SAMPLE EXAM PAPERS

STUDY TIPS

Know what to expect

- **Find out** from your teacher exactly what topics will be assessed in the exam.
- Yearly exams normally test **all work covered** throughout the year.
- **Determine what format** is being used for the exam. Are there multiple-choice questions? Will calculators be allowed in any part of the examination?

Show all working

- **Read the instructions** on the exam paper—marks may be allocated for working.
- Even when **marks are not allocated**, working is important and it is rewarded, even though your answer may be incorrect.
- **Never use correction fluid**—just put a line through any incorrect working. Let your teacher see all your work.

Allocate your time

- Ensure you are **working through the paper efficiently** and not spending too much time on each question, only to find you run out of time at the end of the exam.

Understand mark allocations

- **Questions worth more marks** are often more complex and difficult. Working is crucial in these types of questions.

Re-read and check

- Once you have **completed a question**, rather than moving on to the next, re-read that question to make sure you have in fact answered the correct question. This will only take a second or two.
- If you **finish the exam** with time still remaining, check your answers—often mistakes are found and can be corrected.

Use quality diagrams

- Diagrams should be of **good quality, large and drawn with a lead pencil**. Use an eraser, not correction fluid, to delete mistakes.

Be ready

- Finally, you need to be **prepared for the exam**.
- Always **study mathematics actively**. Active study means using pen and paper to make notes, writing down difficult questions and their solutions, and recording rules to learn.

SAMPLE EXAM PAPERS

Level of Difficulty—Average

60 MINUTES

PAPER 1

Part A Multiple Choice (30 marks)

1 What is the area of this shape?

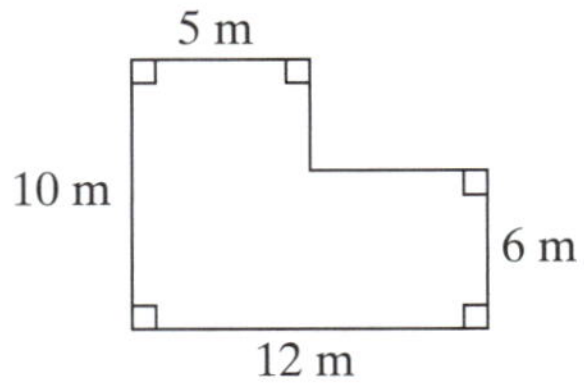

A 92 m^2
B 122 m^2
C 3600 m^2
D There is not enough information to determine the area.

(1 mark)

2 If $k = -3$, $2k^2 = ?$

A 18 **B** 36 **C** –18 **D** –36 (1 mark)

3 Which is closest to the value of x?

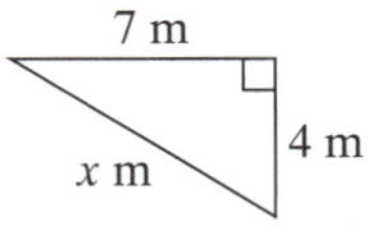

A 8 **B** 9 **C** 10 **D** 11 (1 mark)

4 When written in scientific notation, 0.000 476 = ?

A 476×10^{-6} **B** 4.76×10^{-4} **C** 4.76×10^{-3} **D** 4.76×10^{-6} (1 mark)

5 In the diagram, $\sin\theta = ?$

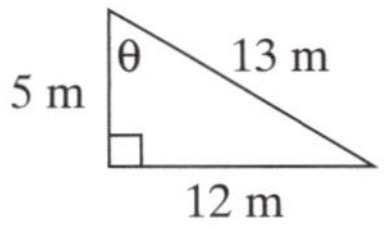

A $\frac{5}{12}$ **B** $\frac{12}{5}$ **C** $\frac{5}{13}$ **D** $\frac{12}{13}$ (1 mark)

6 Which solid is NOT a prism?

A
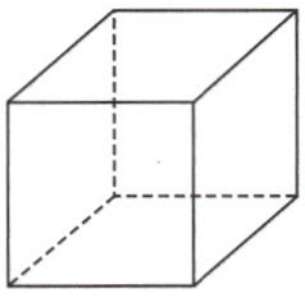
B
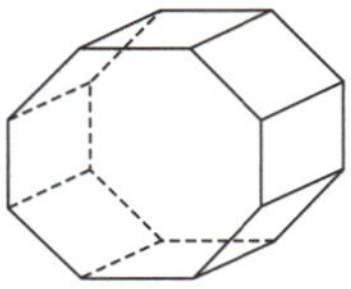
C
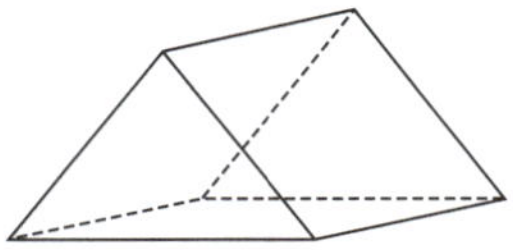
D
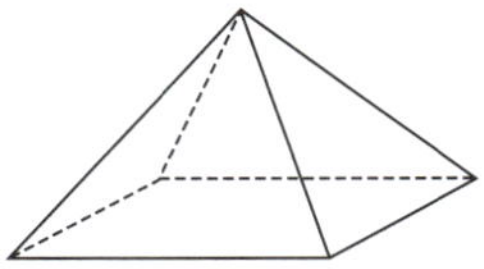

(1 mark)

7 When written correct to two significant figures, 835 = ?

A 83 **B** 84 **C** 830 **D** 840 (1 mark)

8 Melanie is paid \$18.60 per hour. How much does Melanie earn for working 6 hours overtime at 'time-and-a-half'?

A \$111.60 **B** \$120.90 **C** \$167.40 **D** \$279.00 (1 mark)

9 What fraction of one second is a microsecond?

A $\frac{1}{1000}$ **B** $\frac{1}{10\,000}$ **C** $\frac{1}{100\,000}$ **D** $\frac{1}{1\,000\,000}$ (1 mark)

10 $7 - 2(x - 3) = ?$

A $-2x + 4$ **B** $-2x - 1$ **C** $13 - 2x$ **D** $10 - 2x$ (1 mark)

11 What is the volume of this solid?

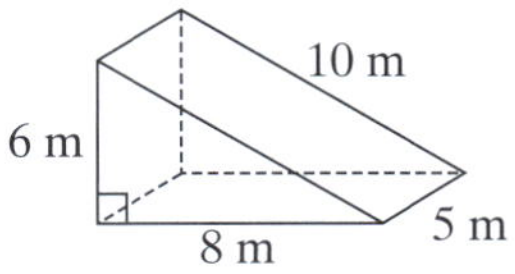

A 120 m^3 B 240 m^3
C 600 m^3 D 1200 m^3 (1 mark)

12 What is the surface area of a cube of side length 5 m?

A 100 m^2 B 125 m^2 C 150 m^2 D 175 m^2 (1 mark)

13 What is the value of $\dfrac{392}{21.3 \times 4.8}$ correct to one decimal place?

A 3.8 B 3.9 C 88.3 D 88.4 (1 mark)

14 Which is a sketch of $y = 4 - x^2$?

A
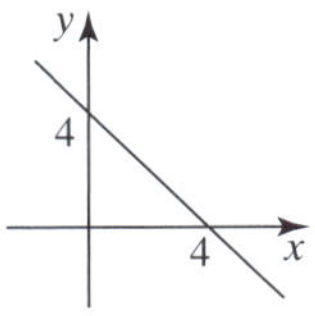

B
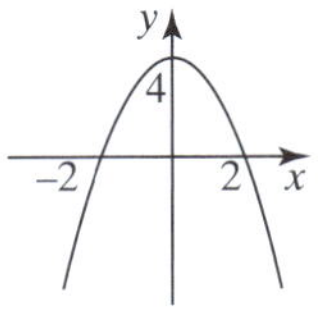

C
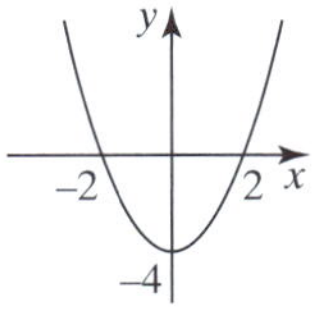

D
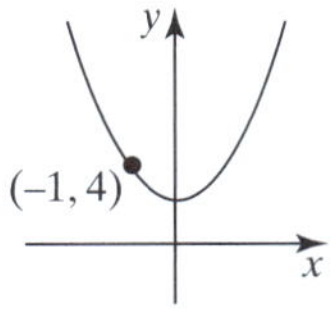

(1 mark)

15 $6^0 + 2x^0 = ?$

A 0 B 1 C 2 D 3 (1 mark)

16 Jason ran 100 m in 15 seconds. What was his speed in kilometres per hour?

A 24 B 30 C 36 D 54 (1 mark)

17 What is the median?

11	3	5	5			
12	0	4	6	8	9	
13	1	2	3	5	7	9
14	0	1	4	7	8	

A 115 B 35
C 132 D 130 (1 mark)

18 P is the point (1, 4) and Q is the point (5, 1). What is the gradient of PQ?

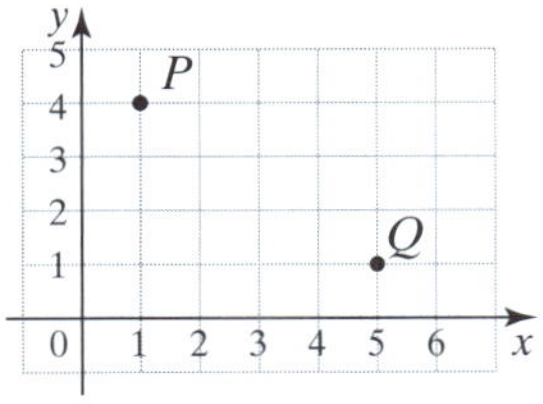

A $\dfrac{3}{4}$ B $-\dfrac{3}{4}$ C $\dfrac{4}{3}$ D $-\dfrac{4}{3}$ (1 mark)

19 P is the point (1, 4) and Q is the point (5, 1).
What is the distance between P and Q?

A 3 units B 4 units C 5 units D 7 units (1 mark)

20 8, 7, 9, 9 and 6 are the marks Ryan scored in his last five class quizzes. 8 is the

A mean B median C mode D range (1 mark)

21 If $\cos \theta = 0.758$, then θ, to the nearest degree, is

A 99° B 42° C 41° D 40° (1 mark)

22 Twins Sandra and Alexander made the following statements:
Sandra: 'When a fair coin is tossed the probability of getting a head is one-half.'
Alexander: 'When a fair coin is tossed the probability of getting a tail is 0.5.'
Who made a correct statement?

A Sandra only B Alexander only
C neither Sandra nor Alexander D both Sandra and Alexander (1 mark)

23 Mitchell measured the distance from Jackson to Red Hill on this map and found it to be 6 cm.
What is the real distance from Jackson to Red Hill?

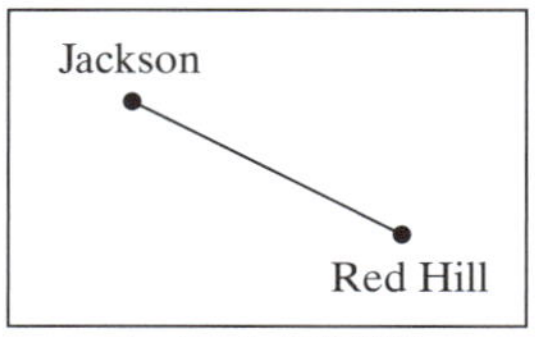

Scale 1:200 000

A 1200 m **B** 12 km
C 120 km **D** 1200 km (1 mark)

24 P is the point $(-2, 8)$ and Q is the point $(4, 2)$. What is the midpoint of PQ?
A $(1, 3)$ **B** $(1, 5)$ **C** $(3, 3)$ **D** $(3, 5)$ (1 mark)

25 $(2x^3)^2 = ?$
A $2x^5$ **B** $2x^6$ **C** $4x^5$ **D** $4x^6$ (1 mark)

26 There are 14 girls and 16 boys in a class. If one pupil is chosen at random from the class, what is the probability that the pupil is a girl?
A $\frac{7}{15}$ **B** $\frac{7}{8}$ **C** $\frac{8}{15}$ **D** $\frac{7}{22}$ (1 mark)

27 A coin was tossed a number of times and the results recorded. The coin showed heads 20 times and tails 30 times. What is the relative frequency of heads?
A $\frac{2}{3}$ **B** $\frac{2}{5}$ **C** $\frac{3}{5}$ **D** $\frac{1}{2}$ (1 mark)

28 What is the approximate volume of a cylinder of radius 4 cm and height 7 cm?
A 117 cm³ **B** 352 cm³ **C** 226 cm³ **D** 176 cm³ (1 mark)

29 Which shows a display that is skewed?

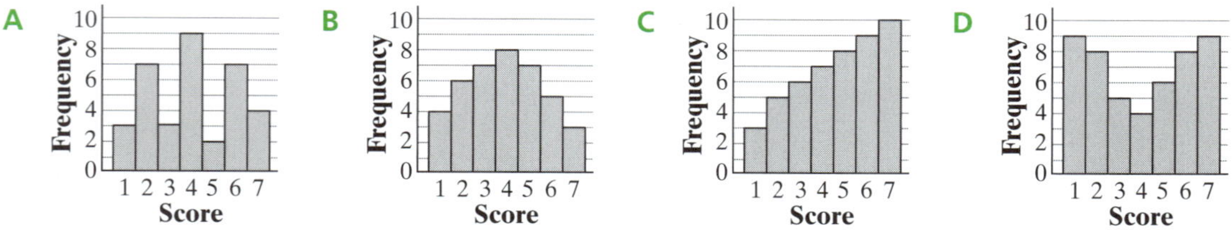

(1 mark)

30 These two triangles are similar. What is the length of side QR?

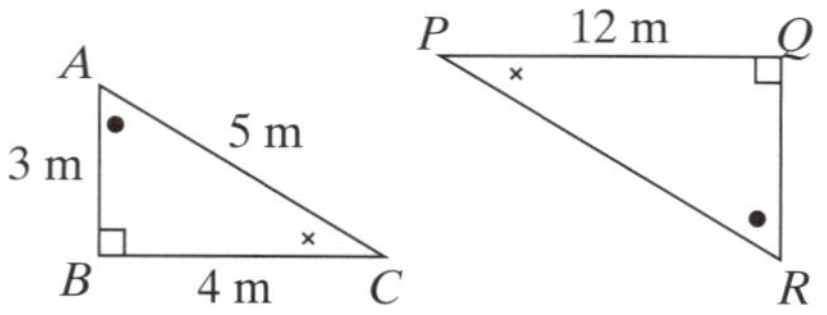

A 9 m **B** 15 m **C** 16 m **D** 25 m (1 mark)

Your Score
/ 30

PAGES 190–191

Part B Short Answer (20 marks)

31 Nico is buying a ride-on mower. The cash price is \$3600. Nico agrees to pay 30% deposit and pay the mower off in monthly instalments over 2 years. Simple interest of 5% p.a. is charged on the balance owing.

- **a** How much is the deposit? (1 mark)
- **b** What is the balance owing after the deposit has been paid? (1 mark)
- **c** How much interest will Nico pay? (1 mark)
- **d** How much is each monthly instalment? (2 marks)

32 Find the value of x. (Give the answer correct to one decimal place.) (2 marks)

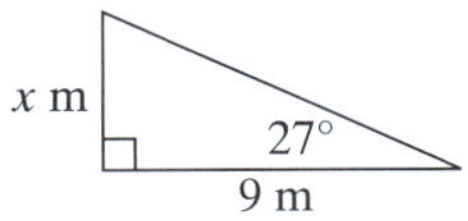

33 A census of students at a school was conducted to see how many computers students had in their homes.

Number of computers	0	1	2	3	4	5
Number of students	13	94	215	76	38	4

- **a** How many students are at the school? (1 mark)
- **b** How many computers in total did the students have in their homes? (1 mark)
- **c** What is the mean number of computers per student? (1 mark)

34 The amount Jim earns in a week is directly proportional to the number of hours he works. The equation $A = kh$ can be used to find the amount he earns, where A is the amount earned in dollars, h is the number of hours and k is a constant. Last week Jim worked 38 hours and earned \$1425.

- **a** What is the value of k? (1 mark)
- **b** What does this value of k represent? (1 mark)
- **c** What will Jim earn in a week where he works 35 hours? (1 mark)
- **d** One week last month Jim earned \$1575. How many more hours did he work that week than last week? (2 marks)

35 The diagram shows a triangular prism.

- **a** Show that $h = 1.25$. (1 mark)
- **b** What is the area of the triangular base of the prism? (1 mark)
- **c** What is the surface area of the prism? (1 mark)
- **d** The prism is an underground tank used to store water. How much water will it hold when full? ($1\text{ m}^3 = 1000\text{ L}$) (2 marks)

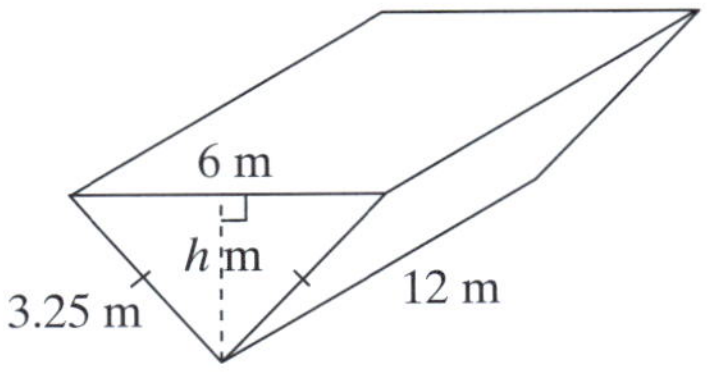

Your Score

____ / 20

PAGES 191–192

Your Feedback

Part A + Part B = ____ / 50 = ____ %

PAGE 198

SAMPLE EXAM PAPERS

Level of Difficulty—Above Average

60 MINUTES

PAPER 2

Part A Multiple Choice **(30 marks)**

1 If $a = 3$ and $b = -5$, then $ab^2 = ?$

A -125 **B** -75 **C** 125 **D** 75 (1 mark)

2 Which statement is correct?

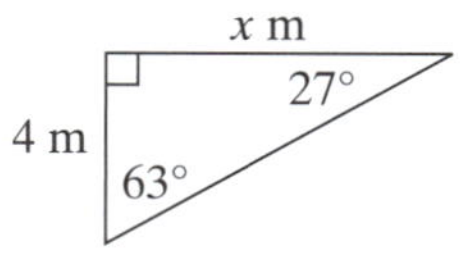

A $x = 4\sin 63°$ **B** $x = 4\cos 27°$ **C** $x = 4\tan 27°$ **D** $x = 4\tan 63°$ (1 mark)

3 $3^{-4} = ?$

A $\frac{1}{81}$ **B** $\frac{1}{64}$ **C** -64 **D** -81 (1 mark)

4 A is the point $(2, 1)$ and B is the point $(5, 3)$.
What is the gradient of the line joining A and B?

A $\frac{2}{3}$ **B** $\frac{3}{2}$

C $-\frac{2}{3}$ **D** $-\frac{3}{2}$

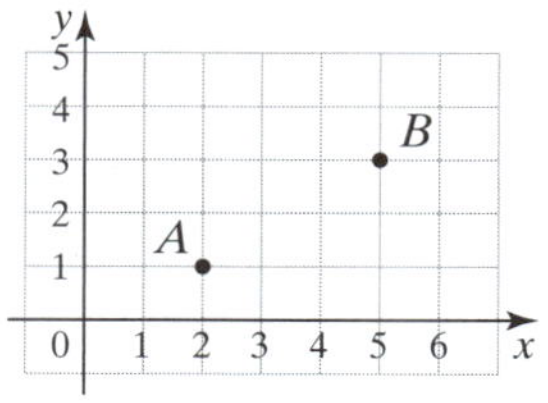

(1 mark)

5 $y = -\frac{x}{6}$ is the equation of

A a circle **B** a parabola **C** a straight line **D** none of these (1 mark)

6 When written correct to 3 significant figures 6.2037 becomes

A 6.204 **B** 6.24 **C** 6.20 **D** 6.237 (1 mark)

7 A speed of 72 kilometres per hour is how many metres per second?

A 20 **B** 24 **C** 36 **D** 93 (1 mark)

8 A survey was taken of the ages of supporters of two hockey teams (the red and blue teams).
The results are shown in the table below.

	Red	Blue
Under 16	12	20
16 or over	18	25

What is the probability that a supporter of the red team is under 16?

A $\frac{2}{3}$ **B** $\frac{2}{5}$ **C** $\frac{3}{5}$ **D** $\frac{4}{25}$ (1 mark)

9 What is the gradient of the line $y = 4 - 3x$?

A 4 **B** 3 **C** -4 **D** -3 (1 mark)

10 The diagram shows a prism. The shaded face is a trapezium with perpendicular height 4 m. What is the volume of this prism?

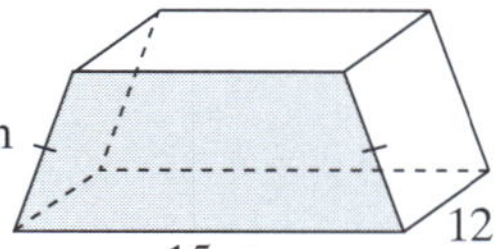

A 240 m^3 B 576 m^3
C 720 m^3 D 8100 m^3 (1 mark)

11 P is the point (3, 5) and Q is the point (–2, 5). What is the length of interval PQ?
A 1 unit B 5 units C 10 units D $\sqrt{50}$ units (1 mark)

12 Expand and simplify $2(2a + 3) + 4(a - 2)$.
A $8a + 1$ B $8a + 4$ C $8a + 14$ D $8a - 2$ (1 mark)

13 What is the value of θ to the nearest whole degree?

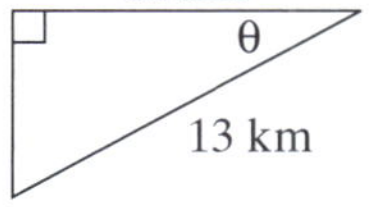

A 23° B 32° C 43° D 67° (1 mark)

14 $16m^{-6} \div 8m^{-2} = ?$
A $2m^{-4}$ B $2m^{-8}$ C $2m^{-3}$ D $2m^{3}$ (1 mark)

15 The diagram shows a sector of a circle of radius 15 cm. What is the area of the sector closest to?

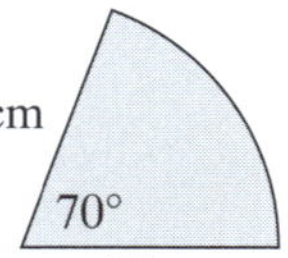

A 18 cm^2 B 137 cm^2
C 247 cm^2 D 275 cm^2 (1 mark)

16 What is the solution of the equation $\frac{n}{4} = \frac{2}{3}$?
A $n = \frac{1}{6}$ B $n = 6$ C $n = 2\frac{2}{3}$ D $n = 4\frac{2}{3}$ (1 mark)

17 A cube has surface area of 541.5 cm^2. What is the length of each side of the cube?
A 3.9 cm B 8.2 cm C 9.5 cm D 23.3 cm (1 mark)

18 For the set of scores 8, 7, 2, 5, 8 and 11, which has the greater value?
A mean B median C mode D range (1 mark)

19 If $A = 2\pi r^2 + 2\pi rh$, find the value of A, to the nearest whole number, when $r = 9$ and $h = 7$.
A 2330 B 905 C 516 D 452 (1 mark)

20 Brian is paid \$800 per week, plus a commission of 3.5% on all sales. What does Brian earn in a week where his sales total \$7000?
A \$828 B \$1013.50 C \$1045 D \$1073 (1 mark)

21 These two triangles are similar. $x = ?$

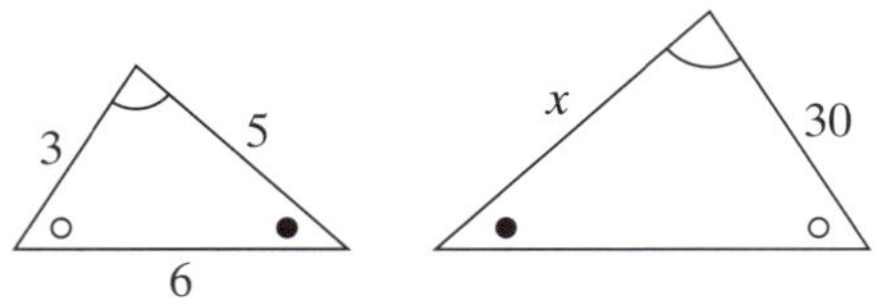

A 15 **B** 18 **C** 50 **D** 60 (1 mark)

22 Which point does NOT lie on the line $x - y = 4$?

A (5, 1) **B** (–1, 5) **C** (4, 0) **D** (2, –2) (1 mark)

23 The graph at right shows the water level in a container over time as it is being filled. Which of the drawings below could represent the container?

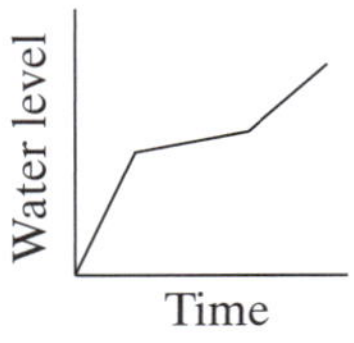

A
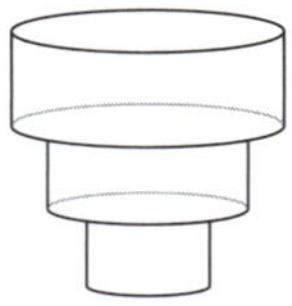

B
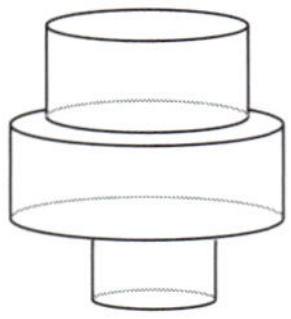

C
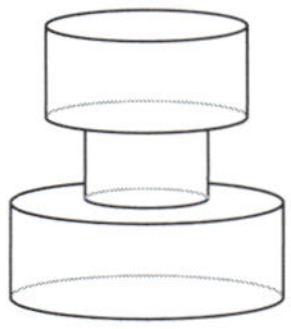

D

(1 mark)

24 A triangular prism has height 8 cm and volume 36 cm^3. What is the area of its base?

A 2.25 cm^2 **B** 4.5 cm^2 **C** 9 cm^2 **D** 12 cm^2 (1 mark)

25 Which triangle is right-angled? (Note: the triangles are not drawn to scale.)

A
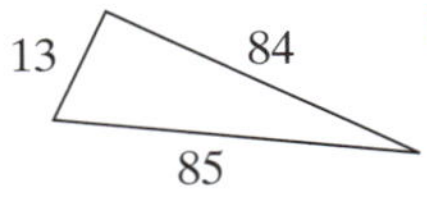

B
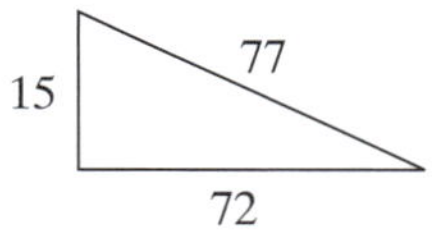

C
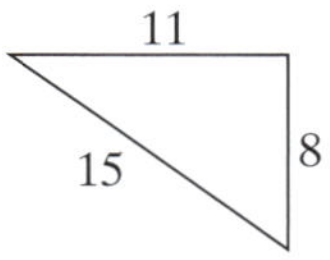

D
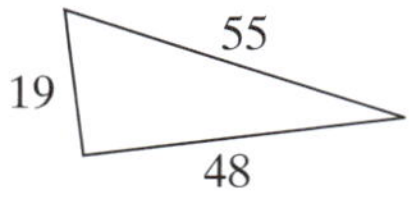

(1 mark)

26 Which number is the smallest?

A 0.7% **B** 6×10^{-2} **C** $\frac{1}{50}$ **D** 0.04 (1 mark)

27 A circle has equation $x^2 + y^2 = 100$. Where does the point (6, 8) lie?

A inside the circle **B** on the circle

C outside the circle **D** There is not enough information. (1 mark)

28 A nanosecond is what fraction of a second?

A $\frac{1}{1000}$ **B** $\frac{1}{100\,000}$ **C** $\frac{1}{1\,000\,000}$ **D** $\frac{1}{100\,000\,000}$ (1 mark)

29 What could the equation of this graph be?

A $y = x^2$ **B** $y = x^2 - 2$

C $y = 2x^2$ **D** $y = \frac{x^2}{2}$

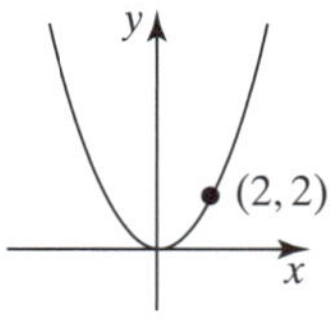

(1 mark)

30 A cylindrical tank has diameter 5 m and height 3 m. Given that 1 m^3 = 1000 L, which of these is closest to the capacity of the tank?
A 6 ML **B** 23.5 ML **C** 60 kL **D** 235 kL (1 mark)

Your Score
/ 30

PAGES 192–194

Part B Short Answer (20 marks)

31 a Expand and simplify $3x(2x - 5) - 4(2x - 7)$. (1 mark)
b Simplify $\dfrac{36a^5bc^2}{(3abc)^2}$ (1 mark)

32 a The Sun is 150 million kilometres from Earth. Express this distance in metres in scientific notation. (1 mark)
b Light travels at the speed of 3×10^8 metres per second. How long will light from the Sun take to reach Earth? (Give the answer in minutes and seconds.) (2 marks)

33 Petria borrows \$9000. Her repayments are \$298.75 per month for 3 years.
a How much will Petria repay in total? (1 mark)
b How much interest does Petria pay? (1 mark)
c What annual rate of simple interest is Petria charged on the loan? (2 marks)

34 Two classes sat for a test and their scores are shown in the back-to-back stem-and-leaf plot.

Class 9H		Class 9R
7 4 3	5	2 8
9 6 2 2 1 0	6	1 1 3 5
9 8 7 5 4 2 1	7	0 2 6 7 9
8 6 5 3 1 0	8	1 2 4 5 7 7 7 9
6 5 4 1	9	2 3 3 4 5 5

a How much greater is the median score of class 9R than that of 9H? (1 mark)
b For which class is the range greater? (1 mark)
c Briefly comment on any similarities and differences between the performance of the two classes. (2 marks)

35 There are some pencils in a box: two are blue, one is red and the other is green. Without looking, Charlotte takes two pencils from the box, one after the other without replacement.
a Draw a tree diagram to show the possible arrangements of pencils. (1 mark)
b What is the probability that one is blue and one is red? (1 mark)

36 A cylindrical tank has a diameter of 3 m and height 2.2 m.

a What is the volume of the tank? (1 mark)

b What is the capacity of the tank in litres to the nearest 100 litres. (1 mark)

37

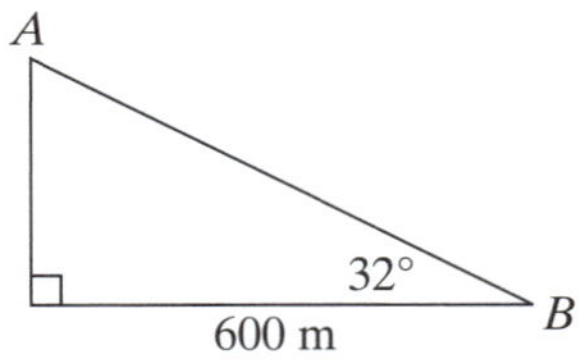

a Find the distance, to the nearest 10 metres, from *A* to *B*. (2 marks)

b If Cressida walks at 5 km/h, about how long will it take her to walk from *A* to *B*? (1 mark)

Your Score

$\frac{\square}{20}$

PAGES 194–195

Your Feedback

$\frac{\square}{\text{Part A}} + \frac{\square}{\text{Part B}} = \frac{\square}{50} = \square\%$

PAGE 198

SAMPLE EXAM PAPERS

Level of Difficulty—Difficult

60 MINUTES

PAPER 3

Part A Multiple Choice **(20 marks)**

1 Correct to three significant figures, $0.395 \div 48.95 =$

A 0.008 **B** 0.0081 **C** 0.008 069 **D** 0.008 07 (1 mark)

2 $-3(2x - 9) - 5(4x - 1) =$

A $-26x + 22$ **B** $-26x + 32$ **C** $14x + 22$ **D** $14x + 32$ (1 mark)

3 What is the equation of the line through $(-2, 5)$ with gradient 3?

A $y = 3x - 3$ **B** $y = 3x + 1$ **C** $y = 3x + 7$ **D** $y = 3x + 11$ (1 mark)

4 Which graph shows P increasing at a decreasing rate?

A

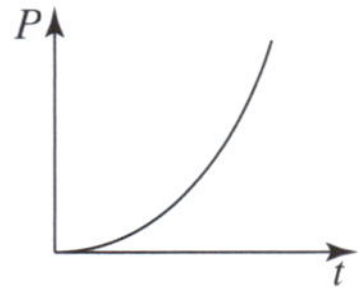

B

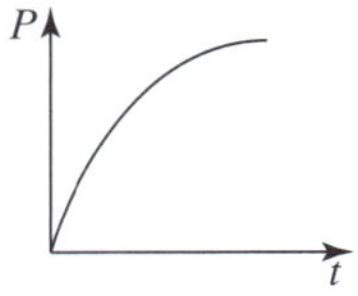

C

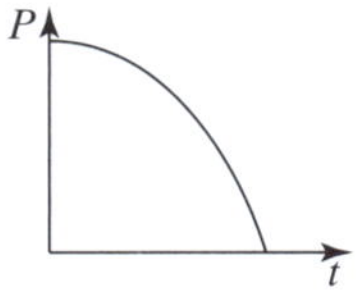

D

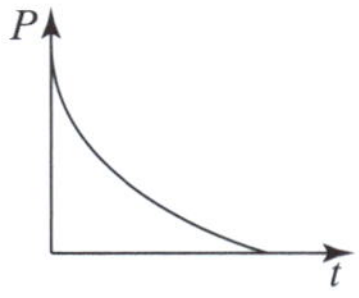

(1 mark)

5 There are five red and four blue pens in a bowl. Without looking, George takes two pens from the bowl. What is the probability that both pens are blue?

A $\frac{4}{9}$ **B** $\frac{16}{81}$ **C** $\frac{1}{6}$ **D** $\frac{2}{9}$ (1 mark)

6 What is the value of θ to the nearest minute?

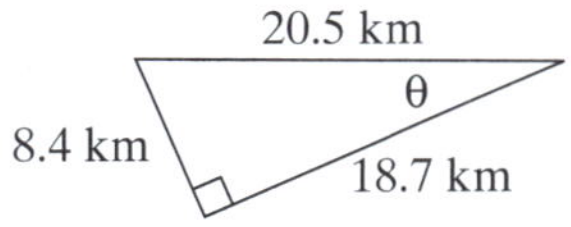

A 24°11′ **B** 24°12′ **C** 24°22′ **D** 24°23′ (1 mark)

7 $(6x^8 \div 3x^3)^2 = ?$

A $2x^{25}$ **B** $2x^{10}$ **C** $4x^{25}$ **D** $4x^{10}$ (1 mark)

8 A square is inscribed in a circle of diameter 12 cm. What is the shaded area closest to?

A 41 cm² **B** 77 cm²

C 31 cm² **D** 380 cm²

(1 mark)

9 Which line will pass through the midpoint of $A(-2, 6)$ and $B(6, 2)$?

A $y = 2$ **B** $x = 4$ **C** $y = x + 2$ **D** $y = 2x - 4$ (1 mark)

10 Which could be the equation of this parabola?

A $y = x^2 - 2$ **B** $y = 2x^2 - 2$

C $y = 2 - 2x^2$ **D** $y = \frac{1}{2}x^2 - 2$

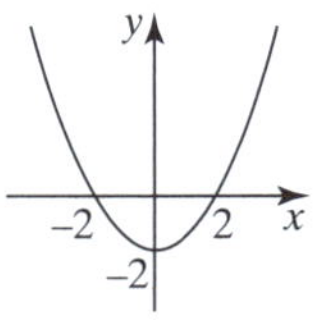

(1 mark)

11 Which is NOT correct?

A 0.06 GL = 600 ML
B 30 000 mg = 0.03 kg
C 2.25 min = 135 s
D 3×10^{-5} km = 3 cm
(1 mark)

12

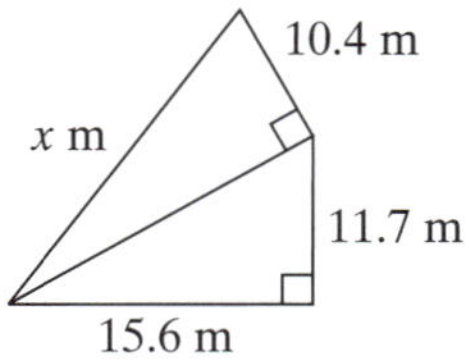

$x = ?$

A 19.1
B 22.1
C 15.7
D 27.1
(1 mark)

13 What annual rate of simple interest is charged if a loan of $24 000 is paid off with 60 equal monthly payments of $530?

A 5%
B 6.25%
C 6.5%
D 7.25%
(1 mark)

14 $2^{-5} \div 4^{-8} = ?$

A 2^3
B 4^3
C 2^9
D 2^{11}
(1 mark)

15 The range of a set of 4 scores is 14. The lowest score is 10 and the mode and median are both 13. What is the mean?

A 13
B 14
C 15
D 16
(1 mark)

16 What is the value of x?

A $11\frac{1}{4}$
B $3\frac{1}{3}$
C $7\frac{1}{2}$
D $10\frac{1}{2}$

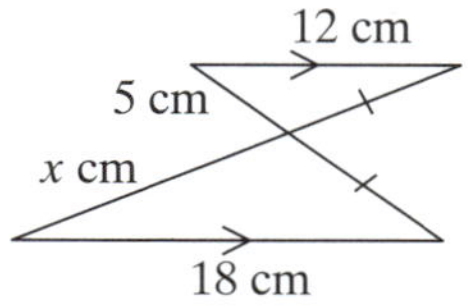

(1 mark)

17 What is the gradient of the line joining (–5, 4) to (7, –2)?

A $\frac{1}{2}$
B $-\frac{1}{2}$
C 2
D –2
(1 mark)

18 Which is closest to the distance a car will travel in 5 seconds at 80 km/h?

A 80 m
B 95 m
C 110 m
D 125 m
(1 mark)

19 The surface area of an open cylinder of radius 35 cm is 12755 cm^2. What is its volume?

A 88 514 cm^3
B 155 862 cm^3
C 434 636 cm^3
D 223 210 cm^3
(1 mark)

20 What is the surface area of this prism?

A 34 680 mm^2
B 27 540 mm^2
C 26 230 mm^2
D 357 000 mm^2

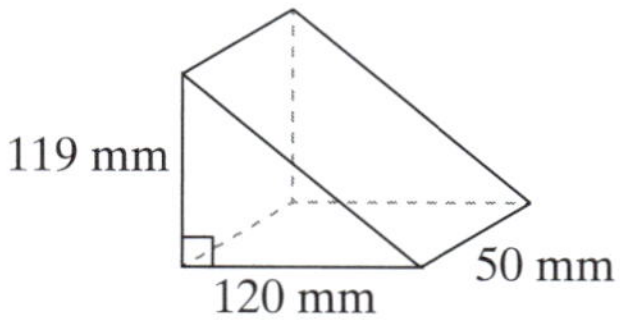

(1 mark)

Your Score

/ 20

PAGES 195–196

Part B Short Answer (30 marks)

21 Find $0.08 \div 126\,000$, expressing the answer in scientific notation correct to three significant figures. (1 mark)

22 Expand and simplify $2xy^2(3x^2 - 2y) - 9x^2y(2 - xy)$. (1 mark)

23 This is a scale drawing of a paddock. The longest boundary is actually 900 m long.

a What scale is used? (1 mark)
b By measurement and calculation find the area of the paddock in hectares. (3 marks)

24 Line l has x-intercept 2 and y-intercept 4.
a What is the gradient of line l? (1 mark)
b What is the equation of line l? (1 mark)
c Q is the point of intersection of line l and the line $y = -6$. What are the coordinates of Q? (1 mark)
d Find the distance from O, the origin, to Q. (1 mark)
e Write down the equation of the circle that has its centre at O and which passes through Q. (1 mark)

25 The monthly repayments on a loan taken over 8 years are \$131.25. Simple interest at 5% p.a. is charged on the loan.
a How much in total is repaid? (1 mark)
b How much was borrowed? (2 marks)

26 Joe rolled two dice several times and drew a graph of the results.

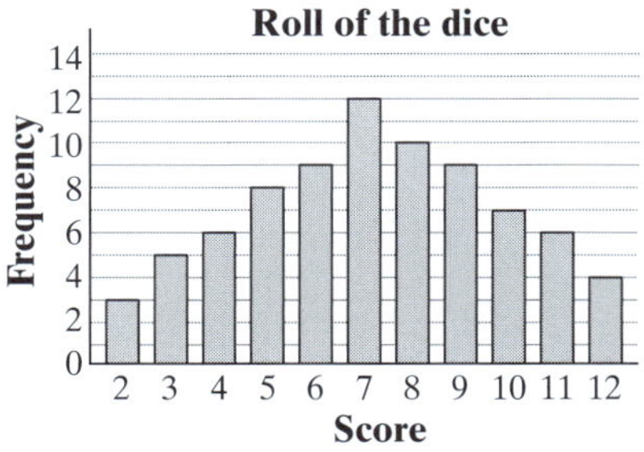

a How many times did Joe roll the dice? (1 mark)
b What it the relative frequency of 7? (1 mark)
c What is the mean? (1 mark)
d What is the median? (1 mark)
e Joe commented that the display was positively skewed. Do you agree? Justify your answer. (1 mark)
f Did a score of 5 occur more or less times than you would expect? (1 mark)
g What is the probability of getting a double when rolling two dice? (1 mark)

27 The area bounded by the parabola $y = x^2 + 3$ and the line $y = 7$ (shown hatched on the diagram) is $10\frac{2}{3}$ square units. Find the shaded area. (3 marks)

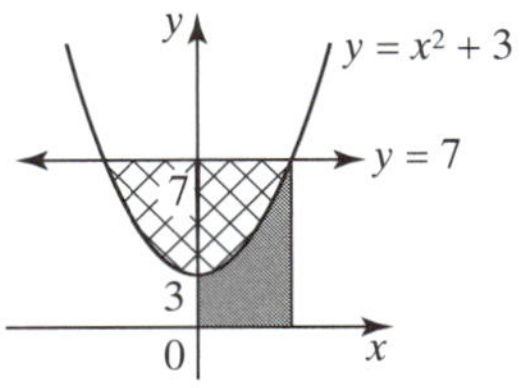

28 The figure consists of a semicircle and a right-angled triangle.

- **a** Use trigonometry to find the diameter of the semicircle. (1 mark)
- **b** Find the perimeter of the figure. (3 marks)
- **c** Find the area of the figure, to the nearest square centimetre. (2 marks)

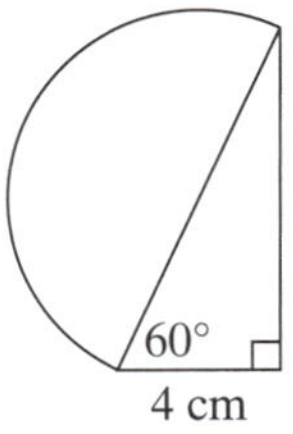

Your Score

___ / 30

PAGES 196–197

Your Feedback

Part A + Part B = ___ / 50 = ___%

PAGE 198

WORKED SOLUTIONS

CHECK YOUR SOLUTIONS

REAL NUMBERS
SKILLS CHECK PAGE 2

1 a $\frac{12}{16} = \frac{3}{4}$
[dividing both numerator and denominator by 4]

b $2\frac{5}{7} = \frac{19}{7}$
$[2 \times 7 + 5 = 19]$

c $\frac{15}{6} = 2\frac{3}{6} = 2\frac{1}{2}$

2 a $4 = \frac{4}{1}$
The reciprocal of 4 is $\frac{1}{4}$.

b The reciprocal of $\frac{3}{4}$ is $\frac{4}{3}$ $\left[= 1\frac{1}{3}\right]$.

c $3\frac{2}{5} = \frac{17}{5}$
The reciprocal of $3\frac{2}{5}$ is $\frac{5}{17}$.

3 a $\frac{2}{3}$ of $24 = 16$
$[24 \div 3 = 8,\ 2 \times 8 = 16]$

b $1\frac{1}{2} + 2\frac{3}{4} = 1\frac{2}{4} + 2\frac{3}{4}$
$= 3\frac{5}{4}$
$= 4\frac{1}{4}$

c $5 - 2\frac{3}{5} = 3 - \frac{3}{5}$
$= 2\frac{2}{5}$

d $\frac{3}{7} \times \frac{2}{5} = \frac{6}{35}$

e $1\frac{2}{3} \times 2\frac{1}{2} = \frac{5}{3} \times \frac{5}{2}$
$= \frac{25}{6}$
$= 4\frac{1}{6}$

f $8 \div \frac{1}{2} = 8 \times 2$
$= 16$

g $2\frac{3}{5} \div 1\frac{1}{4} = \frac{13}{5} \div \frac{5}{4}$
$= \frac{13}{5} \times \frac{4}{5}$
$= \frac{52}{25}$
$= 2\frac{2}{25}$

4 a $0.8 = \frac{8}{10} = \frac{4}{5}$

b $3.5 = 3\frac{1}{2}$

c $0.35 = \frac{35}{100} = \frac{7}{20}$

5 a $0.7 + 0.12 = 0.82$
$[0.70 + 0.12]$

b $0.35 - 0.2 = 0.15$
$[0.35 - 0.20]$

c $0.2 \times 0.4 = 0.08$

d $1.6 \times 1000 = 1600$
[Move the decimal point 3 places to the right.]

e $3.5 \div 100 = 0.035$
[Move the decimal point 2 places to the left.]

f $0.06 \div 3 = 0.02$

g $0.08 \div 0.2 = 0.8 \div 2$
$= 0.4$

6 a $-2 - 5 = -7$

b $7 - 11 = -4$

c $-5 + 8 = 3$

d $-7 + 4 = -3$

e $-2 \times 5 = -10$

f $-12 \div -3 = 4$

g $-4 \times -6 \times -2 = -48$
$[-4 \times -6 = 24,$
$24 \times -2 = -48]$

7 a $\frac{3}{4} = 0.75$ $[3 \div 4]$

b $\frac{5}{8} = 0.625$ $[5 \div 8]$

c $\frac{17}{1000} = 0.017$

8 a $\frac{1}{3} = 0.33333\ldots$ $[1 \div 3]$
$= 0.\dot{3}$

b $\frac{5}{12} = 0.416666666\ldots$
$= 0.41\dot{6}$

c $1\frac{1}{6} = 1.166666666\ldots$
$= 1.1\dot{6}$

9 a $\frac{2}{5} = 0.4$ terminating

b $\frac{2}{3} = 0.66666666\ldots$ recurring

c $\frac{3}{7} = 0.428571428\ldots$ recurring

d $\frac{7}{8} = 0.875$ terminating

10 a $60\% = \frac{60}{100}$
$= \frac{3}{5}$

b $8\% = 0.08$ $[8 \div 100]$

c $\frac{3}{4} = \frac{3}{4} \times 100\%$
$= 75\%$

d $0.3 = 30\%$ $[0.3 \times 100\%]$

11 a $35\% \times \$9000 = \frac{35}{100} \times \$9000$
$= \$3150$

b $\frac{18}{25} \times 100\% = 72\%$

12 15% is 45
1% is 3
100% is 300

REAL NUMBERS
INTERMEDIATE TEST PAGE 3

1 [divide the shape into equal parts]

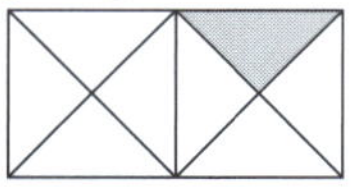

One part of eight is shaded. $\frac{1}{8}$
∴ [D] ✓ (1 mark)

2 The numbers are:
0.2080 0.2800 0.0208 0.0280
The smallest is 0.0208.
∴ [C] ✓ (1 mark)

3 $8\% \text{ of } 800 = \frac{8}{100} \times \frac{800}{1}$
$= 64$
$800 - 64 = 736$
$\therefore$ **[B]** ✓ (1 mark)
[Or: 8% less than 100% is 92%
$92\% \text{ of } 800 = \frac{92}{100} \times \frac{800}{1}$
$= 736$]

4 $10\% = 0.1$
$\frac{1}{10} = 0.1$
All are equal.
$\therefore$ **[D]** ✓ (1 mark)

5 $\frac{1}{3} = 0.333\,33\ldots$ $\frac{1}{4} = 0.25$
[We need a decimal between 0.25 and 0.333 33 ...]
$\frac{3}{10} = 0.3$
$\frac{3}{7} = 0.428\,571\,428\ldots$
$\frac{2}{9} = 0.222\,2222\ldots$
$\frac{5}{12} = 0.416\,6666\ldots$
$\therefore$ **[A]** ✓ (1 mark)

6 $\frac{2}{3} = 0.666\,6666\ldots$
$= 0.7$ (1 d.p.)
$\therefore$ **[D]** ✓ (1 mark)

7 $\frac{1}{3} = 0.333\,333\ldots$
$\frac{5}{6} = 0.833\,3333\ldots$
$\frac{7}{8} = 0.875$
$\frac{7}{9} = 0.777\,777\,77\ldots$
$\therefore$ **[C]** ✓ (1 mark)
[Divide the numerator by the denominator to change a fraction to a decimal.]

8 $-17 + 9 = -8$
The temperature was −8 °C. ✓ (1 mark)

9 $\frac{2.5 + 1.5}{5 \times 0.4} = \frac{4}{2}$
$= 2$ ✓ (1 mark)

10 Percentage mark $= \frac{68}{80} \times 100\%$
$= 85\%$ ✓ (1 mark)

11 $\frac{4}{9}$ is 36
$\frac{1}{9}$ is 9 [dividing by 4]
$\frac{9}{9}$ is 81 [multiplying by 9]
The number is 81.
[$\frac{9}{9} = 1$ whole] ✓ (1 mark)

12 $\left(3\frac{1}{2}\right)^2 = \left(\frac{7}{2}\right)^2$
$= \frac{49}{4}$
$= 12\frac{1}{4}$ ✓ (1 mark)

13 **a** Discount = 15% of \$60
$= \frac{15}{100} \times \frac{\$60}{1}$
$= \$9$
New price = \$60 − \$9
= \$51 ✓ (1 mark)

b Second discount
= 5% of \$51
$= \frac{5}{100} \times \frac{\$51}{1}$
= \$2.55 ✓ (1 mark)
Total savings
= \$9 + \$2.55
= \$11.55 ✓ (2 marks)
(Total: 15 marks)

REAL NUMBERS ADVANCED TEST PAGE 4

1 **a** $2\frac{1}{3} + 3\frac{3}{4} = 5 + \frac{4}{12} + \frac{9}{12}$
$= 5\frac{13}{12}$
$= 6\frac{1}{12}$ ✓ (1 mark)

b $5\frac{2}{7} - 2\frac{3}{5} = 3\frac{10}{35} - \frac{21}{35}$
$= 3 - \frac{11}{35}$
$= 2\frac{24}{35}$ ✓ (1 mark)

c $4\frac{1}{4} \times 3\frac{1}{5} = \frac{17}{4} \times \frac{16}{5}$
$= \frac{17}{1} \times \frac{4}{5}$
$= \frac{68}{5}$
$= 13\frac{3}{5}$ ✓ (1 mark)

d $30 \div \frac{1}{3} = 30 \times 3$
$= 90$ ✓ (1 mark)

e $\left(2\frac{1}{2}\right)^2 = \left(\frac{5}{2}\right)^2$
$= \frac{25}{4}$
$= 6\frac{1}{4}$ ✓ (1 mark)

f $7\frac{1}{2} \div \frac{3}{5} = \frac{15}{2} \times \frac{5}{3}$
$= \frac{5}{2} \times \frac{5}{1}$ [after cancelling]
$= \frac{25}{2}$
$= 12\frac{1}{2}$ ✓ (1 mark)

2 **a** $\frac{3}{5} + \frac{2}{3} \times \frac{1}{8} = \frac{3}{5} + \frac{1}{12}$ ✓
$= \frac{36}{60} + \frac{5}{60}$
$= \frac{41}{60}$ ✓ (2 marks)

b $6\frac{4}{5} \times 7\frac{1}{2} \div 5\frac{2}{3}$
$= \frac{34}{5} \times \frac{15}{2} \times \frac{3}{17}$ ✓
$= 9$ ✓ (2 marks)

3 **a** $-5 \times -4 \times 7 = 140$ ✓ (1 mark)
b $1.8 + 0.25 = 2.05$
[1.80 + 0.25] ✓ (1 mark)
c $0.2 \times 0.3 = 0.06$ ✓ (1 mark)
d $-2 + 3 \times -5 - 8 \times 2$
$= -2 - 15 - 16$
$= -33$ ✓ (1 mark)
e $8 - 2 \times 0.35$
$= 8 - 0.7$
$= 7.3$ ✓ (1 mark)
f $(-0.9)^2 = 0.81$ ✓ (1 mark)

4 **a** $25 \times 37 \times 4$
$= 100 \times 37$
$= 3700$ ✓ (1 mark)

b $39 \times 56 + 39 \times 44$
$= 39 \times (56 + 44)$
$= 39 \times 100$
$= 3900$ ✓ (1 mark)

5 **a** 15% of $\$8420$
$= \$842 + \$421$
$= \$1263$
$[15\% = 10\% + 5\%]$ ✓
(1 mark)

b 67% of 3 kg
$= \frac{67}{100} \times 3000$ g
$= 67 \times 30$ g
$= 2010$ g [or 2.01 kg] ✓
(1 mark)

6 **a** $\frac{52}{80} \times 100\% = 13 \times 5\%$
[after cancelling]
$= 65\%$ ✓
(1 mark)

b 3 cm out of 6 m
= 3 cm out of 600 cm
= 1 out of 200
= 0.5 out of 100
$= 0.5\%$ ✓ (1 mark)

7 **a** $\frac{5}{6} = 0.8\dot{3}$ ✓ (1 mark)

$$\left(6 \overline{)5.{}^{5}0{}^{2}0{}^{2}0{}^{2}0{}^{2}0\ldots} \quad = 0.83333\ldots \right)$$

b $4\frac{2}{3} = \frac{14}{3}$
The reciprocal is $\frac{3}{14}$. ✓
(1 mark)

c $\sqrt{5\frac{4}{9}} = \sqrt{\frac{49}{9}}$ ✓
$= \frac{7}{3}$
$= 2\frac{1}{3}$ ✓ (2 marks)

d $0.2\dot{3} = 0.233\,333\ldots$
Let $x = 0.233\,333\ldots$
$10x = 2.333\,333\ldots$
$9x = 2.1$ ✓
[After subtracting]
$x = \frac{2.1}{9}$
$= \frac{21}{90}$
$= \frac{7}{30}$ ✓ (2 marks)

8 $\frac{1}{4}$ of $240 = 60$
So 60 Year 9 students were on the excursion.
$40\% = 60$ ✓
$10\% = 15$
$100\% = 150$
There were 150 students in Year 9. ✓ (2 marks)
(Total: 30 marks)

ALGEBRAIC TECHNIQUES
SKILLS CHECK PAGE 6

1
a $3x + 5x = 8x$
b $2a^2 + 3a^2 = 5a^2$
c $7pq - 5pq = 2pq$
d $8m + m = 9m$ $[8m + 1m]$
e $6e - e = 5e$ $[6e - 1e]$
f $9y - 8y = y$ $[1y]$
g $5a + 7a - 3a = 9a$
h $-3p - 7p = -10p$
i $6m^3 - 4m^3 - 8m^3 = -6m^3$

2
a $9x + 4y + 2x = 11x + 4y$
b $7p - 2q - p + 8q = 6p + 6q$
$[7p - p = 6p, -2q + 8q = 6q]$
c $4a + 5b + 2a + 7 = 6a + 5b + 7$
d $6m + 5n - 4m + 2n = 2m + 7n$
e $-2k + 3 - 4k - 5 = -6k - 2$
$[-2k - 4k = -6k, 3 - 5 = -2]$
f $4p^2 + 3p + 2p = 4p^2 + 5p$
g $3p^2 - 4q^2 + 5p - 2q^2 - 3q + 4p$
$= 3p^2 - 6q^2 + 9p - 3q$
h $9x^3 - 5x^2 + 8x - 3x^2 + 2x - x^2$
$= 9x^3 - 9x^2 + 10x$

3
a $x \times x = x^2$
b $7 \times a = 7a$
c $2 \times 4y = 8y$
d $k \times 6 = 6k$
e $3 \times 5x^2 = 15x^2$
f $2t \times 4 = 8t$
g $4m \times 2n = 8mn$
h $5a \times b = 5ab$
i $3xy \times 3z = 9xyz$

4
a $7x \div x = 7$
b $8a \div 2a = 4$
c $4g \div 4 = g$ $[1g]$
d $12m \div 3 = 4m$
e $9y^3 \div 3 = 3y^3$
f $20xy \div 4y = 5x$
g $\frac{4p}{2} = 2p$
h $\frac{6x}{12x} = \frac{1}{2}$
[dividing both numerator and denominator by $6x$]
i $\frac{9}{6b} = \frac{3}{2b}$ [dividing by 3]

5
a If $x = 5$, $x + 3 = 5 + 3$
$= 8$
b If $x = 5$, $4x = 4 \times 5$
$= 20$
c If $x = 5$, $x^2 = 5^2$
$= 25$
d If $x = 5$, $3x - 1 = 3 \times 5 - 1$
$= 14$
e If $x = 5$, $15 \div x = 15 \div 5$
$= 3$
f If $x = 5$, $8 - 2x = 8 - 2 \times 5$
$= 8 - 10$
$= -2$

6
a If $a = 2$ and $b = 3$,
$2a + b = 2 \times 2 + 3$
$= 7$
b If $a = 2$ and $b = 3$,
$ab = 2 \times 3$
$= 6$
c If $a = 2$ and $c = 7$,
$c^2 - a = 7^2 - 2$
$= 49 - 2$
$= 47$
d If $b = 3$ and $d = -2$,
$2b^2 + d = 2 \times 3^2 + (-2)$
$= 2 \times 9 - 2$
$= 16$
e If $a = 2$ and $d = -2$,
$ad^2 = 2 \times (-2)^2$
$= 2 \times 4$
$= 8$
f If $a = 2$, $b = 3$ and $c = 7$,
$4bc - 2a = 4 \times 3 \times 7 - 2 \times 2$
$= 80$

7 **a** $2x + 9 = 17$
$2x = 8$
[taking 9 from both sides]
$x = 4$
[dividing both sides by 2]

b $\frac{d-4}{3} = -9$
$d - 4 = -27$
[multiplying by 3]
$d = -23$
[adding 4 to both sides]

c $5p + 8 = 29 - 2p$
$7p + 8 = 29$ [adding $2p$]
$7p = 21$ [subtracting 8]
$p = 3$ [dividing by 7]

ALGEBRAIC TECHNIQUES INTERMEDIATE TEST PAGE 7

1 $8x + x = 8x + 1x$
$= 9x$
∴ **[B]** ✓ (1 mark)

2 $2x \times 3x \times 5x = 30x^3$
[$2 \times 3 \times 5 = 30$, $x \times x \times x = x^3$]
∴ **[D]** ✓ (1 mark)

3 $m^4 + m^4 = 2m^4$ [$1m^4 + 1m^4$]
∴ **[C]** ✓ (1 mark)

4 $5a - 2b - 3a + 4b = 2a + 2b$
[$5a - 3a = 2a$, $-2b + 4b = 2b$]
∴ **[A]** ✓ (1 mark)

5 $5pq^2 = 5 \times p \times q \times q$
[the 'squared' only belongs to q]
∴ **[B]** ✓ (1 mark)

6 $9x^2 - 5x + 2x^2 - x = 11x^2 - 6x$
[$9x^2 + 2x^2 = 11x^2$, $-5x - x = -6x$]
∴ **[A]** ✓ (1 mark)

7 If $k = -2$,
$3k^2 = 3 \times (-2)^2$
$= 3 \times 4$
$= 12$
∴ **[C]** ✓ (1 mark)

8 The number 8 more than $2x - 5$
$= 2x - 5 + 8$
$= 2x + 3$
∴ **[A]** ✓ (1 mark)

9 $\frac{8x^2}{4xy} = \frac{2x}{y}$
∴ **[D]** ✓ (1 mark)
[Divide numerator and denominator by $4x$.]

10 $p = 4$, $q = -3$, $r = 5$
$r - pq = 5 - 4 \times -3$
$= 5 + 12$
$= 17$
∴ **[C]** ✓ (1 mark)

11 $3x^2 - 2x \times 4x = 3x^2 - 8x^2$ ✓
$= -5x^2$ ✓
(2 marks)

12 $8a \times 3a + 2 \times 7a^2 = 24a^2 + 14a^2$ ✓
$= 38a^2$ ✓
(2 marks)

13 $\frac{3a}{12ab} = \frac{1}{4b}$ ✓✓ (2 marks)
[dividing both numerator and denominator by $3a$]

14 **a** $8x + 23 = -5$
[Subtract 23 from both sides.]
$8x = -28$ ✓
[Divide both sides by 8.]
$x = -3.5$ ✓ (2 marks)

b $\frac{m}{6} - 32 = 18$
[Add 32 to both sides.]
$\frac{m}{6} = 50$ ✓
[Multiply both sides by 6.]
$m = 300$ ✓ (2 marks)

15 $7pq + 8p + 2pq - 5p - 6q + 3p - q$
$= 9pq + 6p - 7q$ ✓
When $p = 5$ and $q = 7$,
$9pq + 6p - 7q$
$= 9 \times 5 \times 7 + 6 \times 5 - 7 \times 7$
$= 315 + 30 - 49$
$= 296$ ✓ (2 marks)

16 Jon charges \$5 + \$8 × number of hours
$= \$5 + \$8 \times n$
$= \$(5 + 8n)$ ✓ (1 mark)
[Or $\$5 + \$8n$ Or $\$(8n + 5)$
Or $\$8n + \$5$]

17 $(9x - 7x) \div (5x - 3x)$
$= 2x \div 2x$ ✓
$= 1$ ✓ (2 marks)
(Total: 25 marks)

ALGEBRAIC TECHNIQUES ADVANCED TEST PAGE 8

1 **a** $-9x + x = -8x$ ✓ (1 mark)
b $-9x - x = -10x$ ✓ (1 mark)
c $-9x \times x = -9x^2$ ✓ (1 mark)
d $-9x \div x = -9$ ✓ (1 mark)
e $5x + 2y = 5x + 2y$ ✓ (1 mark)
f $5x \times 2y = 10xy$ ✓ (1 mark)
g $\frac{5ab}{10bc} = \frac{a}{2c}$ ✓ (1 mark)
h $\frac{p^2q^2}{(pq)^2} = \frac{p^2q^2}{p^2q^2}$
$= 1$ ✓ (1 mark)

2 **a** $2x + 3x \times 4x$
$= 2x + 12x^2$ ✓ (1 mark)
b $12k \div 3k - 2k$
$= 4 - 2k$ ✓ (1 mark)
c $4n \times -2n \times n \times -3$
$= 24n^3$ ✓ (1 mark)

3 **a** $5x^2 + 2x - x^2 + 7x$
$= 4x^2 + 9x$ ✓ (1 mark)
b $-6a + 8b - 9a - 3b - 4$
$= -15a + 5b - 4$ ✓ (1 mark)
c $4k^3 - 3k^2 + 2k - 5k^2 + 8k$
$= 4k^3 - 8k^2 + 10k$ ✓ (1 mark)
d $-3 - 9 - 4m + 7m + 2$
$= -10 + 3m$ ✓ (1 mark)
e $8e + 9f - 4f + 3e + f - 5e$
$= 6e + 6f$ ✓ (1 mark)
f $m^5 - 3m^5 - 4n^4 + 2n^2 - 6n^4$
$= -2m^5 - 10n^4 + 2n^2$ ✓ (1 mark)

4 **a** $4a \times 5a - 2a \times 8a$
$= 20a^2 - 16a^2$ ✓
$= 4a^2$ ✓ (2 marks)
b $12k^2 \div 6 + 3k \times 4k$
$= 2k^2 + 12k^2$ ✓
$= 14k^2$ ✓ (2 marks)
c $(x^2 + x^2) \times (7x^2 \div x^2)$
$= 2x^2 \times 7$ ✓
$= 14x^2$ ✓ (2 marks)
d $(7x - 4x) \div (5x + x)$
$= 3x \div 6x$ ✓
$= \frac{1}{2}$ ✓ (2 marks)

5 a $y - x = 5 - (-2)$
$= 7$ ✓ (1 mark)

b $x^2y = (-2)^2 \times 5$
$= 4 \times 5$
$= 20$ ✓ (1 mark)

c $5x - 2y = 5 \times -2 - 2 \times 5$
$= -20$ ✓ (1 mark)

d $3xy + x = 3 \times -2 \times 5 + (-2)$
$= -32$ ✓ (1 mark)

6 a $5x + 3$ ✓ (1 mark)

b $6x - 7$ ✓ (1 mark)

c $2x + x^2$ ✓ (1 mark)

7 a $6a + 9 - a = 8 - 3a + 1$
$5a + 9 = 9 - 3a$ ✓
$8a + 9 = 9$
$8a = 0$
$a = 0$ ✓ (2 marks)

b $\frac{2t+4}{7} - 3 = 5$
$\frac{2t+4}{7} = 8$
$2t + 4 = 56$ ✓
$2t = 52$
$t = 26$ ✓ (2 marks)
[Or multiply each term by 7 first.
$2t + 4 - 21 = 35$
$2t - 17 = 35$
$2t = 52$
$t = 26$]

(Total: 36 marks)

INDICES AND SCIENTIFIC NOTATION
SKILLS CHECK PAGE 10

1 $7 \times 7 \times 7 \times 7 \times 7 \times 7 = 7^6$

2 $3^4 = 3 \times 3 \times 3 \times 3$

3 a $2^3 = 8$ $[2 \times 2 \times 2]$

b $3^4 = 81$

c $5^6 = 15\,625$

d $10^5 = 100\,000$

e $6^0 = 1$

f $9^1 = 9$

g $37^2 = 1369$

h $7^9 = 40\,353\,607$

4 a $3^4 \times 3^3 = 3^{4+3}$
$= 3^7$

b $2^5 \times 2 = 2^{5+1}$
$= 2^6$

c $10^2 \times 10^8 = 10^{2+8}$
$= 10^{10}$

5 a $7^8 \div 7^2 = 7^{8-2}$
$= 7^6$

b $3^6 \div 3^3 = 3^{6-3}$
$= 3^3$

c $2^9 \div 2 = 2^{9-1}$
$= 2^8$

6 a $(5^2)^3 = 5^{2 \times 3}$
$= 5^6$

b $(7^7)^2 = 7^{7 \times 2}$
$= 7^{14}$

c $(2^4)^4 = 2^{4 \times 4}$
$= 2^{16}$

7 a $2^{-1} = \frac{1}{2}$

b $6^{-1} = \frac{1}{6}$

c $15^{-1} = \frac{1}{15}$

d $5^{-2} = \frac{1}{5^2} = \frac{1}{25}$

e $2^{-3} = \frac{1}{2^3} = \frac{1}{8}$

f $10^{-6} = \frac{1}{10^6} = \frac{1}{1\,000\,000}$

8 a $5^{-1} = \frac{1}{5} = 0.2$

b $10^{-4} = \frac{1}{10^4} = 0.0001$

c $2^{-2} = \frac{1}{2^2} = \frac{1}{4} = 0.25$

9 a $\frac{1}{3} = 3^{-1}$

b $\frac{1}{49} = \frac{1}{7^2} = 7^{-2}$

c $\frac{1}{32} = \frac{1}{2^5} = 2^{-5}$

10 a $3^4 = 3 \times 3 \times 3 \times 3$
$= 81$ $(\neq 64)$ False

b $5^3 = 5 \times 5 \times 5$
$= 125$ $(\neq 15)$ False

c $2^5 = 2 \times 2 \times 2 \times 2 \times 2$
$= 32$ True

d $7^{-1} = \frac{1}{7}$ True

e $9^{-2} = \frac{1}{9^2}$
$= \frac{1}{81}$ $(\neq \frac{1}{3})$ False

f $4^{-3} = \frac{1}{4^3}$
$= \frac{1}{64}$ $(\neq \frac{1}{81})$ False

g $10^7 = 10 \times 10 \times 10 \times 10 \times 10 \times 10 \times 10$
$= 10\,000\,000$
$(\neq 1\,000\,000)$ False

h $10^{-4} = \frac{1}{10^4}$
$= \frac{1}{10\,000}$
$= 0.0001$
$(\neq 0.000\,01)$ False

i $10^{-2} = \frac{1}{100}$
$= 0.01$ True

j $30^2 = 900$
$3 \times 10^2 = 3 \times 100$
$= 300$ $(\neq 30^2)$ False

k $10 \times 10^{-1} = 10^{1+(-1)}$
$= 10^0$
$= 1$ True

11 a $1000 = 10^3$

b $10\,000\,000 = 10^7$

c $0.01 = \frac{1}{100} = 10^{-2}$

d $0.000\,01 = 10^{-5}$

12 a $720\,000 = 7.2 \times 10^5$

b $9000 = 9 \times 10^3$

c $576 = 5.76 \times 10^2$

d $289\,000\,000 = 2.89 \times 10^8$

e $4300 = 4.3 \times 10^3$

f $26\,540\,000 = 2.654 \times 10^7$

g $0.027 = 2.7 \times 10^{-2}$

h $0.000\,05 = 5 \times 10^{-5}$

i $0.001\,376 = 1.376 \times 10^{-3}$

13 a $4 \times 10^2 = 400$

b $9.1 \times 10^3 = 9100$

c $1.827 \times 10^5 = 182\,700$
d $3.7 \times 10^{-4} = 0.000\,37$
e $8 \times 10^7 = 80\,000\,000$
f $7 \times 10^{-3} = 0.007$
g $1.2 \times 10^{-6} = 0.000\,0012$
h $6.45 \times 10^{-5} = 0.000\,0645$
i $1 \times 10^{-2} = 0.01$

INDICES AND SCIENTIFIC NOTATION
INTERMEDIATE TEST PAGE 11

1 $3^2 \times 3^4 = 3^{2+4}$
$= 3^6$
$\therefore$ [A] ✓ (1 mark)

2 $8^{12} \div 8^4 = 8^{12-4}$
$= 8^8$
$\therefore$ [D] ✓ (1 mark)

3 $(2^2)^5 = 2^{2\times5}$
$= 2^{10}$
$\therefore$ [B] ✓ (1 mark)

4 $3^0 + 2^0 = 1 + 1$
$= 2$
$\therefore$ [B] ✓ (1 mark)

5 $4^{-2} = \frac{1}{4^2}$
$= \frac{1}{16}$
$\therefore$ [D] ✓ (1 mark)

6 $3^2 \times 2^4 = 9 \times 16$
$= 144$
$= 12^2$
$\therefore$ [A] ✓ (1 mark)
[Or $3^2 \times 2^4 = 3^2 \times (2^2)^2$
$= 3^2 \times 4^2$
$= (3 \times 4)^2$
$= 12^2$]

7 $8^2 \times 4^3 = (2^3)^2 \times (2^2)^3$
$= 2^6 \times 2^6$
$= 2^{12}$
$\therefore$ [C] ✓ (1 mark)

8 $2^3 \div 2^{-2} = 2^{3-(-2)}$
$= 2^5$
$\therefore$ [C] ✓ (1 mark)

9 $\frac{3^2 \times 2^3}{2^2 \times 3^3} = \frac{\cancel{3}^1 \times \cancel{3}^1 \times \cancel{2}^1 \times \cancel{2}^1 \times 2}{\cancel{2}^1 \times \cancel{2}^1 \times \cancel{3}^1 \times \cancel{3}^1 \times 3}$
$= \frac{2}{3}$
$\therefore$ [B] ✓ (1 mark)

10 $0.0008 = 8 \times 10^{-4}$
$\therefore$ [D] ✓ (1 mark)
$[0.0008 = 8 \times \frac{1}{10\,000}]$

11 $2.3 \times 10^6 = 2\,300\,000$
Two million three hundred thousand
$\therefore$ [B] ✓ (1 mark)

12 $5 \times 10^{-2} + 2.6 \times 10^{-3}$
$= 0.05 + 0.0026$
$= 0.0526$
$= 5.26 \times 10^{-2}$
$\therefore$ [C] ✓ (1 mark)

13 $\frac{32 \times 500}{0.8 \div 0.05} = \frac{16\,000}{16}$
$= 1000$ ✓
$= 1 \times 10^3$ ✓
(2 marks)

14 $\frac{(3.2 \times 10^7) \times (5.6 \times 10^3)}{2.8 \times 10^{-5}}$
$= \frac{1.792 \times 10^{11}}{2.8 \times 10^{-5}}$
$= 6.4 \times 10^{15}$ ✓ (1 mark)

15 $2.87 \times 10^{-2} = 0.0287$
$6 \times 10^{-2} = 0.0600$
$5 \times 10^{-3} = 0.0050$ ✓
In ascending order:
5×10^{-3}, 2.87×10^{-2},
6×10^{-2} ✓ (2 marks)

16 $(2^5 \times 2^3)^2 \div (2^4 \div 2)^3$
$= (2^8)^2 \div (2^3)^3$ ✓
$= 2^{16} \div 2^9$ ✓
$= 2^7$ ✓ (3 marks)
(Total: 20 marks)

INDICES AND SCIENTIFIC NOTATION
ADVANCED TEST PAGE 12

1 a $10^{-4} = \frac{1}{10^4}$
$= \frac{1}{10\,000}$
$= 0.0001$ ✓ (1 mark)

b $2^{-3} = \frac{1}{2^3}$
$= \frac{1}{8}$
$= 0.125$ ✓ (1 mark)

c $5^{-2} = \frac{1}{5^2}$
$= \frac{1}{25}$
$= 0.04$ ✓ (1 mark)

2 a $7^{-3} = \frac{1}{7^3}$
$= \frac{1}{343}$ ✓ (1 mark)

b $\left(\frac{1}{2}\right)^{-1} = 2$ ✓ (1 mark)

c $\left(\frac{2}{3}\right)^{-2} = \left(\frac{3}{2}\right)^2$
$= \frac{9}{4}$
$= 2\frac{1}{4}$ ✓ (1 mark)

3 a $2^x = 1024$
$2^x = 2^{10}$
$x = 10$ ✓ (1 mark)

b $7^x = 1$
$7^x = 7^0$
$x = 0$ ✓ (1 mark)

c $4^x = 2^{12}$
$(2^2)^x = 2^{12}$
$2^{2x} = 2^{12}$
$2x = 12$
$x = 6$ ✓ (1 mark)

d $10^x = \frac{1}{1000}$
$= \frac{1}{10^3}$
$10^x = 10^{-3}$
$x = -3$ ✓ (1 mark)

e $2^x = \frac{1}{64}$
$= \frac{1}{2^6}$
$= 2^{-6}$
$x = -6$ ✓ (1 mark)

f $x^{-2} = \frac{25}{16}$
$x^2 = \frac{16}{25}$
$x = \pm\sqrt{\frac{16}{25}}$
$= \pm\frac{4}{5}$ ✓ (1 mark)

4 **a** $5^8 \times 5^{-3} = 5^5$ ✓ (1 mark)

b $7^4 \div 7^9 = 7^{-5}$ ✓ (1 mark)

c $6^{-3} \times 6^{-5} = 6^{-8}$ ✓ (1 mark)

d $2^{-4} \div 2 = 2^{-5}$ ✓ (1 mark)

e $(3^4)^2 \div (3^5)^3$
$= 3^8 \div 3^{15}$
$= 3^{-7}$ ✓ (1 mark)

f $(10^{-3})^{-2} \times (10^{-4})^4$
$= 10^6 \times 10^{-16}$
$= 10^{-10}$ ✓ (1 mark)

g $(5^{-5})^2 \div (5^0)^{-2}$
$= 5^{-10} \div 1$
$= 5^{-10}$ ✓ (1 mark)

h $2^{-7} \times 4^{-3}$
$= 2^{-7} \times (2^2)^{-3}$
$= 2^{-7} \times 2^{-6}$
$= 2^{-13}$ ✓ (1 mark)

5 **a** 32×10^7
$= 3.2 \times 10^8$ ✓ (1 mark)

b 0.35×10^5
$= 3.5 \times 10^4$ ✓ (1 mark)

c 560×10^{-4}
$= 5.6 \times 10^{-2}$ ✓ (1 mark)

d 7×100^2
$= 7 \times (10^2)^2$
$= 7 \times 10^4$ ✓ (1 mark)

e 0.06×10^{-5}
$= 6 \times 10^{-7}$ ✓ (1 mark)

f 97×10^{-9}
$= 9.7 \times 10^{-8}$ ✓ (1 mark)

6 **a** $E = mc^2$
$= 0.8 \times (3 \times 10^5)^2$
$= 7.2 \times 10^{10}$ ✓ (1 mark)

b $E = mc^2$
$5.76 \times 10^{13} = m \times (3 \times 10^5)^2$
$m = \dfrac{5.76 \times 10^{13}}{(3 \times 10^5)^2}$
$= 640$ ✓
(1 mark)

7 **a** $(7.5 \times 10^8) \times (4.8 \times 10^7)$
$= 3.6 \times 10^{16}$ ✓ (1 mark)

b $(3.7 \times 10^6)^2$
$= 1.369 \times 10^{13}$ ✓
(1 mark)

c $6.5 \times 10^{-8} + 5.3 \times 10^{-7}$
$= 0.000\,000\,595$
$= 5.95 \times 10^{-7}$ ✓ (1 mark)

d $8.7 \times 10^{-4} - 3.9 \times 10^{-6}$
$= 0.000\,8661$
$= 8.661 \times 10^{-4}$ ✓ (1 mark)

e $\sqrt{9.604 \times 10^{-5}}$
$= 0.0098$
$= 9.8 \times 10^{-3}$ ✓ (1 mark)

8 **a** Distance from Earth to Mercury
$= (1.49 \times 10^8 - 5.8 \times 10^7)$ km
$= 91\,000\,000$ km
$= 9.1 \times 10^7$ km ✓
(1 mark)

b Distance from Venus to Earth
$= (1.49 \times 10^8 - 1.08 \times 10^8)$ km
$= 41\,000\,000$ km
Venus is closer to Earth than Mercury because the distance from Earth to Venus is less than half of the distance from Earth to Mercury. ✓ (1 mark)

(Total: 35 marks)

PYTHAGORAS' THEOREM

SKILLS CHECK PAGE 14

1 **a** $13^2 = 169$

b $8^2 + 6^2 = 64 + 36$
$= 100$

c $37^2 - 24^2 = 1369 - 576$
$= 793$

2 **a** $\sqrt{5625} = 75$

b $\sqrt{25 + 144} = \sqrt{169}$
$= 13$

c $\sqrt{25 - 9} = \sqrt{16}$
$= 4$

3 **a** $\sqrt{527} = 22.956\,480\ldots$
$= 23.0$ (1 d.p.)

b $\sqrt{49 + 16} = \sqrt{65}$
$= 8.062\,2577\ldots$
$= 8.1$ (1 d.p.)

c $\sqrt{3.2^2 - 1.7^2} = \sqrt{7.35}$
$= 2.711\,088\ldots$
$= 2.7$ (1 d.p.)

4 **a** XY

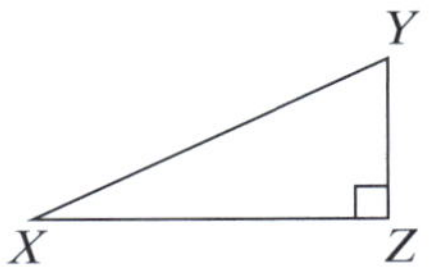

b QR

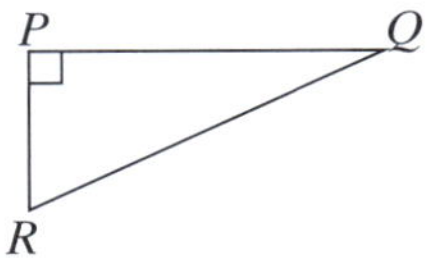

c DF

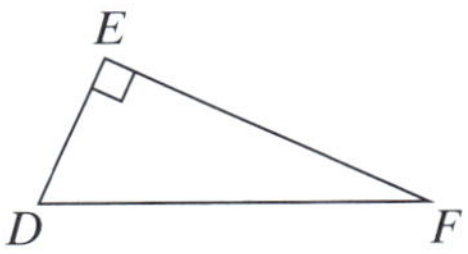

5 **a**

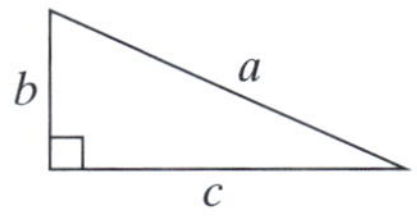

$a^2 = b^2 + c^2$

b

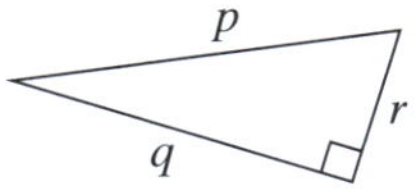

$p^2 = q^2 + r^2$

6 **a**

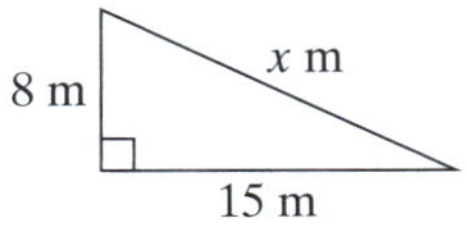

$x^2 = 8^2 + 15^2$
$= 64 + 225$
$= 289$
$x = \sqrt{289}$
$= 17$
∴ the length of the hypotenuse is 17 m.

b

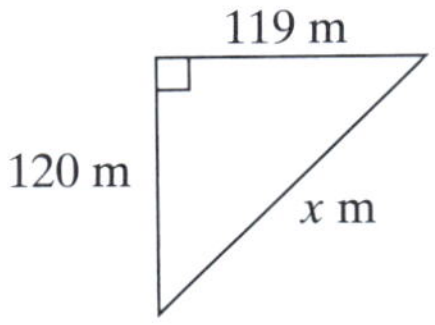

$x^2 = 119^2 + 120^2$
$= 14\,161 + 14\,400$
$= 28\,561$
$x = \sqrt{28\,561}$
$= 169$
∴ the length of the hypotenuse is 169 m.

7 **a**

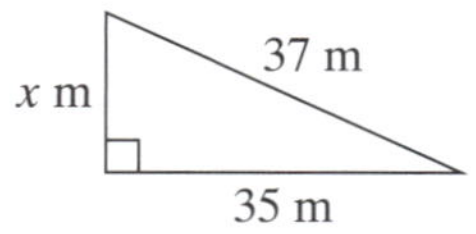

$37^2 = x^2 + 35^2$
$x^2 = 37^2 - 35^2$
$= 1369 - 1225$
$= 144$
$x = \sqrt{144}$
$= 12$
∴ the side is 12 m long.

b

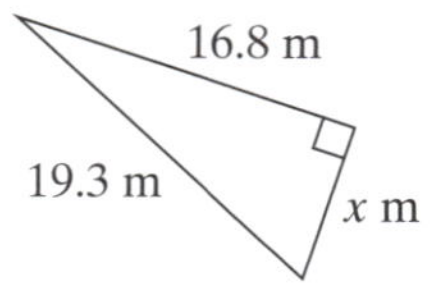

$x^2 = 19.3^2 - 16.8^2$
$= 372.49 - 282.24$
$= 90.25$
$x = \sqrt{90.25}$
$= 9.5$
∴ the side is 9.5 m long.

8 **a**

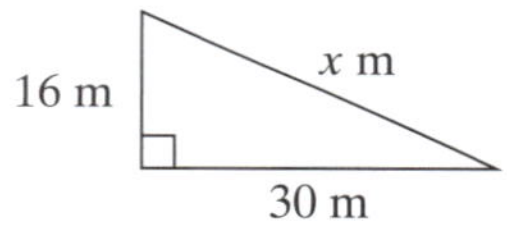

$x^2 = 16^2 + 30^2$
$= 256 + 900$
$= 1156$
$x = \sqrt{1156}$
$= 34$
∴ the side is 34 m long.

b

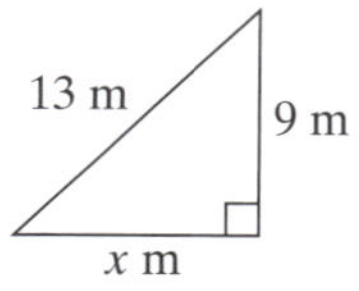

$13^2 = x^2 + 9^2$
$x^2 = 13^2 - 9^2$
$= 169 - 81$
$= 88$
$x = \sqrt{88}$
$= 9.380\,831\,52\ldots$
∴ the side is 9.4 m long (1 d.p.).

9 **a** $11^2 + 48^2 = 55^2$?
$11^2 + 48^2 = 121 + 2304$
$= 2425$
$55^2 = 3025$
False

b $225^2 + 272^2 = 353^2$?
$225^2 + 272^2 = 50\,625 + 73\,984$
$= 124\,609$
$353^2 = 124\,609$
True

c $21.9^2 + 29.2^2 = 36.5^2$?
$21.9^2 + 29.2^2 = 479.61 + 852.64$
$= 1332.25$
$36.5^2 = 1332.25$
True

PYTHAGORAS' THEOREM
INTERMEDIATE TEST PAGE 15

1

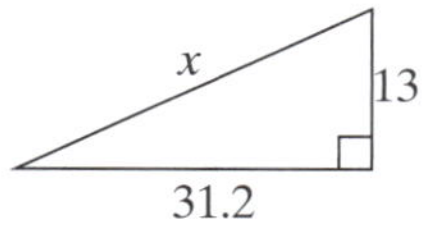

$x^2 = 13^2 + 31.2^2$
$= 1142.44$
$x = \sqrt{1142.44}$
$= 33.8$
∴ [B] ✓ (1 mark)

2

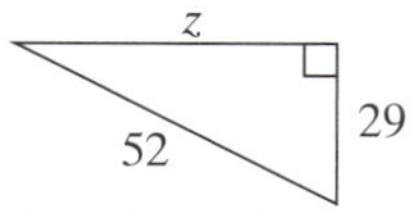

$z^2 = 52^2 - 29^2$
$= 1863$
$z = \sqrt{1863}$
$= 43.162\,48\ldots$
$= 43.2$ (1 d.p.)
∴ A ✓ (1 mark)

3 Try each option.
A: $23.3^2 = 542.89$
$10.5^2 + 20.8^2 = 542.89$
So triangle A is right-angled.
B: $18.5^2 = 342.25$
$10.4^2 + 15.3^2 = 342.25$
So triangle B is right-angled.
C: $20.5^2 = 420.25$
$13.3^2 + 15.6^2 = 420.25$
So triangle C is right-angled.
D: $35.3^2 = 1246.09$
$20.5^2 + 25.8^2 = 1085.89$
So triangle D is not right-angled.
∴ [D] ✓ (1 mark)

4 Let the length of the stay be x m.

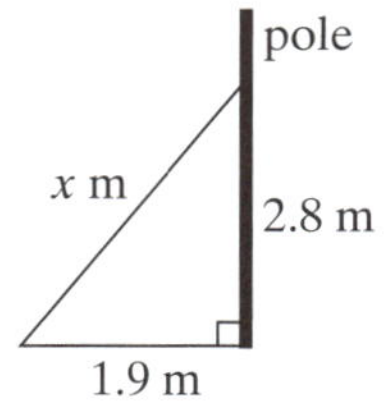

$x^2 = 2.8^2 + 1.9^2$
$= 11.45$
$x = \sqrt{11.45}$
$= 3.383\,7848\ldots$ ✓
$= 3.4$ (1 d.p.)
The length of the stay is 3.4 m to one decimal place. ✓
(2 marks)

5

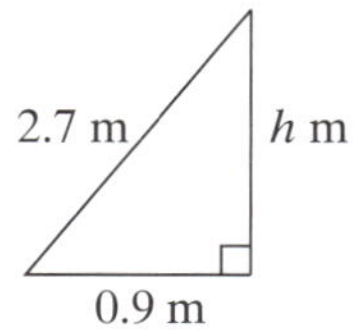

Let h m be the height the ladder reaches up the wall.
$h^2 = 2.7^2 - 0.9^2$
$= 6.48$
$h = \sqrt{6.48}$
$= 2.545\,5844\ldots$ ✓
$= 2.5$ (1 d.p.)
The height of the ladder up the wall is 2.5 m to one decimal place. ✓ (2 marks)

6

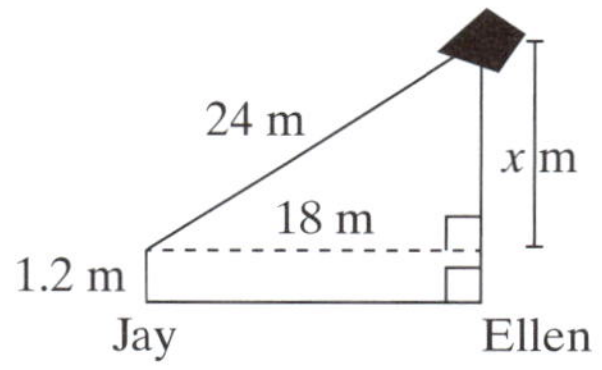

Let x m be the height of the kite above Jay's arm.
$x^2 = 24^2 - 18^2$
$= 252$
$x = \sqrt{252}$ ✓
$= 15.874\,507\ldots$
$= 15.9$ (1 d.p.) ✓
Height $= (15.9 + 1.2)$ m
$= 17.1$ m
The height of the kite is 17.1 m above ground to one decimal place. ✓ (3 marks)

7 Let the length of the brace be x m.

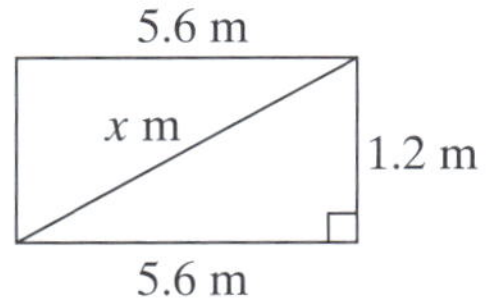

$x^2 = 1.2^2 + 5.6^2$
$= 32.8$
$x = \sqrt{32.8}$
$= 5.727\,128\,425\ldots$ ✓
$= 5.73$ (2 d.p.)

The length of the brace is 5.73 m or 573 cm to the nearest centimetre. ✓ (2 marks)

8

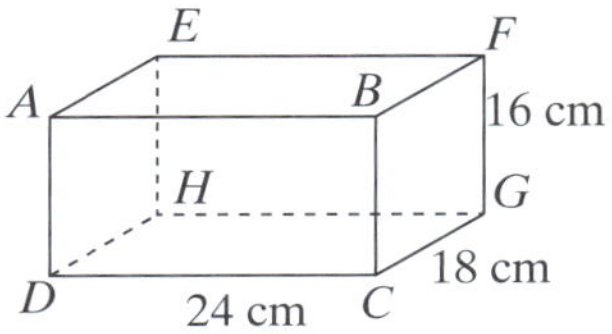

a $DG^2 = DC^2 + CG^2$
$= 24^2 + 18^2$
$= 900$
$DG = \sqrt{900}$ ✓
$= 30$

The length of DG is 30 cm. ✓ (2 marks)

b $AG^2 = AD^2 + DG^2$
$= 16^2 + 30^2$ ✓
$= 1156$
$AG = \sqrt{1156}$ ✓
$= 34$

The length of AG is 34 cm. ✓ (3 marks)

9 Let h cm be the height of the triangle.

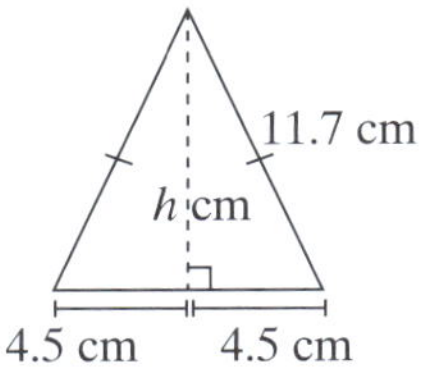

$h^2 = 11.7^2 - 4.5^2$ ✓
$= 116.64$
$h = \sqrt{116.64}$
$= 10.8$ ✓

$A = \frac{1}{2}bh$
$= \frac{1}{2} \times 9 \times 10.8$
$= 48.6$

The area is 48.6 cm². ✓

(3 marks)

(Total: 20 marks)

PYTHAGORAS' THEOREM
ADVANCED TEST PAGE 16

1 **a** Let the length of side AB be x km.

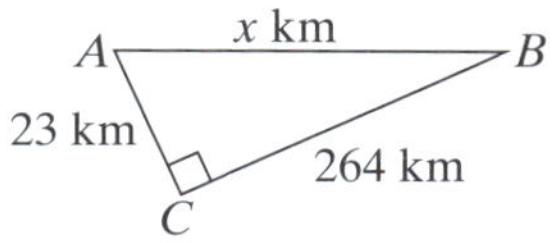

$x^2 = 23^2 + 264^2$
$= 70225$ ✓
$x = \sqrt{70\,225}$ $(x > 0)$
$= 265$

The length of AB is 265 km. ✓ (2 marks)

b Let the length of side AB be x cm.

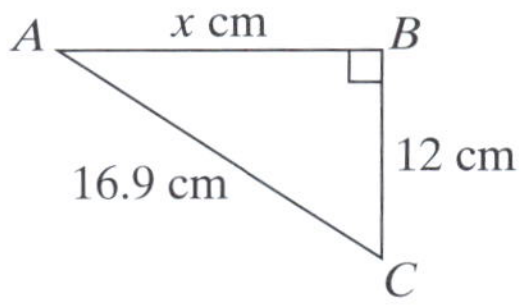

$x^2 + 12^2 = 16.9^2$
$x^2 = 16.9^2 - 12^2$
$= 141.61$ ✓
$x = \sqrt{141.61}$ $(x > 0)$
$= 11.9$

The length of AB is 11.9 cm. ✓ (2 marks)

c Let the length of side AB be x m.

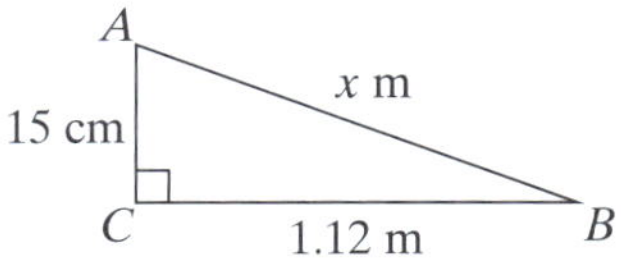

$AC = 15$ cm $= 0.15$ m
$x^2 = 1.12^2 + 0.15^2$
$= 1.2769$ ✓
$x = \sqrt{1.2769}$ $(x > 0)$
$= 1.13$

The length of AB is 1.13 m. ✓ (2 marks)

2 **a** $73^2 = 5329$
$48^2 + 55^2 = 5329 = 73^2$

The numbers do form a Pythagorean triad. ✓

(1 mark)

b $15.17^2 = 230.1289$
$7.95^2 + 12.92^2 = 230.1289$
$= 15.17^2$

The numbers do form a Pythagorean triad. ✓

(1 mark)

3 **a** Let x cm be the length of the hypotenuse.

$x^2 = 11^2 + 25^2$
$= 746$
$x = \sqrt{746}$ $(x > 0)$ ✓
$= 27.313\,0005\ldots$
$= 27.3$ (1 d.p.)

The length of the hypotenuse is 27.3 cm. ✓

(2 marks)

b $x^2 = 12^2 + 25^2$
$= 769$
$x = \sqrt{769}$ $(x > 0)$
$= 27.730\,849\ldots$
$= 27.7$ (1 d.p.)

The length of the hypotenuse is 27.7 cm. ✓ The incorrect measurement meant that the hypotenuse was thought to be 4 mm shorter than it actually is. ✓ (2 marks)

4 Let the base of the triangle be x cm.

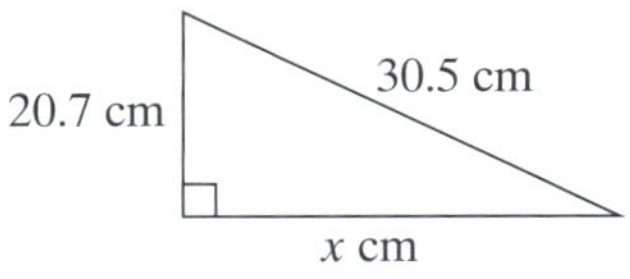

$x^2 = 30.5^2 - 20.7^2$
$= 501.76$ ✓
$x = \sqrt{501.76}$ $(x > 0)$
$= 22.4$

The base is 22.4 cm long. ✓

$A = \frac{1}{2}bh$
$= \frac{1}{2} \times 22.4 \times 20.7$
$= 231.84$

The area is 231.84 cm². ✓

(3 marks)

5 Let the length of each side be x cm.

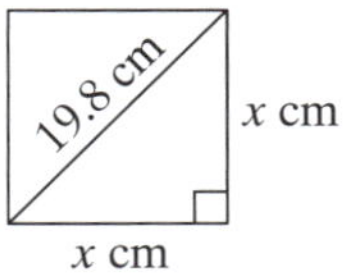

$x^2 + x^2 = 19.8^2$ ✓
$2x^2 = 392.04$
$x^2 = 196.02$ ✓
$x = \sqrt{196.02}$ $(x > 0)$
$= 14.000714\ldots$
The length of each side is 14 cm. ✓ (3 marks)

6 **a** Let h m be the height of the pole.

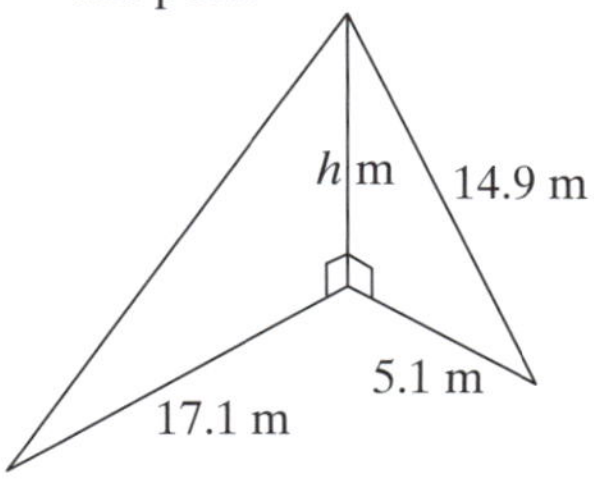

$h^2 = 14.9^2 - 5.1^2$
$= 196$ ✓
$h = \sqrt{196}$ $(h > 0)$
$= 14$
The pole is 14 m high. ✓
(2 marks)

b Let x m be the length of the second wire.
$x^2 = 14^2 + 17.1^2$
$= 488.41$ ✓
$x = \sqrt{448.41}$ $(x > 0)$
$= 22.1$
The second wire is 22.1 m long. ✓ (2 marks)

7 The javelin would need to fit diagonally across the case.

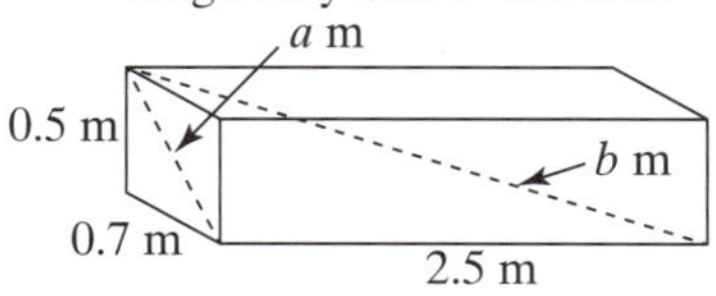

$a^2 = 0.5^2 + 0.7^2$
$= 0.74$
$b^2 = a^2 + 2.5^2$
$= 0.74 + 6.25$
$= 6.99$ ✓
$b = \sqrt{6.99}$ $(b > 0)$
$= 2.64386\ldots$

So, the length of the diagonal of the case is 2.64 m, to the nearest centimetre. ✓
The javelin should just fit in the case. (But this doesn't allow for the thickness of the javelin and it assumes that the javelin is exactly 2.6 m long and not slightly longer, but rounded to 2.6 m. Ben should be prepared to find the javelin might not fit.) ✓ (3 marks)
(Total: 25 marks)

UNITS OF MEASUREMENT AND TIME SCALES
SKILLS CHECK PAGE 18

1 **a** 3 m = 300 cm
b 5 kg = 5000 g
c 16 L = 16 000 mL
d 7 cm = 70 mm
e 9000 L = 9 kL
f 23 000 mg = 23 g

2 **a** 5 min = 5×60 s
= 300 s
b 24 h = 24×60 min
= 1440 min
c 480 s = $(480 \div 60)$ min
= 8 min
d 3 days = 3×24 h
= 72 h
e 7 decades = 7×10 years
= 70 years
f 2½ centuries
= 2.5×100 years
= 250 years

3 **a** 3 h after 7:20 am is 10:20 am. 15 min after that is 10:35 am.
b 9 h after 8:52 pm is 5:52 am. Another 8 min after that is 6:00 am. 40 min after 6:00 am is 6:40 am.

4 **a** From 6:35 am until 6:35 pm is 12 hours. It is another hour until 7:35 pm. From 7:35 pm until 7:45 pm is 10 minutes.
Total time = 13 h 10 min
b From 0850 to 1350 is 5 hours. From 1350 to 1400 is 10 minutes. From 1400 to 1435 is 35 minutes.
Total time = 5 h 45 min
c There are 31 days in May. From 12 noon 19 May until 12 noon 31 May is 12 days. There are 30 days in June. From 12 noon 31 May until 12 noon 30 June is 30 days. From 12 noon 30 June until 12 noon 5 July is 5 days.
Total time = 47 days

5 There are 30 days in September. So, 30 days after 13 September is 13 October. Another 15 days after this is 28 October.

6 1 megalitre = 1 000 000 litres
1 gigalitre = 1 000 000 000 litres
1 kilolitre = 1000 litres
So, kilolitre is the smallest, followed by megalitre and gigalitre.

7 micro means one-millionth
milli means one-thousandth
nano is one-thousand-millionth
So nanosecond is shortest, followed by microsecond and millisecond.

8 A millennium is 1000 years.

9 **a** $624 \approx 600$
b $578 \approx 600$
c $1295 \approx 1300$
d $45\,378\,632 \approx 45\,378\,600$
e $453\,720 \approx 453\,700$
f $63\,999 \approx 64\,000$

10 **a** $0.47 = 0.5$ (1 d.p.)
b $2.54 = 2.5$ (1 d.p.)
c $8.15 = 8.2$ (1 d.p.)
d $24.6378 = 24.6$ (1 d.p.)
e $1.0365 = 1.0$ (1 d.p.)
f $0.98 = 1.0$ (1 d.p.)

11 **a** $0.376 = 0.38$ (2 d.p.)
b $1.2345 = 1.23$ (2 d.p.)

c $576.399 = 576.40$ (2 d.p.)

12 a 21 576
The first significant figure is 2.
b 0.005 78
The first significant figure is 5.
c 1.027 65
The first significant figure is 1.
d 0.020 79
The first significant figure is 2.
e 700 000
The first significant figure is 7.
f 0.800
The first significant figure is 8.

13 a $5\,876\,235 = 5\,880\,000$
(3 significant figures)
b $562\,877 = 563\,000$
(3 significant figures)
c $9461 = 9460$
(3 significant figures)
d $0.058\,76 = 0.0588$
(3 significant figures)
e $0.003\,0854 = 0.003\,09$
(3 significant figures)
f $0.000\,047\,239 = 0.000\,0472$
(3 significant figures)

UNITS OF MEASUREMENT AND TIME SCALES
INTERMEDIATE TEST PAGE 19

1 The length of a car would be measured in metres.
$\therefore$ [B] ✓ (1 mark)

2 0.789 is closest to 0.8
$\therefore$ [B] ✓ (1 mark)
[$0.8 = 0.800$, $0.7 = 0.700$, $0.77 = 0.770$, $0.83 = 0.830$]

3 $1\,000\,000\,000$ ns = 1 s
A nanosecond is one-thousand-millionth of a second.
$\therefore$ [D] ✓ (1 mark)

4 1 MB $\approx$ 1000 kB
1 GB $\approx$ 1000 MB
$\approx 1000 \times 1000$ kB
$\approx 1\,000\,000$ kB
$\therefore$ [C] ✓ (1 mark)

5 $5378 = 5400$ (2 significant figures)
$\therefore$ [D] ✓ (1 mark)
[The value of the number cannot change. It must be about 5000 so answers of 50 cannot be right.]

6 Error $= \pm 0.5$ cm
$= \pm 5$ mm
Length of pencil was measured to be 70 mm. Best interval is between 65 mm and 75 mm.
$\therefore$ [A] ✓ (1 mark)

7 $0.028\,19 = 0.0282$ (3 significant figures)
$\therefore$ [D] ✓ (1 mark)
[The first significant figure is 2.]

8 The digit must be 5.
$\therefore$ [B] ✓ (1 mark)
[Only 2.563 will be equal to 2.6 when written correct to one decimal place.]

9 micro means one-millionth
$1\mu g = 0.000\,001$ g
$\therefore$ [C] ✓ (1 mark)

10 Speed $= 3 \times 10^5$ km/s
$= 300\,000$ km/s
Time $= (1 \div 300\,000)$ s
$= 0.000\,003\,333\ldots$ s
$\approx 0.000\,003$ s
≈ 3 microseconds
$\therefore$ [D] ✓ (1 mark)

11 a Number of doses
$= 700 \div 15$
$= 46.666\,6666\ldots$ ✓
The bottle holds 46 15 mL doses. ✓ (2 marks)
[The bottle holds an amount closer to 47 doses than 46, but it only actually holds 46 doses.]
b Number of doses
$= 700 \div 12$
$= 58.333\,33\ldots$
The bottle holds 58 12 mL doses. ✓ (1 mark)
c 3×58 doses $= 174$ doses ✓
Three bottles hold
3×700 mL $= 2100$ mL
Number of doses
$= 2100 \div 12$
$= 175$
Three bottles hold 175 doses. ✓
No, three bottles hold one more dose than three times the answer to part **b**. ✓ (3 marks)
[This is because there is one-third of a dose left over. Three times one-third is one whole. So three bottles give one more dose.]

12 $20\,000\,000 \div 2.6 = 7\,692\,307.692\ldots$
Henry is correct with his calculation but be cannot claim that the area is correct to this degree of accuracy. ✓
The population is only approximately 20 million and the density approximately 2.6 so the best Henry can say is the area is approximately 7.7 million square km. ✓
(2 marks)

13 a Length of time
$= (245 - 65)$ million years
$= 180$ million years ✓
b For an age of 14 years:
Number of times longer
$= 208$ million $\div 14$
$= 14.857\,142\ldots$ million
≈ 15 million ✓ (2 marks)
(Total: 20 marks)

UNITS OF MEASUREMENT AND TIME SCALES
ADVANCED TEST PAGE 20

1 a 0.065×18.235
$= 1.185\,275$
$= 1.19$ (2 d.p.) ✓ (1 mark)

b $\sqrt{912} = 30.1993377\ldots$
$= 30.20$ (2 d.p.) ✓ (1 mark)

c $\frac{16.35 + 2.97}{4.85 - 3.2} = 11.7090909\ldots$
$= 11.71$ (2 d.p.) ✓ (1 mark)

2 **a** $95 \div 0.875 = 108.571428\ldots$
$= 109$ (3 sig.fig.) ✓ (1 mark)

b $(0.246)^2 = 0.060516$
$= 0.0605$ (3 sig.fig.) ✓ (1 mark)

c $563.7 \div 0.006$
$= 93950$
$= 94000$ (3 sig.fig) ✓ (1 mark)

3 **a** 1 t = 1000 kg
= 1000 000 g
= 1 000 000 000 mg
= 1 000 000 000 000 µg ✓ (1 mark)

b 1 GL = 1 000 000 000 L
= 1 000 000 000 000 mL ✓ (1 mark)

c 1 h = 60 min
= 3600 s
= 3 600 000 000 000 ns ✓ (1 mark)

4 **a** 255 min = 4 h 15 min ✓ (1 mark)

b 0.08 g = (0.08 × 1000) mg
= 80 mg ✓ (1 mark)

c 43.5 ML = 43.5 × 1 000 000 L
= 43 500 000 L ✓ (1 mark)

d 720 µg = (720 ÷ 1000) mg
= 0.72 mg ✓ (1 mark)

e 0.0087 s = 0.0087 × 1000 ms
= 8.7 ms ✓ (1 mark)

f 8.5 millennia
= 8.5 × 1000 years
= 8500 years
= 85 centuries ✓ (1 mark)

g 12 288 GB
= (12 288 ÷ 1024) TB
= 12 TB ✓ (1 mark)

h 29 420 ns
= 29.42 µs
= 0.029 42 ms
= 0.000 029 42 s ✓ (1 mark)

i 5.6×10^{-7} kg
= 0.000 000 56 kg
= 0.000 56 g
= 0.56 mg
= 560 µg ✓ (1 mark)

j 9.4×10^8 kL
$= 9.4 \times 10^5$ ML
$= 9.4 \times 10^2$ GL
= 940 GL ✓ (1 mark)

5 **a** 700 MB = 700 × 1024 kB
= 716 800 kB
Number of files
= 716 800 ÷ 1240
= 578.0645…
= 578
(nearest whole number) ✓ (1 mark)

b 4 GB = 4 × 1024 MB
= 4096 MB
Number of disks
= 4096 ÷ 700
= 5.851 4285 …
= 6
(nearest whole number)
The device can hold as much information as about 6 disks, but it cannot hold all the information that can be stored on 6 disks. ✓ (1 mark)

6 **a** Speed $= 3 \times 10^5$ km/s
$\text{Time} = \frac{\text{distance}}{\text{speed}}$
$= \frac{3}{3 \times 10^5}$
= 0.000 01 s
= 0.01 ms
= 10 µs ✓ (1 mark)

b Speed = 1200 km/h
= 20 km/min
$= \frac{1}{3}$ km/s
So sound travels 1 km in 3 seconds.
It will take 9 seconds to travel 3 km. ✓ (1 mark)

c $T = \frac{4 \times 10^{13}}{3 \times 10^5}$
= 133 333 333.3… s ✓
= 37 037.037… h
= 1543.209… days
= 4.2279… years
The time is about 4 years and 3 months. ✓ (2 marks)
(Total: 25 marks)

EARNING MONEY
SKILLS CHECK — PAGE 22

1 **a** Pay = 8 × \$37.65
= \$301.20

b Pay = 40 × \$23.50
= \$940

c Pay = 35 × \$41.70
= \$1459.50

d Pay = 37.5 × \$38.76
= \$1453.50

2 **a** Weekly pay = \$41 028 ÷ 52
= \$789

b Weekly pay = \$612.50 ÷ 2
= \$306.25

3 **a** Annual income
= \$752 × 52
= \$39 104

b Annual income
= \$1825 × 26
= \$47 450

c Annual income
= \$3015 × 12
= \$36 180

4 **a** Weekly pay = \$54 288 ÷ 52
= \$1044

b Fortnightly pay
= \$54 288 ÷ 26
= \$2088
[Or \$1044 × 2 = \$2088]

c Monthly pay = \$54 288 ÷ 12
= \$4524
[This is not \$1044 × 4.]

5 **a** Hourly rate
= \$1274.80 ÷ 40
= \$31.87 per hour

b Hourly rate
= \$1438.50 ÷ 35
= \$41.10 per hour

c Hourly rate = \$912 ÷ 38
= \$24 per hour

6 8 hours on Monday
7 hours on Tuesday
3.5 hours on Wednesday
8 hours on Thursday
9 hours on Friday
Total hours = 8 + 7 + 3.5 + 8 + 9
= 35.5

7 Overtime hours: 0.5 hours on Tuesday, 3.5 hours on Thursday and 1.5 hours on Friday.
Total overtime = 0.5 + 3.5 + 1.5
= 5.5 hours

8 **a** Commission = 7% of $8000
$= \frac{7}{100} \times \$8000$
= $560

b Commission
= 2½% of $15 000
$= \frac{2.5}{100} \times \$15\,000$
= $375

9 **a** Earnings = $0.80 × 56
= $44.80

b Earnings = $5 × 15
= $75

10 **a** Overtime = 1 × 1.5 × $21.60
= $32.40

b Overtime = 3 × 2 × $21.60
= $129.60

c Overtime = 8 × 2.5 × $21.60
= $432.00

11 **a** Earnings = 4 × $895
= $3580

b 17½% of $2872
$= \frac{175}{100} \times \$2872$
= $502.60

12 Deductions
= $209 + $68 + $29.50 + $15
= $321.50

13 Net earnings = $876 – $277
= $599

EARNING MONEY
INTERMEDIATE TEST PAGE 23

1 Pay = 12 × 1.5 × $24.38
= $438.84
∴ [D] ✓ (1 mark)

2 Royalty = 0.125 × $19.95 × 1500
= $3740.625
∴ [D] ✓ (1 mark)

3 Try each option.
$53 872 ÷ 52 = $1036
$2052.50 ÷ 2 = $1026.25
$30.75 × 35 = $1076.25
$26.58 × 40 = $1063.20
$1076.25 is the highest.
∴ [C] ✓ (1 mark)

4 **a** Weekly pay = $28.40 × 35
= $994.00 ✓ (1 mark)

b Overtime hours = 42 – 35
= 7 ✓
Overtime pay
= 7 × 1.5 × $28.40
= $298.20 ✓
Total weekly pay
= $994.00 + $298.20
= $1292.20 ✓ (3 marks)

5 **a** Fortnightly pay
= $78 000 ÷ 26
= $3000 ✓ (1 mark)

b Total deductions
= $215 + $892 + $39.80 + $20
= $1166.80 ✓
Take-home pay
= $3000 – $1166.80
= $1833.20 ✓ (2 marks)

6 **a** Earnings for the day
= 24 × $7.50
= $180 ✓ (1 mark)

b Number of surveys
= $1290 ÷ $7.50
= 172
Eric will have completed 172 surveys. ✓ (1 mark)

7 Commission on first $100 000
= 4% of $100 000
$= \frac{4}{100} \times \frac{\$100\,000}{1}$
= $4000 ✓
Remainder = $315 000 – $100 000
= $215 000 ✓
Commission on remainder
= 3½% of $215 000
= 0.035 × $215 000
= $7525 ✓
Total commission
= $4000 + $7525
= $11 525 ✓ (4 marks)

8 **a** Pay rate = $1808.00 ÷ 40
= $45.20
The hourly rate of pay is $45.20 per hour. ✓ (1 mark)

b Saturday: 6 hours at double time
Pay for Saturday
= 6 × 2 × $45.20
= $542.40 ✓
Sunday: 4.5 hours at double time and a half
Pay for Sunday
= 4.5 × 2.5 × $45.20
= $508.50 ✓
Total for weekend
= $542.40 + $508.50
= $1050.90 ✓ (3 marks)

c 4 weeks pay = 4 × $1808.00
= $7232.00 ✓
Holiday loading
= 17½% of $7232.00
= 0.175 × $7232
= $1265.60 ✓
Total holiday pay
= $7232.00 + $1265.60
= $8497.60 ✓ (3 marks)
(Total: 23 marks)

EARNING MONEY
ADVANCED TEST PAGE 24

1 **a** B: Annual pay = 26 × $3250
= $84 500
Difference = $87 512 – $84 500
= $3012
Sam will receive an extra $3012 per year. ✓ (1 mark)

b Some possible answers:
Position B might have fewer hours.
Position B might have better conditions.
Position B might be a more secure position.
Position B might have better prospects for promotion and/or improved pay conditions in the future.
✓✓✓ (3 marks)

2 Yearly income = $876 × 52
= $45 552 ✓
Monthly income = $45 552 ÷ 12
= $3796 ✓
(2 marks)

WORKED SOLUTIONS

CHECK YOUR SOLUTIONS

3 **a** 9:00 a.m. until 5:00 p.m.
= 8 hours
8:00 a.m. until 4:30 p.m.
= 8½ hours
8:30 a.m. until 5:00 p.m.
= 8½ hours
11:00 a.m. until 9:00 p.m.
= 10 hours
11:00 a.m. until 4:00 p.m.
= 5 hours
Total hours
= 8 + 8.5 + 8.5 + 10 + 5
= 40 ✓ (1 mark)

b 8 hours on Monday
Earnings = 8 × \$19.70
= \$157.60 ✓
(1 mark)

c Kendall worked 10 hours on Thursday, 6 hours at normal time and 4 hours at time-and-a-half.
Normal pay = 6 × \$19.70
= \$118.20
Overtime = 4 × 1.5 × \$19.70
= \$118.20 ✓
Total for Thursday
= \$118.20 + \$118.20
= \$236.40 ✓ (2 marks)

d 22 hours worked on Tuesday, Wednesday and Friday
Earnings = 22 × \$19.70
= \$433.40 ✓
Weekly wage
= \$157.60 + \$236.40 + \$433.40
= \$827.40 ✓ (2 marks)

4 **a** Taxable income
= \$92 560 – \$2595
= \$89 965 ✓ (1 mark)

b Tax = \$17 547 + \$0.37 × (89 965 – 80 000) ✓
= \$21 234.05 ✓
(2 marks)

c Medicare levy
= 1.5% of \$89 965
= 0.015 × \$89 965
= \$1349.475
= \$1349.48
(nearest cent) ✓ (1 mark)

d Total tax payable
= \$21 234.05 + \$1349.48
= \$22 583.53

Total tax paid
= 52 × \$453.50
= \$23 582 ✓
Difference
= \$23 582 – \$22 583.53
= \$998.47
Margaret will receive a refund, (because she has paid more tax than is required) of \$998.47. ✓
(2 marks)

5 Sales exceeding \$5000
= \$27 000 – \$5000
= \$22 000
Commission = 0.0325 × \$22 000
= \$715 ✓
Total earnings
= \$320 + \$715
= \$1035 ✓ (2 marks)

6 **a** 4 weeks pay = 4 × \$995.40
= \$3981.60
Leave loading
= 0.175 × \$3981.60
= \$696.78 ✓
Total holiday pay
= \$3981.60 + \$696.78
= \$4678.38 ✓ (2 marks)

b Normal hourly rate of pay
= \$995.40 ÷ 36
= \$27.65
Overtime rate
= 1.5 × \$27.65
= \$41.475 ✓
Now \$696.78 ÷ \$41.475
= 16.8 ✓
16.8 hours = 16 h 48 min
So Celine would need to work for 16 hours and 48 minutes overtime. ✓
(3 marks)
(Total: 25 marks)

SIMPLE INTEREST
SKILLS CHECK PAGE 26

1 **a** 3 years = 3 × 12 months
= 36 months

b 8 years = 8 × 12 months
= 96 months

c 25 years = 25 × 12 months
= 300 months

d 5½ years
= (5 × 12 + 6) months
= 66 months

2 **a** 10% of \$900 = $\frac{10}{100}$ × \$900
= \$90

b 5% of \$8000 = $\frac{5}{100}$ × \$8000
= \$400

c 3% of \$250 000
= $\frac{3}{100}$ × \$250 000
= \$7500

3 **a** $\frac{20}{200} \times 100\% = 10\%$

b $\frac{180}{3000} \times 100\% = 6\%$

c $\frac{1380}{18\,400} \times 100\% = 7.5\%$

4 **a** 5% = \$9000
1% = \$1800
100% = \$180 000

b 12% = \$75 000
1% = \$6250
100% = \$625 000

5 **a** $I = PRN$
= \$500 × $\frac{6}{100}$ × 1
= \$30

b $I = PRN$
= \$2000 × $\frac{4}{100}$ × 2
= \$160

c $I = PRN$
= \$8000 × $\frac{5}{100}$ × 6
= \$2400

d $I = PRN$
= \$950 × $\frac{7.5}{100}$ × 4
= \$285

6 **a** $I = PRN$
= \$600 × $\frac{5}{100}$ × 1
= \$30

b Total amount = \$600 + \$30
= \$630

c $I = PRN$
= \$630 × $\frac{5}{100}$ × 1
= \$31.50

7 **a** Amount owing
= \$5000 – \$1500
= \$3500

b Deposit = 20% of \$5000

$= \frac{20}{100} \times \$5000$

$= \$1000$

Amount owing

$= \$5000 - \$1000$

$= \$4000$

8 **a** Total paid = \$125 × 12

= \$1500

b 2 years = 24 months

Total paid = \$78 × 24

= \$1872

c Total paid

= \$1000 + \$321.70 × 18

= \$6790.60

d 25 years = 25 × 12 months

= 300 months

Total paid

= \$30 000 + \$813.47 × 300

= \$274 041

9 **a** Monthly instalment

= \$2724 ÷ 12

= \$227

b 3 years = 36 months

Monthly instalment

= \$11 311.20 ÷ 36

= \$314.20

10 **a** Interest = \$3212.50 – \$2500

= \$712.50

b 4 years = 48 months

Total paid = \$220 × 48

= \$10 560

Interest = \$10 560 – \$8000

= \$2560

SIMPLE INTEREST
INTERMEDIATE TEST PAGE 27

1 $P = \$2750,\ R = 4\%,\ N = 3$

$I = PRN$

$= \$2750 \times \frac{4}{100} \times 3$

$= \$330$

∴ **[B]** ✓ (1 mark)

2 $P = \$60\,000,\ R = 9\%,\ N = 15$

$I = PRN$

$= \$60\,000 \times \frac{9}{100} \times 15$

$= \$81\,000$

∴ **[D]** ✓ (1 mark)

3 $P = \$25\,000,\ R = 6\%,\ N = 1.5$

$I = PRN$

$= \$25\,000 \times \frac{6}{100} \times 1.5$

$= \$2250$

∴ **[A]** ✓ (1 mark)

4 $I = 100,\ P = 500,\ N = 4$

$I = PRN$

$100 = 500 \times R \times 4$

$= 2000R$

$R = \frac{100}{2000}$

$= \frac{5}{100}$

The rate is 5%.

∴ **[B]** ✓ (1 mark)

5 $I = 5400,\ P = 12\,000,\ R = 7.5\%$

$I = PRN$

$5400 = 12\,000 \times \frac{7.5}{100} \times N$

$= 900N$

$N = \frac{5400}{900}$

$= 6$

The loan was for 6 years. ✓

(1 mark)

6 **a** $P = \$2000,\ R = 3\%,\ N = 1$

$I = PRN$

$= \$2000 \times \frac{3}{100} \times 1$

$= \$60$ ✓ (1 mark)

b New balance = \$2000 + \$60

= \$2060 ✓

(1 mark)

c $P = \$2060,\ R = 3\%,\ N = 1$

$I = PRN$

$= \$2060 \times \frac{3}{100} \times 1$

$= \$61.80$ ✓ (1 mark)

d After two years:

Balance = \$2060 + \$61.80

= \$2121.80

$P = \$2121.80,\ R = 3\%,\ N = 1$

$I = PRN$

$= \$2121.80 \times \frac{3}{100} \times 1$

$= \$63.654$

$= \$63.65$ (nearest cent) ✓

Final balance

= \$2121.80 + \$63.65

= \$2185.45 ✓ (2 marks)

7 Cash price: \$7800

a Deposit = 15% of \$7800

$= \frac{15}{100} \times \$7800$

$= \$1170$ ✓

(1 mark)

b Balance owing

= \$7800 – \$1170

= \$6630 ✓ (1 mark)

c $P = \$6630,\ R = 8\%,\ N = 3$

$I = PRN$

$= \$6630 \times \frac{8}{100} \times 3$ ✓

$= \$1591.20$ ✓ (2 marks)

[Or: interest per year

= \$530.40 ✓; for 3 years

= \$1591.20 ✓]

d Total to repay

= \$6630 + \$1591.20

= \$8221.20 ✓ (1 mark)

e 3 years = 36 months

36 monthly repayments ✓

Each payment

= \$8221.20 ÷ 36

= \$228.37 (nearest cent) ✓

(2 marks)

8 $I = 2400,\ R = 5\%,\ N = 4$

$I = PRN$

$2400 = P \times \frac{5}{100} \times 4$ ✓

$= 0.2P$

$P = 2400 \div 0.2$

$= 12\,000$

The amount of money was \$12 000. ✓ (2 marks)

9 **a** 2 years = 24 months

Total paid

= \$700 + 24 × \$240 ✓

= \$6460 ✓ (2 marks)

b Interest = \$6460 – \$5569

= \$891 ✓ (1 mark)

c Because \$700 deposit was paid, Rosa has only borrowed \$4869.

Yearly interest

= \$891 ÷ 2

= \$445.50 ✓

Percentage interest

$= \frac{\$445.50}{\$4869} \times 100\%$

$= 9.1497227...\%$

Rosa pays 9.15% interest (to 2 d.p.) ✓ (2 marks)

10 $P = \$1840,\ R = 0.0445\%,$
$N = 30$
$[0.0445\% = 0.0445 \div 100$
$= 0.000\,445]$
$I = PRN$
$= \$1840 \times 0.000\,445 \times 30$
$= \$24.564$
$= \$24.56$ (nearest cent)
Fiona would pay \$24.56 interest. ✓ (1 mark)
(Total: 25 marks)

SIMPLE INTEREST ADVANCED TEST PAGE 28

1 18 months = 1.5 years
$I = PRN$
$= \$6400 \times 0.048 \times 1.5$
$= \$460.80$ ✓ (1 mark)

2 25 years = 25 × 12 months
= 300 months ✓
$I = PRN$
$= \$145\,000 \times 0.008 \times 300$
$= \$348\,000$ ✓ (2 marks)

3 $I = PRN$
$\$588 = P \times 0.07 \times 3$
$= 0.21P$ ✓
$P = \$588 \div 0.21$
$= \$2800$
The amount invested was \$2800. ✓ (2 marks)

4 $I = PRN$
$\$6750 = \$12\,500 \times 0.06 \times N$
$= \$750N$ ✓
$N = \$6750 \div \$750$
$= 9$
The amount would need to be invested for 9 years. ✓ (2 marks)

5 Annual interest $= \$2494 \div 4$
$= \$623.50$ ✓
Rate $= \dfrac{\$623.50}{\$8600} \times 100\%$
$= 7.25\%$ ✓ (2 marks)

6 $I = PRN$
$= \$16\,000 \times 0.075 \times 6$
$= \$7200$ ✓
Total $= \$16\,000 + \$7200$
$= \$23\,200$ ✓ (2 marks)

7 From 18 Aug until 9 Oct
= 14 + 30 + 9 days
= 53 days ✓
$I = PRN$
$= \$3750 \times 0.0006 \times 53$
$= \$119.25$
May paid a total of \$119.25 in interest. ✓ (2 marks)

8 Deposit $= 0.15 \times \$26\,400$
$= \$3960$
Balance owing
$= \$26\,400 - \$3960$
$= \$22\,440$ ✓
$I = PRN$
$= \$22\,440 \times 0.085 \times 5$
$= \$9537$ ✓
Total to be repaid
$= \$22\,440 + \$9537$
$= \$31\,977$
Now 5 years = 5 × 12 months
= 60 months
Repayments
$= \$31\,977 \div 60$
$= \$532.95$ ✓ (3 marks)

9 Total repaid $= 4 \times 12 \times \$450$
$= \$21\,600$ ✓
Interest $= \$21\,600 - \$16\,000$
$= \$5600$ ✓
$I = PRN$
$\$5600 = \$16\,000 \times R \times 4$
$= \$64\,000R$
$R = \$5600 \div \$64\,000$
$= 0.0875$
The interest rate is 8.75%.
$[0.0875 \times 100\%]$ ✓ (3 marks)

10 **a** Interest $= 0.05 \times \$30\,000$
$= \$1500$ ✓
(1 mark)

b 2nd year: Amount invested
$= \$30\,000 + \$1500$
$= \$31\,500$
Interest $= 0.05 \times \$31\,500$
$= \$1575$ ✓
(1 mark)

c 3rd year: Amount invested
$= \$31\,500 + \$1575$
$= \$33\,075$
Interest $= 0.05 \times \$33\,075$
$= \$1653.75$ ✓
Total interest
$= \$1500 + \$1575 + \$1653.75$
$= \$4728.75$ ✓ (2 marks)

d $I = PRN$
$= \$30\,000 \times 0.0525 \times 3$
$= \$4725$ ✓
For a period of 3 years the option of having the interest added to the investment produces a slightly higher amount.
This amount, \$3.75, is small so there is not much difference over 3 years. If the investment continued the first option would produce a better result as time goes on. ✓ (2 marks)
(Total: 25 marks)

RATES AND PROPORTION SKILLS CHECK PAGE 30

1 **a** 80 km in 2 h = 40 km in 1 h
= 40 km/h

b 200 m in 10 s = 20 m in 1 s
= 20 m/s

c \$8 for 4 L = \$2 for 1 L
= \$2/L

d 189 km in 3 h
= 63 km in 1 h
= 63 km/h

e \$86 for 4 people
= \$21.50 for 1 person
= \$21.50 per person

f \$120 for 5 h
= \$24 for 1 hour's work
= \$24 per hour

2 **a** 140 km in 2 h = 70 km in 1 h
= 70 km/h

b 720 km in 9 h = 80 km/h
[divide by 9]

c 420 km in 5 h = 84 km/h

d 348 km in 6 h = 58 km/h
[divide by 6]

e 180 km in $2\frac{1}{2}$ h = 72 km/h
[divide by 2.5]

f 27 km in 20 minutes
= 81 km in 60 minutes
= 81 km/h

3 Distance = 3 × 75 km
= 225 km

4 **a** 720 [multiply by 60]
b 1 [divide by 60]

c 16 000 [multiply by 1000]

d 25 [divide by 1000]

5 a direct [as radius increases so does circumference]

b inverse [as age increases value decreases]

c inverse [as length increases strength decreases]

d direct [as speed increases so does distance]

6 a 400 L ≈ 88 gallons

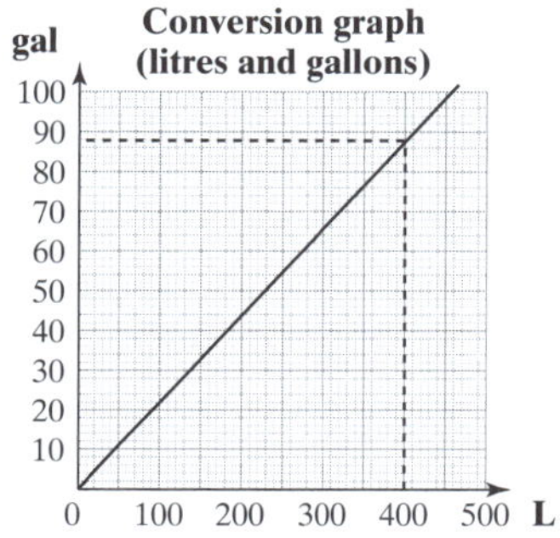

b 70 gallons ≈ 320 L

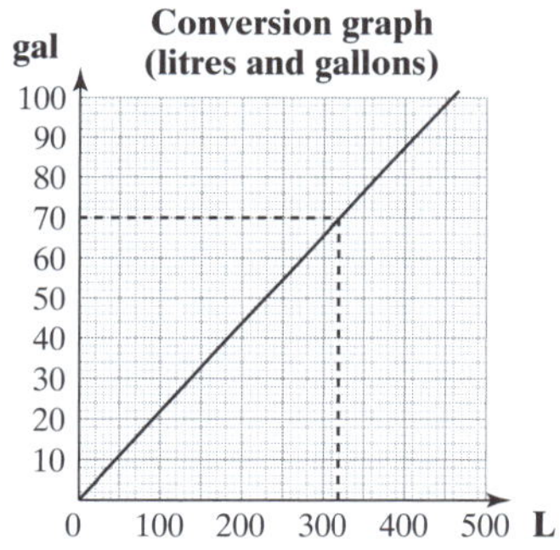

7 $y = 700x$

a When $x = 6$, $y = 700 \times 6$
$= 4200$

b When $y = 3150$,
$3150 = 700x$
$x = 3150 \div 700$
$= 4.5$

8 a $y = kx$
When $x = 15$, $y = 72$
$72 = k \times 15$
$k = 72 \div 15$
$= 4.8$

b $y = 4.8x$
When $x = 25$, $y = 4.8 \times 25$
$= 120$

c $y = 4.8x$
When $y = 96$, $96 = 4.8x$
$x = 96 \div 4.8$
$= 20$

9 i a increasing

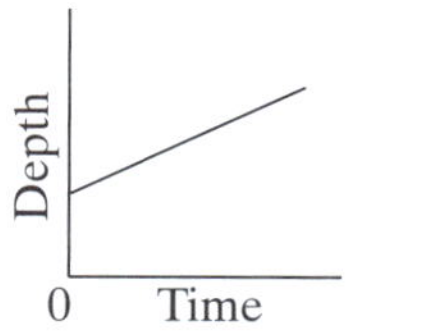

[The depth is getting larger over time.]

b decreasing

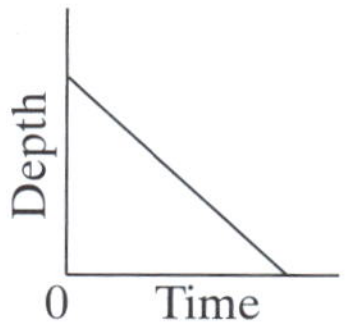

[The depth is getting smaller over time.]

c increasing

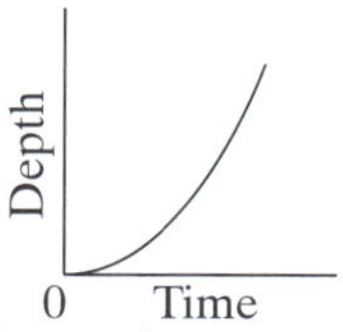

d increasing

ii a constant [straight line]

b constant

c variable [curved line]

d variable

RATES AND PROPORTION
INTERMEDIATE TEST PAGE 31

1 Flow rate = 40 L in 5 min
= 8 L/min
Volume in 12 min
$= 12 \times 8$ L
= 96 L
∴ [C] ✓ (1 mark)

2 Time $= (378 \div 72)$ h
= 5.25 h
= 5 h 15 min
∴ [A] ✓ (1 mark)

3 1 hour and 20 minutes
= 80 minutes
75 runs in 80 minutes
$= (75 \div 80)$ runs per minute
= 0.9375 runs per minute
∴ [B] ✓ (1 mark)

4 The water level will be shown by the graph in A.

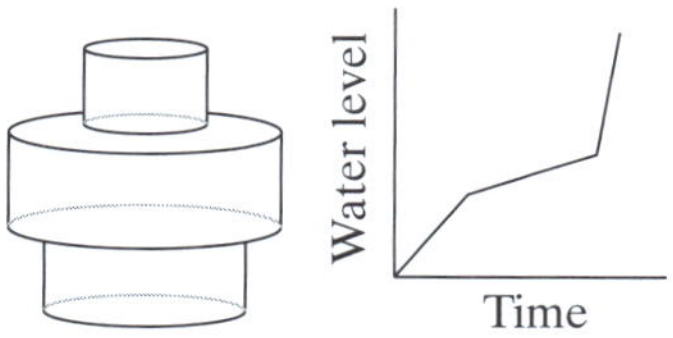

[The water level will rise at a constant rate at first, then at a slower rate for the larger middle section, then at the fastest rate for the top (smallest) section.]
∴ [A] ✓ (1 mark)

5 Let the length of the second rectangle be x mm.

$$\frac{x}{69} = \frac{91}{39}$$

$$x = \frac{91}{39} \times 69$$

$= 161$

The second rectangle is 161 mm long. ✓ (1 mark)

6 a Rate = 50 words/min
In 5 minutes,
number of words $= 50 \times 5$
$= 250$ ✓
(1 mark)

b Time required
$= (2000 \div 50)$ min
= 40 min ✓ (1 mark)

c For 3000 words,
Shelley's time
$= (3000 \div 50)$ min
= 60 min
Jeremy's time
$= (3000 \div 60)$ min
= 50 min ✓
Extra time $= (60 - 50)$ min
= 10 min
Jeremy can type a 3000-word essay 10 minutes more quickly than Shelley can. ✓
(2 marks)

WORKED SOLUTIONS

CHECK YOUR SOLUTIONS

7 100 m in 10 seconds
= 600 m in 60 seconds
= 600 m/minute
= (600 × 60) m in 1 hour
= 36 000 m/h ✓
= (36 000 ÷ 1000) km/h
= 36 km/h ✓ (2 marks)

8 **a** 5 mL in 10 L
= 50 mL in 100 L
= 250 mL in 500 L ✓
(1 mark)

b 250 mL for 2 ha
= 125 mL for 1 ha ✓
= (125 × 15) mL for 15 ha
= 1875 mL for 15 ha ✓
= 1.875 L for 15 ha ✓
(3 marks)

9 78 km/h = (78 × 1000) m/h
= 78 000 m/h
= (78 000 ÷ 60) m/minute
= 1300 m/minute ✓
= (1300 ÷ 60) m/s
= $21.\dot{6}$ m/s ✓
(2 marks)

10

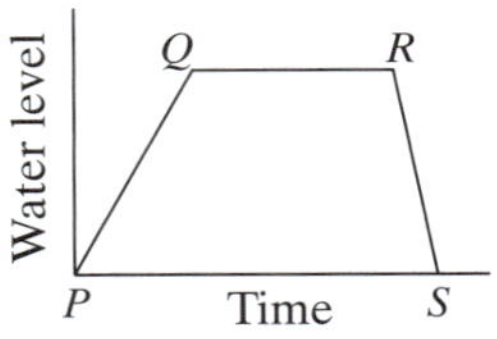

a **i** Between P and Q, the level of water in the machine is increasing constantly. ✓ (1 mark)

ii Between Q and R the level of water remains the same. It is constant. ✓ (1 mark)

iii Between R and S the level of water is decreasing constantly. ✓
(1 mark)

b The washing machine empties quicker than it fills. ✓
The line between P and Q is not as steep as the line between R and S. ✓
(2 marks)
[It takes less time for the water level to change by the same amount.]

11 $d = kS$
When $S = 80$, $d = 12$
$12 = k \times 80$
$k = 12 \div 80$
$= 0.15$ ✓
So $d = 0.15S$ ✓
When $S = 100$
$d = 0.15 \times 100$
$= 15$
The distance travelled at 100 km/h is 15 km. ✓ (3 marks)
(Total: 25 marks)

RATES AND PROPORTION
ADVANCED TEST PAGE 32

1 **a** Amount of petrol
= (65.52 ÷ 1.56) L
= 42 L ✓ (1 mark)

b Fuel consumption
= 42 L per 450 km
= $9\,^{1}/_{3}$ L per 100 km
[42 ÷ 4.5] ✓ (1 mark)

c Distance
= $(60 \div 9\,^{1}/_{3}) \times 100$ km
= 642.857 … km ✓
So, Jim should be able to travel 625 km if he maintains the same rate of fuel consumption. But, if he uses slightly more petrol for every 100 km travelled he will not have enough. Jim should plan to fill his fuel tank at some time on the journey to be certain that he has enough petrol. ✓
(2 marks)

2

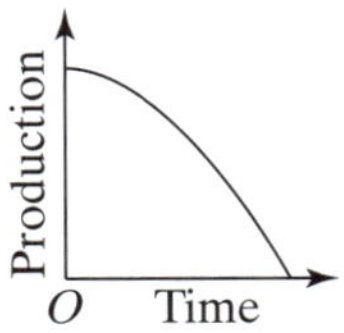

✓ (1 mark)

3

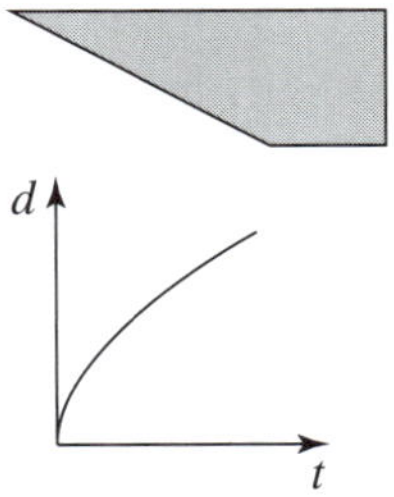

✓✓

[The depth increases at a variable rate so the graph is curved. The depth increases quickly at first but more slowly as time goes on.] (2 marks)

4

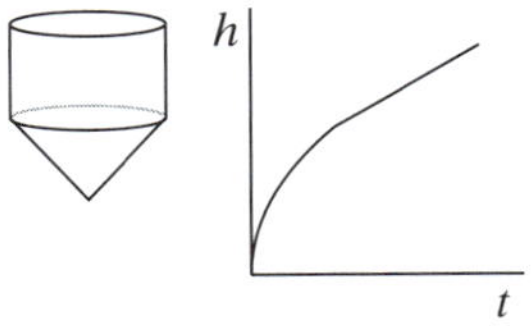

✓✓ (2 marks)
[The depth will increase quickly at first but then more slowly as the conical section is filled. The cylindrical section will fill at a constant rate.]

5

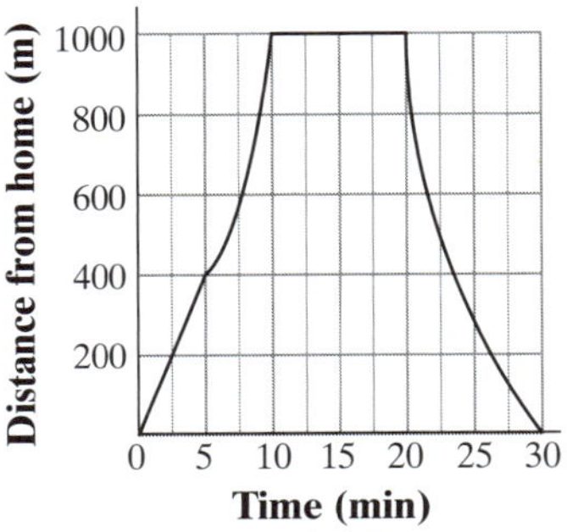

✓✓✓ (3 marks)
[Between 5 and 10 minutes the curve increases at an increasing rate and between 20 and 30 minutes the curve decreases at a decreasing rate. The exact path of the graph between these times cannot be determined.]

6

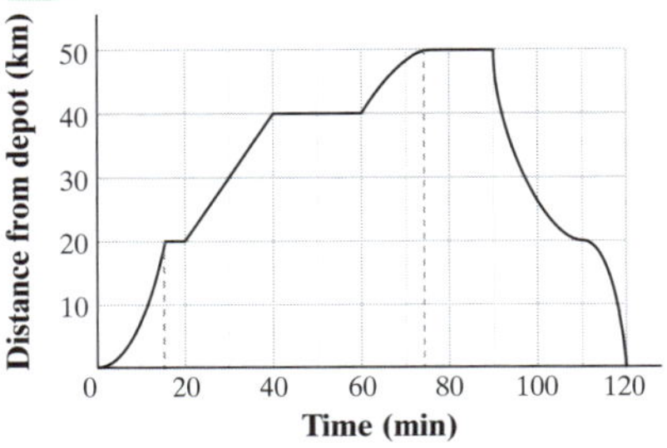

a The journey took two hours. ✓ (1 mark)

b The bus leaves the depot and travels slowly at first gradually increasing speed for the first 15 minutes.

It is stopped for 5 minutes. ✓ It then travels at a constant speed (60 km/h) for the next 20 minutes. ✓ (2 marks)

c The bus travels 50 km in 30 minutes. ✓ At this rate it would travel 100 km in 60 minutes so the average speed is 100 km/h. ✓ (2 marks)

d The bus travelled 10 km. ✓ (1 mark)

e In the last 10 minutes the bus travels 20 km. This is an average speed of 120 km/h. ✓ But it is travelling at a variable rate, slower at first and then faster. So the bus would need to be travelling faster than 120 km/h for part of that time in order to average that speed. ✓ (2 marks)

(Total: 20 marks)

THE INDEX LAWS SKILLS CHECK PAGE 34

1 **a** $2^4 \times 2^5 = 2^{4+5}$
$= 2^9$

b $3^2 \times 3^3 = 3^{2+3}$
$= 3^5$

c $7^{12} \times 7^2 = 7^{12+2}$
$= 7^{14}$

2 **a** $x^4 \times x^5 = x^{4+5}$
$= x^9$

b $a^2 \times a^3 = a^{2+3}$
$= a^5$

c $t^{12} \times t^2 = t^{12+2}$
$= t^{14}$

d $2e^7 \times e = 2e^{7+1}$
$= 2e^8$

e $8m^4 \times 2m^2 = 16m^6$
[$8 \times 2 = 16$, $m^4 \times m^2 = m^{4+2}$]

f $4p^7 \times 3p^7 = 12p^{14}$

g $b^4 \times 5b^2 = 5b^6$
[$1 \times 5 = 5$, $b^4 \times b^2 = b^{4+2}$]

h $4a^2 \times 3a^7 \times 2a^3 = 24a^{12}$
[$4 \times 3 \times 2 = 24$, $a^2 \times a^7 \times a^3 = a^{2+7+3}$]

i $6x^2y^3 \times 2x^3y^5 = 12x^5y^8$
[$6 \times 2 = 12$, $x^2 \times x^3 = x^{2+3}$, $y^3 \times y^5 = y^{3+5}$]

j $2a^4 \times 3a^2b = 6a^6b$

k $4t^3u^7 \times 4tu^2 = 16t^4u^9$
[$4 \times 4 = 16$, $t^3 \times t = t^{3+1}$, $u^7 \times u^2 = u^{7+2}$]

l $3p^2qr^3 \times 5pqr^4 = 15p^3q^2r^7$
[$p^2 \times p = p^{2+1}$, $q \times q = q^2$]

3 **a** $2^{12} \div 2^2 = 2^{12-2}$
$= 2^{10}$

b $5^8 \div 5^3 = 5^{8-3}$
$= 5^5$

c $6^9 \div 6^6 = 6^3$

4 **a** $x^{12} \div x^2 = x^{12-2}$
$= x^{10}$

b $a^8 \div a^3 = a^{8-3}$
$= a^5$

c $h^9 \div h^6 = h^3$

d $12m^{12} \div 2m^2 = 6m^{10}$
[$12 \div 2 = 6$, $m^{12} \div m^2 = m^{12-10}$]

e $16x^2 \div 4x = 4x$
[$x^2 \div x = x^{2-1} = x^1$]

f $15p^7 \div 5p^6 = 3p$ [$3p^1$]

g $x^4y^6 \div x^2y^5 = x^{4-2}y^{6-5}$
$= x^2y$

h $9m^4n^3 \div 3m^2n^2 = 3m^2n$

i $8a^3b^2 \div 2ab^2 = 4a^2$

j $\frac{8x^2}{12x} = \frac{2x}{3}$
[dividing both numerator and denominator by $4x$]

k $\frac{7n^6}{14n^3} = \frac{n^3}{2}$ [dividing by $7n^3$]

l $\frac{9a^2b^7}{6a^2b^3} = \frac{3b^4}{2}$
[dividing by $3a^2b^3$]

5 **a** $(x^4)^2 = x^{4\times 2}$
$= x^8$

b $(a^2)^3 = a^{2\times 3}$
$= a^6$

c $(g^7)^4 = g^{7\times 4}$
$= g^{28}$

d $(m^3)^3 = m^9$

e $(2n)^5 = 2^5n^5$
$= 32n^5$

f $(4k^2)^2 = 16k^4$

g $(3t^6)^3 = 27t^{18}$

h $5(p^4)^9 = 5p^{36}$
[5 is not in the brackets. It does not get raised to the ninth power.]

i $(x^2y^3)^2 = x^{2\times 2}y^{3\times 2}$
$= x^4y^6$

j $(a^3b^2)^4 = a^{12}b^8$

k $(pq^2)^3 = p^3q^6$ [$p^{1\times 3}$, $q^{2\times 3}$]

l $(2m^2n^4)^4 = 16m^8n^{16}$

6 **a** $7^0 = 1$

b $9^0 = 1$

c $x^0 = 1$

d $k^0 = 1$

e $2a^0 = 2 \times a^0$
$= 2 \times 1$
$= 2$

f $(2a)^0 = 1$

g $5e^0 = 5 \times 1$
$= 5$

h $m^0 + n^0 = 1 + 1$
$= 2$

i $7x^0 + 2 = 7 \times 1 + 2$
$= 7 + 2$
$= 9$

j $4m^0 + 2m^0 = 4 \times 1 + 2 \times 1$
$= 6$
[Or $4m^0 + 2m^0 = 6m^0 = 6 \times 1$]

k $9e^0 + 3e = 9 \times 1 + 3e$
$= 9 + 3e$

l $b^0 - c^0 = 1 - 1$
$= 0$

THE INDEX LAWS INTERMEDIATE TEST PAGE 35

1 $3a^3 \times 5a^5 = 15a^8$
∴ [C] ✓ (1 mark)
[$3 \times 5 = 15$, $a^3 \times a^5 = a^{3+5}$]

2 $4n^9 \div 4n^3 = n^6$
∴ [B] ✓ (1 mark)
[$4 \div 4 = 1$, $n^9 \div n^3 = n^{9-3}$]

3 Try each option:
$2x \times 2x = 4x^2$ $8x^2 \div 2x^2 = 4$
$4x \times x = 4x^2$ $3x^2 + x^2 = 4x^2$
$8x^2 \div 2x^2 \neq 4x^2$
∴ [B] ✓ (1 mark)

4 $(2x^3)^3 = 8x^9$
∴ [D] ✓ (1 mark)
[$2^3 = 8$, $(x^3)^3 = x^{3\times 3}$]

5 $7x^0 + 7^0 = 7 \times x^0 + 1$
$= 7 \times 1 + 1$
$= 7 + 1$
$= 8$
$\therefore$ [D] ✓ (1 mark)

6 $8x^2 \times 3x = 24x^{2+1}$
$= 24x^3$
$\therefore$ [B] ✓ (1 mark)

7 $2 + 4p^0 = 2 + 4$
$= 6$
$\therefore$ [D] ✓ (1 mark)

8 $\frac{12m^8}{16m^9} = \frac{3}{4m}$
[dividing both numerator and denominator by $4m^8$]
$\therefore$ [A] ✓ (1 mark)

9 $2n^5 + 3n^2 \times n^3 = 2n^5 + 3n^5$
$= 5n^5$
$\therefore$ [A] ✓ (1 mark)

10 $(x^2y^3)^4 = x^8y^{12}$
$[(x^2)^4 = x^{2\times4}, (y^3)^4 = y^{3\times4}]$
$\therefore$ [B] ✓ (1 mark)

11 $3(m^5)^2 = 3m^{10}$
[3 is not in the bracket so does not get squared, $(m^5)^2 = m^{5\times2}$]
$\therefore$ [D] ✓ (1 mark)

12 $a^2b^3c^4 \times ab^2c^3 = a^{2+1}b^{3+2}c^{4+3}$
$= a^3b^5c^7$
$\therefore$ [B] ✓ (1 mark)

13 $a^{12} \div a^2 \times a^6 = a^{10} \times a^6$
$= a^{16}$ ✓ (1 mark)

14 $12x^8 - 3x^3 \times 5x^5 = 12x^8 - 15x^8$ ✓
$= -3x^8$ ✓
(2 marks)

15 $\frac{6x^7 \div 3x^3}{5x^2 \times 2x^4} = \frac{2x^4}{10x^6}$ ✓✓
$= \frac{1}{5x^2}$ ✓ (3 marks)
[dividing both numerator and denominator by $2x^4$.
$2x^4 \div 2x^4 = 1$ so there is 1 in the numerator.]

16 $(12p^5 \div 6p^2)^2 = (2p^3)^2$ ✓
$= 4p^6$ ✓ (2 marks)

(Total: 20 marks)

THE INDEX LAWS
ADVANCED TEST PAGE 36

1 a $\frac{12a^3b^5c^2}{20ab^3c^2} = \frac{3a^2b^2}{5}$ ✓ (1 mark)

b $-2x^3y^5z^8 \times 5xy^3z^7$
$= -10x^4y^8z^{15}$ ✓ (1 mark)

c $(3x^2y^3)^3 = 27x^6y^9$ ✓ (1 mark)

d $(5x^2y)^0 = 1$ ✓ (1 mark)

e $18p^9q^7r \div -3p^5qr$
$= -6p^4q^6$ ✓ (1 mark)

f $4k^3mn^4 \times 3kmn^2 \times 2m^6n$
$= 24k^4m^8n^7$ ✓ (1 mark)

2 a $5a^3b^2 \times 3a^2b^6 + 9ab^5 \times 2a^2b^4$
$= 15a^5b^8 + 18a^3b^9$ ✓
(1 mark)

b $\frac{2m^5 \times 3m^4}{12m^{12} \div 2m^2} = \frac{6m^9}{6m^{10}}$
$= \frac{1}{m}$ ✓ (1 mark)

3 a $36a^8b^7c^6 \div 3ab^3c^3 \div 3abc^3$
$= 12a^7b^4c^3 \div 3abc^3$ ✓
$= 4a^6b^3$ ✓ (2 marks)

b $(-3x^4y^5 \times 2x^3y^2)^2$
$= (-6x^7y^7)^2$ ✓
$= 36x^{14}y^{14}$ ✓ (2 marks)

c $\frac{9p^5q^6 \times 8p^3q^3}{6pq^4 \times 6p^4q^5} = \frac{72p^8q^9}{36p^5q^9}$ ✓
$= 2p^3$ ✓
(2 marks)

4 a $3x^2 \times 4y^2 = 12x^2y^2$ ✓ (1 mark)

b $((x^2y^3)^2)^2 = (x^2y^3)^4$
$= x^8y^{12}$ ✓ (1 mark)

c $a^m \times a^n = a^{m+n}$ ✓ (1 mark)

d $b^m \div c^n = \frac{b^m}{c^n}$
[It can't be simplified.] ✓
(1 mark)

5 a $\frac{2}{b^2} = 2b^{-2}$ ✓ (1 mark)

b $\frac{1}{5x} = (5x)^{-1}$ ✓ (1 mark)

c $\frac{2}{(3a)^3} = 2(3a)^{-3}$ ✓ (1 mark)

6 a $5a^{-1} = \frac{5}{a}$ ✓ (1 mark)

b $(3t)^{-1} = \frac{1}{3t}$ ✓ (1 mark)

c $6p^{-2} = \frac{6}{p^2}$ ✓ (1 mark)

7 a $20a^7 \div 4a^7 = 5a$?
False
$20a^7 \div 4a^7 = 5$ ✓ (1 mark)

b $(4x^3)^2 = 4x^6$?
False
$(4x^3)^2 = 16x^6$ ✓ (1 mark)

c $3^0 + 3x^0 = 2$?
False
$3^0 + 3x^0 = 1 + 3$
$= 4$ ✓ (1 mark)

d $\left(\frac{2}{x}\right)^{-1} = \frac{x}{2}$?
True ✓ (1 mark)

e $6m^4 \div 2m^7 = 3m^3$?
False
$6m^4 \div 2m^7 = 3m^{-3}$ ✓
(1 mark)

f $\frac{1}{2}n^{-2} = \frac{1}{2n^2}$?
True ✓ (1 mark)

8 a $5x^{-2} \times 3x^{-5} = 15x^{-7}$ ✓ (1 mark)

b $2(x^{-4})^{-3} = 2x^{12}$ ✓ (1 mark)

c $(4k^3)^{-1} = \frac{1}{4k^3}$ ✓ (1 mark)

d $8n^5 \div 4n^{-3} = 2n^8$ ✓ (1 mark)

e $12a^{-2} \div -3a^{-9} = -4a^7$ ✓
(1 mark)

(Total: 35 marks)

THE DISTRIBUTIVE LAW
SKILLS CHECK PAGE 38

1 a $2(x + y)$
$= 2 \times x + 2 \times y$
$= 2x + 2y$

b $5(p - q) = 5 \times p - 5 \times q$
$= 5p - 5q$

c $8(a + b + c)$
$= 8 \times a + 8 \times b + 8 \times c$
$= 8a + 8b + 8c$

d $3(x + 2) = 3 \times x + 3 \times 2$
$= 3x + 6$

e $6(a - 4) = 6 \times a - 6 \times 4$
$= 6a - 24$

f $9(t + 1) = 9 \times t + 9 \times 1$
$= 9t + 9$

g $2(3e + 4) = 2 \times 3e + 2 \times 4$
$= 6e + 8$

h $4(5p-2) = 4 \times 5p - 4 \times 2$
$= 20p - 8$

i $7(2a+3b) = 7 \times 2a + 7 \times 3b$
$= 14a + 21b$

2 **a** $5a + 10 = 5(a+2)$
b $2e - 8 = 2(e-4)$
c $6t + 8u = 2(3t + 4u)$

3 **a** $4(a+3) = 4 \times a + 4 \times 3$
$= 4a + 12$

b $x(x-2) = x \times x - x \times 2$
$= x^2 - 2x$

c $2m(3m-1)$
$= 2m \times 3m - 2m \times 1$
$= 6m^2 - 2m$

d $4(x - 2y + 5)$
$= 4 \times x - 4 \times 2y + 4 \times 5$
$= 4x - 8y + 20$

e $3ab(a^2 - 5b) = 3a^3b - 15ab^2$
f $x^2y(xy - y^4) = x^3y^2 - x^2y^5$

4 **a** $-2(3x+5)$
$= -2 \times 3x + (-2) \times 5$
$= -6x - 10$

b $-5(6-5e) = -30 + 25e$
c $-4(x + y - z) = -4x - 4y + 4z$
d $-3(4a - 5b) = -12a + 15b$
e $-x(x+8) = -x^2 - 8x$
f $-7e(f - 2g) = -7ef + 14eg$

5 **a** $-(3x-4) = -1 \times 3x - (-1) \times 4$
$= -3x + 4$

b $-(a+2b) = -a - 2b$
c $-(4x^2 - 1) = -4x^2 + 1$
d $-(2x - 3y + z) = -2x + 3y - z$
e $-(-2x + 7) = 2x - 7$
f $-(-3ab - 4c) = 3ab + 4c$

6 **a** $6x + 5 + 3x + 8 = 9x + 13$
b $8a - 4 + 3a - 6 = 11a - 10$
c $7p - 2 - 3p + 5 = 4p + 3$
d $x^2 + 3x + 5x + 15 = x^2 + 8x + 15$
e $a^2 - 2a + 3a - 6 = a^2 + a - 6$
f $n^2 - 5n - n + 5 = n^2 - 6n + 5$

7 **a** $12 + 3(4t+1) = 12 + 12t + 3$
$= 12t + 15$

b $6(x-2) + 5 = 6x - 12 + 5$
$= 6x - 7$

c $4(2p-3) + 3p - 2$
$= 8p - 12 + 3p - 2$
$= 11p - 14$

d $8x - (x-3) = 8x - x + 3$
$= 7x + 3$

8 **a** $8(a+2) + 6(b+3)$
$= 8a + 16 + 6b + 18$
$= 8a + 6b + 34$

b $2(3x+5) + 4(2x+7)$
$= 6x + 10 + 8x + 28$
$= 14x + 38$

c $5(x+4) + 3(x-7)$
$= 5x + 20 + 3x - 21$
$= 8x - 1$

d $7(5x-4) - (3x+2)$
$= 35x - 28 - 3x - 2$
$= 32x - 30$

e $4(3p + 5q) - 3(2p + q)$
$= 12p + 20q - 6p - 3q$
$= 6p + 17q$

f $6(2a+b) - 5(3a-b)$
$= 12a + 6b - 15a + 5b$
$= -3a + 11b$

g $-4(5a+2b) - 3(6a+5)$
$= -20a - 8b - 18a - 15$
$= -38a - 8b - 15$

h $-2(8x-y) - (3x+y)$
$= -16x + 2y - 3x - y$
$= -19x + y$

THE DISTRIBUTIVE LAW
INTERMEDIATE TEST PAGE 39

1 $-2(3-7x) = -2 \times 3 - (-2) \times 7x$
$= -6 + 14x$
$= 14x - 6$
∴ [D] ✓ (1 mark)

2 $3x - (2x-1) = 3x - 2x + 1$
$= x + 1$
∴ [A] ✓ (1 mark)

3 $7(x+2) + 3 = 7x + 14 + 3$
$= 7x + 17$
∴ [C] ✓ (1 mark)

4 $5a - 2(3+a) = 5a - 6 - 2a$
$= 3a - 6$
∴ [B] ✓ (1 mark)

5 $3x^2(4x^3 + 2xy) = 12x^5 + 6x^3y$
∴ [A] ✓ (1 mark)

6 Check each option by expanding.
A: $3(x^2 - 2) = 3x^2 - 6$
B: $4(2a - b) = 8a - 4b$
C: $4(3xy + 2y + 1) = 12xy + 8y + 4$
D: $3(5k - 8m) = 15k - 24m$
$\neq 15k - 27m$
∴ [D] ✓ (1 mark)

7 $x(x-1) - 5(x-1)$
$= x^2 - x - 5x + 5$
$= x^2 - 6x + 5$
∴ [A] ✓ (1 mark)

8 $2p^3(p^2 - q) - 2p^3q$
$= 2p^5 - 2p^3q - 2p^3q$
$= 2p^5 - 4p^3q$
∴ [C] ✓ (1 mark)

9 $2a^2(3ab^2 - 5a^4)$
$= 2a^2 \times 3ab^2 - 2a^2 \times 5a^4$
$= 6a^3b^2 - 10a^6$
∴ [C] ✓ (1 mark)
$[2a^2 \times 3ab^2 = 6a^{2+1}b^2,$
$2a^2 \times 5a^4 = 10a^{2+4}]$

10 $a(a+b) + b(a+b)$
$= a^2 + ab + ab + b^2$
$= a^2 + 2ab + b^2$
∴ [B] ✓ (1 mark)

11 $4u^3(u^3 - 1) = 4u^3 \times u^3 - 4u^3 \times 1$
$= 4u^6 - 4u^3$ ✓
(1 mark)

12 $3x^2y^3(x^4 - 5xy^2)$
$= 3x^6y^3 - 15x^3y^5$ ✓ (1 mark)

13 $6(3x-4) - 3(2x+7)$
$= 18x - 24 - 6x - 21$ ✓
$= 12x - 45$ ✓ (2 marks)

14 $5(3a-7) = 2(a+2)$
$15a - 35 = 2a + 4$ ✓
$15a = 2a + 39$
$13a = 39$
$a = 3$ ✓ (2 marks)

15 $a(3a+2b) - 4(2a^2 - 3b)$
$= 3a^2 + 2ab - 8a^2 + 12b$ ✓
$= -5a^2 + 2ab + 12b$ ✓
(2 marks)

16 $2(3x-1) - (x-2)$
$= 6x - 2 - x + 2$ ✓
$= 5x$ ✓ (2 marks)
(Total: 20 marks)

THE DISTRIBUTIVE LAW
ADVANCED TEST PAGE 40

1 a $-3x^2(2x^2 - 5y)$
$= -6x^4 + 15x^2y$ ✓ (1 mark)

b $7a^2b(2ab^2 - 3b^3)$
$= 14a^3b^3 - 21a^2b^4$ ✓ (1 mark)

c $2xy^2z^3(-4xy + 11xz^2)$
$= -8x^2y^3z^3 + 22x^2y^2z^5$ ✓ (1 mark)

d $9p^4q^3(7p^5r^2 - 6q^7)$
$= 63p^9q^3r^2 - 54p^4q^{10}$ ✓ (1 mark)

e $11t^5u^9(3t^3u^5 - 8v^6)$
$= 33t^8u^{14} - 88t^5u^9v^6$ ✓ (1 mark)

f $-5a^3b^7(2a^4 + 9b^5)$
$= -10a^7b^7 - 45a^3b^{12}$ ✓ (1 mark)

g $p^{-3}(p^4 - p^{-2}) = p - p^{-5}$ ✓ (1 mark)

h $5n^4(2n^4 + n^{-4})$
$= 10n^8 + 5$ ✓ (1 mark)

i $6xy\left(\frac{1}{x} + \frac{1}{y}\right) = 6y + 6x$ ✓ (1 mark)

2 a $2 - (5 - x) = 2 - 5 + x$
$= -3 + x$ ✓ (1 mark)

b $3x^2 + x(x - 7)$
$= 3x^2 + x^2 - 7x$
$= 4x^2 - 7x$ ✓ (1 mark)

c $(p + q) - (p - q)$
$= p + q - p + q$
$= 2q$ ✓ (1 mark)

d $9x(3x - 5) - 17x + 2x^2$
$= 27x^2 - 45x - 17x + 2x^2$
$= 29x^2 - 62x$ ✓ (1 mark)

e $a(a - b) + b(a - b)$
$= a^2 - ab + ab - b^2$
$= a^2 - b^2$ ✓ (1 mark)

3 a $3p^2(q - 4) + 5pq(p - 1)$
$= 3p^2q - 12p^2 + 5p^2q - 5pq$ ✓
$= 8p^2q - 12p^2 - 5pq$ ✓ (2 marks)

b $3x^2y^3(5x^3 - 2y) - 7x^4y^2(6xy + 1)$
$= 15x^5y^3 - 6x^2y^4 - 42x^5y^3 - 7x^4y^2$ ✓
$= -27x^5y^3 - 6x^2y^4 - 7x^4y^2$ ✓ (2 marks)

c $12(9 - 4a) + 8(6a - 17)$
$= 108 - 48a + 48a - 136$ ✓
$= -28$ ✓ (2 marks)

d $-6a^2bc^5(2a + 7bc^2)$
$- (6a^3bc^5 - 7a^2b^2c^7)$
$= -12a^3bc^5 - 42a^2b^2c^7$
$-6a^3bc^5 + 7a^2b^2c^7$ ✓
$= -18a^3bc^5 - 35a^2b^2c^7$ ✓ (2 marks)

4 a $7p^3q^5(4p^5q^6 + 3p^6q^7)$
$- 5p^8q^{10}(2q + 5p - 3pq)$
$= 28p^8q^{11} + 21p^9q^{12}$
$- 10p^8q^{11} - 25p^9q^{10}$
$+ 15p^9q^{11}$ ✓
$= 18p^8q^{11} + 21p^9q^{12}$
$- 25p^9q^{10} + 15p^9q^{11}$ ✓ (2 marks)

b $9a^4(3a - 5b^3c^2)$
$+ 2abc^2(7a^3b^2 - 6b)$
$- a(2bc - 12a^4)$
$= 27a^5 - 45a^4b^3c^2$
$+ 14a^4b^3c^2 - 12ab^2c^2$
$- 2abc + 12a^5$ ✓
$= 39a^5 - 31a^4b^3c^2$
$- 12ab^2c^2 - 2abc$ ✓ (2 marks)

5 a $(a^2 \times b^2)(a^2 + b^2)$
$= a^2b^2(a^2 + b^2)$ ✓
$= a^4b^2 + a^2b^4$ ✓ (2 marks)

b $(x^3y^2 \times xy^4)(5x + 3y)$
$= x^4y^6(5x + 3y)$ ✓
$= 5x^5y^6 + 3x^4y^7$ ✓ (2 marks)

c $(2xy^2)^2(3x^2 - 1)$
$= 4x^2y^4(3x^2 - 1)$ ✓
$= 12x^4y^4 - 4x^2y^4$ ✓ (2 marks)

d $(12k^2 \div 3k)(2k + m)$
$= 4k(2k + m)$ ✓
$= 8k^2 + 4km$ ✓ (2 marks)

6 a $4(3x - 2) + 3(2x + 1) = 7$
$12x - 8 + 6x + 3 = 7$ ✓
$18x - 5 = 7$
$18x = 12$
$x = \frac{2}{3}$ ✓ (2 marks)

b $6(4m - 5) - 5(2m + 3) = 9(m - 5)$
$24m - 30 - 10m - 15 = 9m - 45$ ✓
$14m - 45 = 9m - 45$
$14m = 9m$
$5m = 0$
$m = 0$ ✓ (2 marks)

c $7(5k - 1) - 9(3k - 4) = 2(6 - k) - 23$
$35k - 7 - 27k + 36 = 12 - 2k - 23$ ✓
$8k + 29 = -2k - 11$
$10k = -40$
$k = -4$ ✓ (2 marks)

(Total: 40 marks)

DISTANCE BETWEEN TWO POINTS
SKILLS CHECK PAGE 42

1

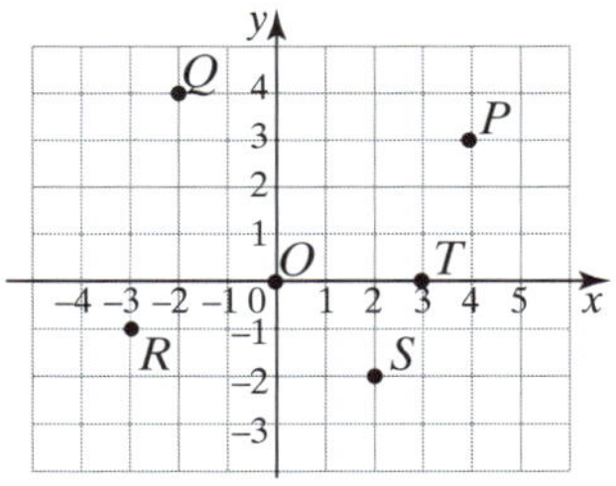

a $P(4, 3)$

b $Q(-2, 4)$

c $R(-3, -1)$

d $S(2, -2)$

e $T(3, 0)$

f $O(0,0)$

2 a

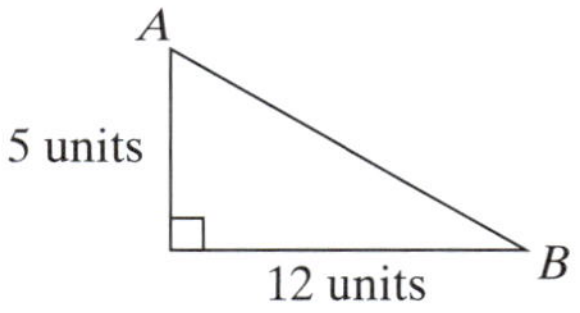

$AB^2 = 5^2 + 12^2$
$= 25 + 144$
$= 169$

$AB = \sqrt{169}$
$= 13$

$\therefore$ the length AB is 13 units

b

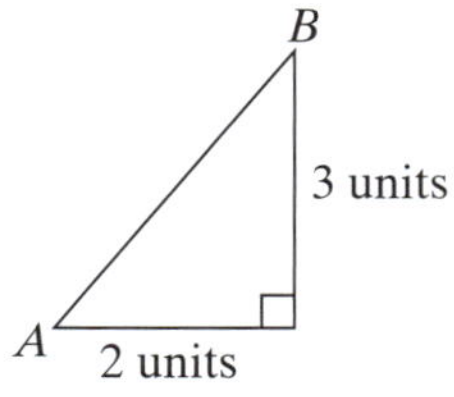

$AB^2 = 2^2 + 3^2$
$= 4 + 9$
$= 13$

$AB = \sqrt{13}$

The length AB is $\sqrt{13}$ units.

3 **a** $AB = 8$ units
[By counting the squares. Or: $(5 - (-3))$ units]

b $BC = 2$ units
[$(5 - 3)$ units]

c $AD = 8$ units
[$(5 - (-3))$ units]

d

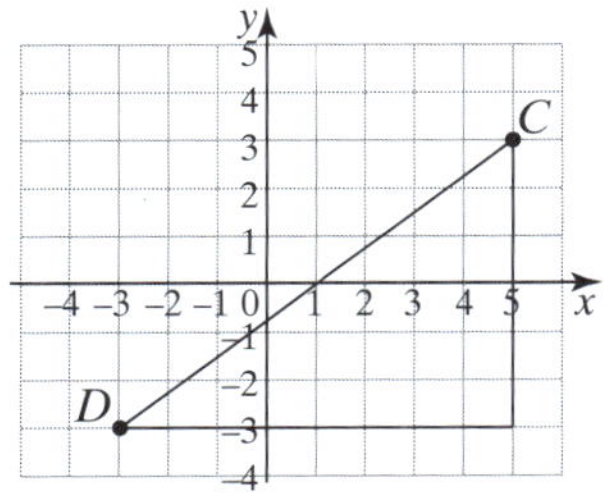

$d^2 = 6^2 + 8^2$
$= 36 + 64$
$= 100$
$d = \sqrt{100}$
$= 10$
The distance is 10 units.

4 **a** (3, 2) and (3, 7)
Distance = $(7 - 2)$ units
= 5 units

b (4, –8) and (4, 1)
Distance = $(1 - (-8))$ units
= 9 units

c (–5, –3) and (–5, –9)
Distance = $(-3 - (-9))$ units
= 6 units

d (5, 0) and (11, 0)
Distance = $(11 - 5)$ units
= 6 units

e (–6, –2) and (12, –2)
Distance = $(12 - (-6))$ units
= 18 units

f (–1, 7) and (–3, 7)
Distance = $(-1 - (-3))$ units
= 2 units

5 **a** $x_1 = -3, x_2 = 5, y_1 = 1, y_2 = 7$

$$d = \sqrt{(x_2 - x_1)^2 + (y_2 - y_1)^2}$$
$$= \sqrt{(5-(-3))^2 + (7-1)^2}$$
$$= \sqrt{64 + 36}$$
$$= \sqrt{100}$$
$$= 10$$

b $x_1 = -9, x_2 = 6, y_1 = -2, y_2 = 6$

$$d = \sqrt{(x_2 - x_1)^2 + (y_2 - y_1)^2}$$
$$= \sqrt{(6-(-9))^2 + (6-(-2))^2}$$
$$= \sqrt{15^2 + 8^2}$$
$$= \sqrt{289}$$
$$= 17$$

6 $x_1 = 2, x_2 = 3, y_1 = -4, y_2 = -2$

$$d = \sqrt{(x_2 - x_1)^2 + (y_2 - y_1)^2}$$
$$= \sqrt{(3-2)^2 + (-2-(-4))^2}$$
$$= \sqrt{1 + 4}$$
$$= \sqrt{5}$$

7 **a** (4, –5) and (–1, 7)
[$x_1 = 4, x_2 = -1, y_1 = -5, y_2 = 7$]

$$d = \sqrt{(x_2 - x_1)^2 + (y_2 - y_1)^2}$$
$$= \sqrt{(-1-4)^2 + (7-(-5))^2}$$
$$= \sqrt{25 + 144}$$
$$= \sqrt{169}$$
$$= 13$$

The distance is 13 units.

b (6, 2) and (4, 5)

$$d = \sqrt{(x_2 - x_1)^2 + (y_2 - y_1)^2}$$
$$= \sqrt{(4-6)^2 + (5-2)^2}$$
$$= \sqrt{4 + 9}$$
$$= \sqrt{13}$$

The distance is $\sqrt{13}$ units.

DISTANCE BETWEEN TWO POINTS
INTERMEDIATE TEST PAGE 43

1 (–2, 5) to (–2, –8)
Distance = $5 - (-8)$ units
= 13 units
∴ [A] ✓ (1 mark)

2 (5, 5) is more than 5 units from the origin.

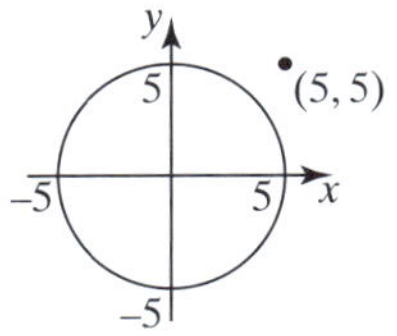

∴ [D] ✓ (1 mark)

3 (–1, –3), (5, 2)

$$d = \sqrt{(x_2 - x_1)^2 + (y_2 - y_1)^2}$$
$$= \sqrt{(5-(-1))^2 + (2-(-3))^2}$$
$$= \sqrt{61}$$

The distance is $\sqrt{61}$ units.
∴ [D] ✓ (1 mark)

4 **a**

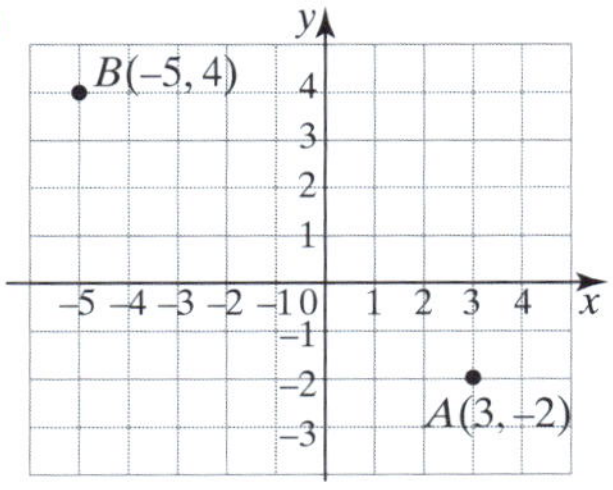

✓ (1 mark)

b $d^2 = 6^2 + 8^2$ ✓
$= 36 + 64$
$= 100$
$d = \sqrt{100}$ $(d > 0)$
$= 10$ units ✓ (2 marks)

5

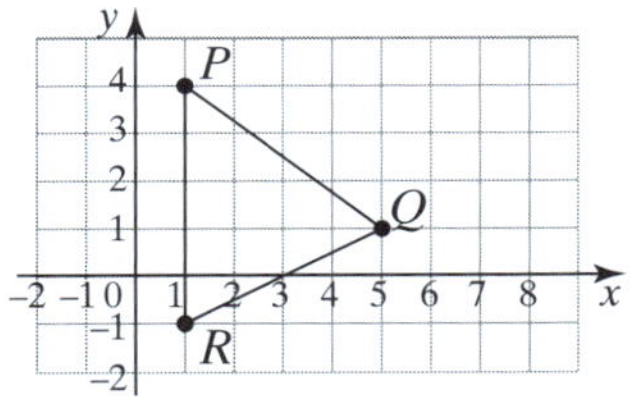

a $PR = 5$ units ✓ (1 mark)

b $d^2 = 3^2 + 4^2$ ✓
$= 9 + 16$
$= 25$
$d = \sqrt{25}$
$= 5$
$PQ = 5$ units ✓ (2 marks)

c $RQ \neq 5$ units
[$d^2 = 2^2 + 4^2$
$= 4 + 16$
$= 20$ $(\neq 25)$]
Triangle PQR is isosceles. ✓
$PQ = PR$ ✓ (2 marks)

d Q is 4 units from PR. ✓
$A = \frac{1}{2}bh$
$= \frac{1}{2} \times 5 \times 4$
$= 10$ units^2 ✓ (2 marks)

6 **a** $P(-11, 7)$, $Q(14, 7)$
$PQ = 25$ units ✓ (1 mark)
[from –11 to 14 along the line $y = 7$]

b $P(-11, 7)$, $R(-2, -5)$

$d = \sqrt{(x_2 - x_1)^2 + (y_2 - y_1)^2}$

$= \sqrt{(-2-(-11))^2 + (-5-7)^2}$ ✓

$= \sqrt{225}$

$= 15$

$\therefore PR = 15$ units ✓ (2 marks)

c $Q(14, 7)$, $R(-2, -5)$

$d = \sqrt{(x_2 - x_1)^2 + (y_2 - y_1)^2}$

$= \sqrt{(-2-14)^2 + (-5-7)^2}$ ✓

$= \sqrt{400}$

$= 20$

$\therefore QR = 20$ units ✓

(2 marks)

d ΔPQR is right-angled ✓

$PR^2 + QR^2 = 15^2 + 20^2$

$= 225 + 400$

$= 625$

$PQ^2 = 25^2$

$= 625$

$\therefore PR^2 + QR^2 = PQ^2$ ✓

(2 marks)

(Total: 20 marks)

DISTANCE BETWEEN TWO POINTS
ADVANCED TEST PAGE 44

1 a (3, 7) and (11, 1)

$d = \sqrt{(x_2 - x_1)^2 + (y_2 - y_1)^2}$

$= \sqrt{(11-3)^2 + (1-7)^2}$ ✓

$= 10$ units ✓ (2 marks)

b (–2, 8) and (–7, –4)

$d = \sqrt{(x_2 - x_1)^2 + (y_2 - y_1)^2}$

$= \sqrt{(-7-(-2))^2 + (-4-8)^2}$ ✓

$= 13$ units ✓ (2 marks)

c (5, –9) and (–7, 7)

$d = \sqrt{(x_2 - x_1)^2 + (y_2 - y_1)^2}$

$= \sqrt{(-7-5)^2 + (7-(-9))^2}$ ✓

$= 20$ units ✓ (2 marks)

2 a (5, 8) and (1, 7)

$d = \sqrt{(x_2 - x_1)^2 + (y_2 - y_1)^2}$

$= \sqrt{(1-5)^2 + (7-8)^2}$ ✓

$= \sqrt{17}$ units ✓ (2 marks)

b (3, –6) and (8, 0)

$d = \sqrt{(x_2 - x_1)^2 + (y_2 - y_1)^2}$

$= \sqrt{(8-3)^2 + (0-(-6))^2}$ ✓

$= \sqrt{61}$ units ✓ (2 marks)

c (–4, –5) and (1, –9)

$d = \sqrt{(x_2 - x_1)^2 + (y_2 - y_1)^2}$

$= \sqrt{(1-(-4))^2 + (-9-(-5))^2}$ ✓

$= \sqrt{41}$ units ✓ (2 marks)

3 a $A(0, 2)$, $B(4, -1)$

$d = \sqrt{(x_2 - x_1)^2 + (y_2 - y_1)^2}$

$= \sqrt{(4-0)^2 + (-1-2)^2}$

$= 5$ units ✓

$B(4, -1)$, $C(9, -1)$

distance $= (9 - 4)$ units

$= 5$ units

So $AB = BC$

ΔABC is isosceles. ✓

(2 marks)

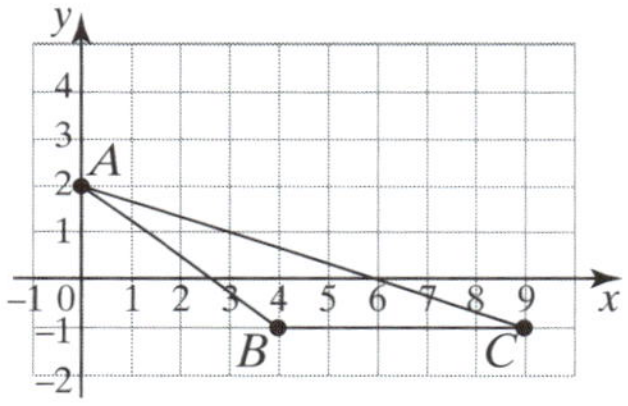

b Perpendicular height

$= 3$ units ✓

$A = \frac{1}{2}bh$

$= \frac{1}{2} \times 5 \times 3$

$= 7.5$ units^2 ✓ (2 marks)

4 a $P(7, 7)$, $Q(8, 0)$

$d = \sqrt{(x_2 - x_1)^2 + (y_2 - y_1)^2}$

$= \sqrt{(8-7)^2 + (0-7)^2}$ ✓

$PQ = \sqrt{50}$ units ✓ (2 marks)

b $Q(8, 0)$, $R(3, -5)$

$d = \sqrt{(x_2 - x_1)^2 + (y_2 - y_1)^2}$

$= \sqrt{(3-8)^2 + (-5-0)^2}$ ✓

$QR = \sqrt{50}$ units

$\therefore QR = PQ$ ✓ (2 marks)

c

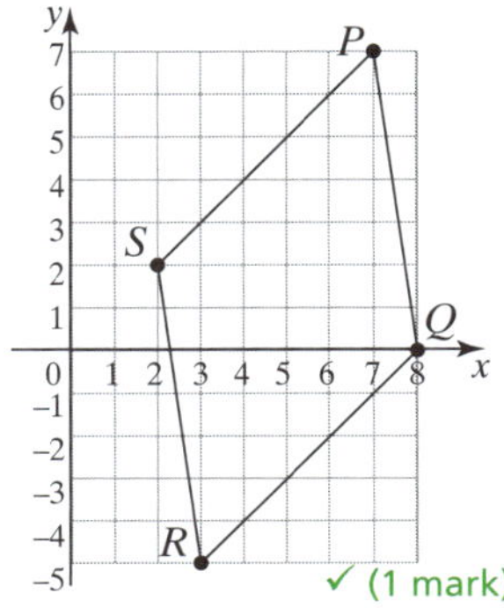

✓ (1 mark)

d $P(7, 7)$, $R(3, -5)$

$d = \sqrt{(x_2 - x_1)^2 + (y_2 - y_1)^2}$

$= \sqrt{(3-7)^2 + (-5-7)^2}$ ✓

$= \sqrt{160}$ units ✓ (2 marks)

e $Q(8, 0)$, $S(2, 2)$

$d = \sqrt{(x_2 - x_1)^2 + (y_2 - y_1)^2}$

$= \sqrt{(2-8)^2 + (2-0)^2}$

$= \sqrt{40}$ units ✓

Rhombus:

$A = \frac{1}{2}xy$

$= \frac{1}{2} \times \sqrt{160} \times \sqrt{40}$ ✓

$= 40$ units^2 ✓ (3 marks)

5 a $P(0, 4)$, $Q(8, 10)$

$d = \sqrt{(x_2 - x_1)^2 + (y_2 - y_1)^2}$

$= \sqrt{(8-0)^2 + (10-4)^2}$ ✓

$= 10$ units ✓ (2 marks)

b R lies on the x-axis so at R, $y = 0$ ✓

Now $PQ = QR$

So $QR = 10$ units

But Q is 10 units above the x-axis.

So, R lies on the x-axis directly below Q.

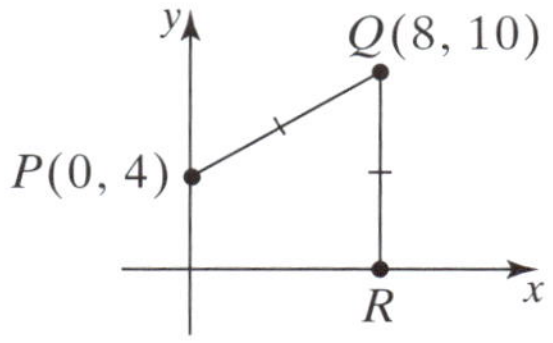

R is the point (8, 0) ✓

(2 marks)

(Total: 30 marks)

WORKED SOLUTIONS

MIDPOINT AND GRADIENT
SKILLS CHECK PAGE 46

1 a Sum = 3 + 9 = 12
Average = 12 ÷ 2 = 6

b Average $= \dfrac{-3+7}{2}$
$= 2$

c Average $= \dfrac{-5-4}{2}$
$= -4\frac{1}{2}$

2

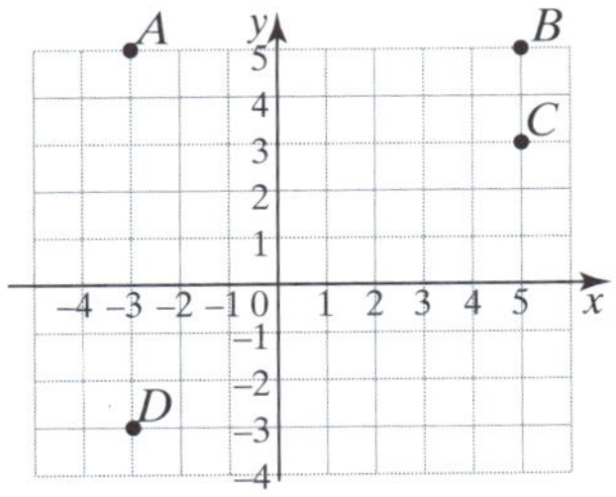

a $A(-3, 5)$, $B(5, 5)$
Midpoint (1, 5) [Both points have y-coordinate 5. The midpoint must also have y-coordinate 5. The x-coordinate is halfway between –3 and 5.]

b $B(5, 5)$, $C(5, 3)$
Midpoint (5, 4)

c $A(-3, 5)$, $D(-3, -3)$
Midpoint (–3, 1)

d $C(5, 3)$, $D(-3, -3)$
Midpoint (1, 0) [The x-coordinate is halfway between 5 and –3. The y-coordinate is halfway between 3 and –3.]

3 a negative [As you move from left to right across the grid the line goes down.]

b positive [As you move from left to right across the grid the line goes up.]

c positive

d negative

4 a

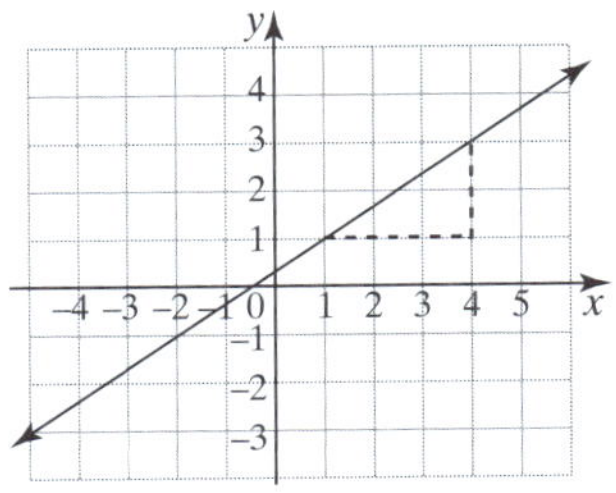

gradient $= \dfrac{\text{rise}}{\text{run}} = \dfrac{2}{3}$

b

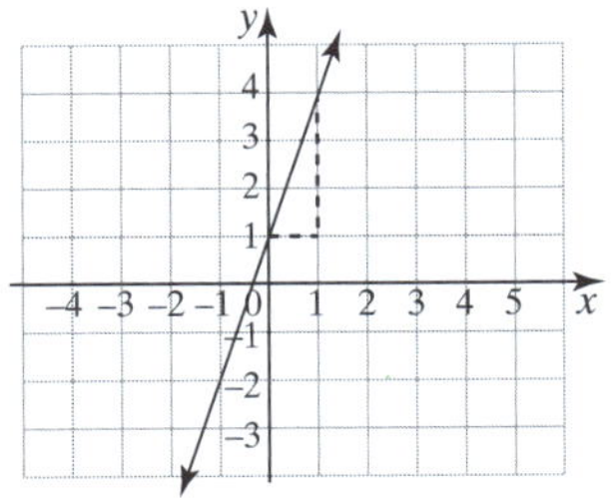

gradient $= \dfrac{\text{rise}}{\text{run}} = \dfrac{3}{1} = 3$

c

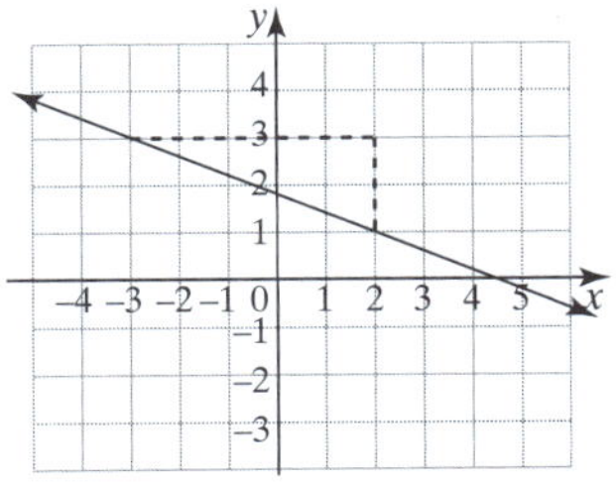

gradient $= \dfrac{\text{rise}}{\text{run}} = -\dfrac{2}{5}$

[As you move from left to right the line goes down so the 'rise' is negative.]

5 a $x_1 = 5, x_2 = 9, y_1 = -2, y_2 = 4$

$M(x, y) = \left(\dfrac{x_1+x_2}{2}, \dfrac{y_1+y_2}{2}\right)$
$= \left(\dfrac{5+9}{2}, \dfrac{-2+4}{2}\right)$
$= (7, 1)$

b $x_1 = -3, x_2 = 8, y_1 = 3, y_2 = -4$

$M(x, y) = \left(\dfrac{x_1+x_2}{2}, \dfrac{y_1+y_2}{2}\right)$
$= \left(\dfrac{-3+8}{2}, \dfrac{3+(-4)}{2}\right)$
$= \left(\dfrac{5}{2}, \dfrac{-1}{2}\right)$
$= \left(2\dfrac{1}{2}, -\dfrac{1}{2}\right)$

6 a (3, 7) and (9, 1)
$[x_1 = 3, x_2 = 9, y_1 = 7, y_2 = 1]$

$M(x, y) = \left(\dfrac{x_1+x_2}{2}, \dfrac{y_1+y_2}{2}\right)$
$= \left(\dfrac{3+9}{2}, \dfrac{7+1}{2}\right)$
$= (6, 4)$

b (4, 0) and (–2, 7)

$M(x, y) = \left(\dfrac{x_1+x_2}{2}, \dfrac{y_1+y_2}{2}\right)$
$= \left(\dfrac{4+(-2)}{2}, \dfrac{0+7}{2}\right)$
$= \left(1, 3\dfrac{1}{2}\right)$

7 a $x_1 = -7, x_2 = -1, y_1 = 5, y_2 = 8$

$m = \dfrac{y_2-y_1}{x_2-x_1}$
$= \dfrac{8-5}{-1-(-7)}$
$= \dfrac{3}{6}$
$= \dfrac{1}{2}$

b $x_1 = 2, x_2 = 3, y_1 = 1, y_2 = -5$

$m = \dfrac{y_2-y_1}{x_2-x_1}$
$= \dfrac{-5-1}{3-2}$
$= \dfrac{-6}{1}$
$= -6$

8 a (3, 7) and (9, –1)
$[x_1 = 3, x_2 = 9, y_1 = 7, y_2 = -1]$

$m = \dfrac{y_2-y_1}{x_2-x_1}$
$= \dfrac{-1-7}{9-3}$
$= \dfrac{-8}{6}$
$= -\dfrac{4}{3}$

b (–8, –2) and (6, 5)

$m = \dfrac{y_2-y_1}{x_2-x_1}$
$= \dfrac{5-(-2)}{6-(-8)}$
$= \dfrac{7}{14}$
$= \dfrac{1}{2}$

9 a If the gradients are equal the lines are **parallel**.

b If two lines are parallel then their gradients are **equal**.

MIDPOINT AND GRADIENT
INTERMEDIATE TEST PAGE 47

1 (3, 7) and (7, 3)

$\frac{3+7}{2} = 5 \qquad \frac{7+3}{2} = 5$

The midpoint is (5, 5).

∴ **[C]** ✓ (1 mark)

2 Try each of the options.

A $\frac{-1+1}{2} = 0 \neq 3$

B: $\frac{-1+5}{2} = 2 \neq 3$

C: $\frac{-1+5}{2} = 2 \neq 3$

D: $\frac{-1+7}{2} = 3$

$\frac{4+12}{2} = 8$

(7, 12) is the point.

∴ **[D]** ✓ (1 mark)

3

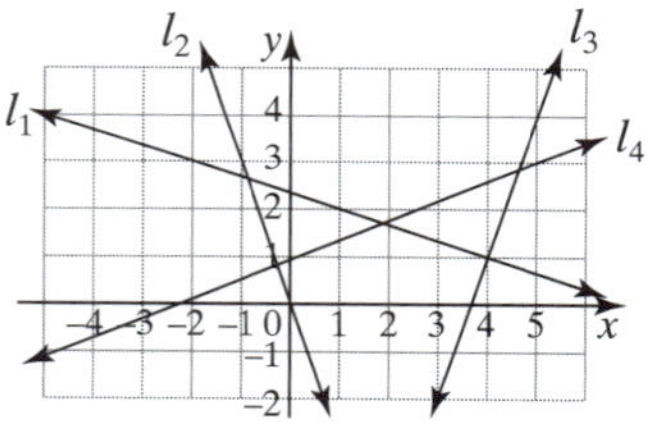

l_1: gradient $= \frac{\text{rise}}{\text{run}} = \frac{-1}{3}$

∴ **[A]** ✓ (1 mark)

4 (3, 7), (8, –5)

$M(x, y) = \left(\frac{x_1 + x_2}{2}, \frac{y_1 + y_2}{2}\right)$

$= \left(\frac{3+8}{2}, \frac{7+(-5)}{2}\right)$

$= (5.5, 1)$

∴ **[B]** ✓ (1 mark)

5 (–2, 4), (5, –10)

$m = \frac{y_2 - y_1}{x_2 - x_1}$

$= \frac{-10-4}{5-(-2)}$

$= \frac{-14}{7}$

$= -2$

∴ **[D]** ✓ (1 mark)

6

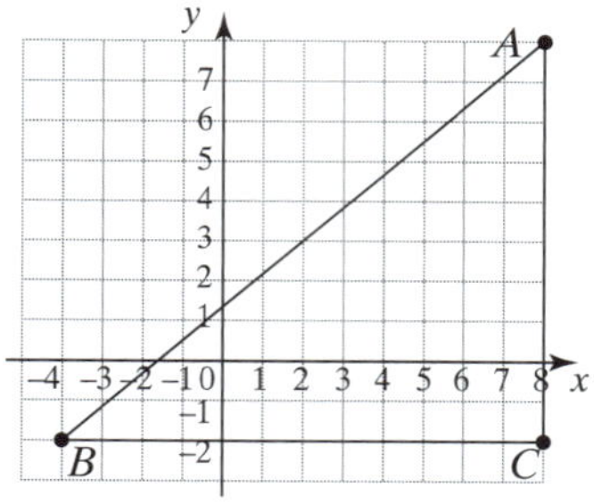

a Midpoint is (2, 2½) ✓✓ (2 marks)

[The x-coordinate of A is $x = 8$ and of B is $x = -4$. Halfway between these points is $x = 2$. The y-coordinate of A is $y = 7$ and of B is $y = -2$. Halfway between these points is $y = 2½$.]

b gradient $= \frac{\text{rise}}{\text{run}}$

$= \frac{9}{12}$ ✓

$= \frac{3}{4}$ ✓ (2 marks)

c BC is a horizontal line. So the gradient is 0. ✓ (1 mark)

7 a

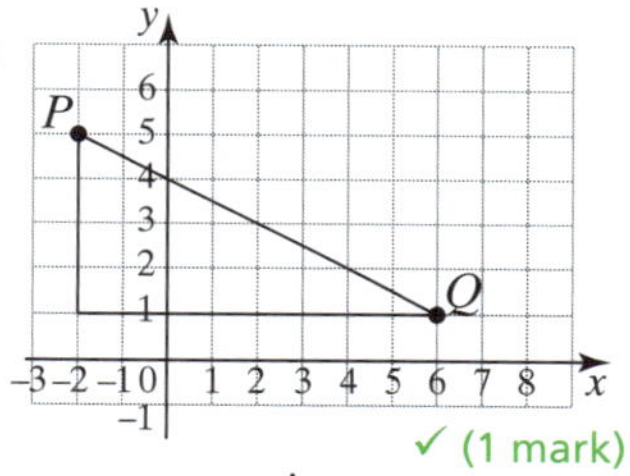

✓ (1 mark)

b gradient $= \frac{\text{rise}}{\text{run}}$

$= \frac{-4}{8}$ ✓

$= -\frac{1}{2}$ ✓ (2 marks)

8

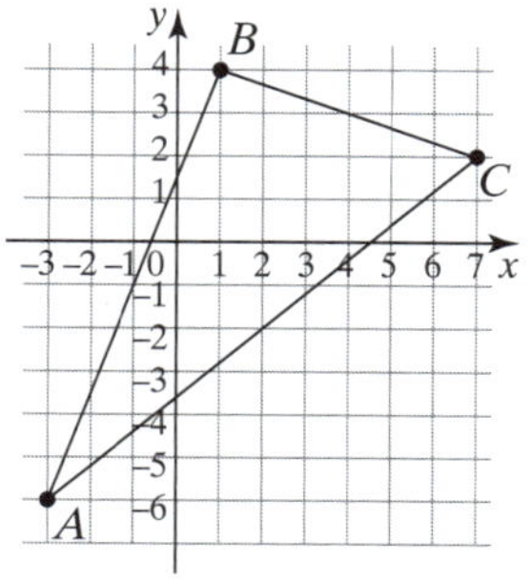

a gradient $m = \frac{\text{rise}}{\text{run}}$

AB: $m = \frac{10}{4}$

$= \frac{5}{2}$ ✓

AC: $m = \frac{8}{10}$

$= \frac{4}{5}$ ✓

BC: $m = \frac{-2}{6}$

$= -\frac{1}{3}$ ✓ (3 marks)

b From the diagram:

P is the point (–1, –1) ✓

Q is the point (2, –2) ✓

R is the point (4, 3) ✓

(3 marks)

c

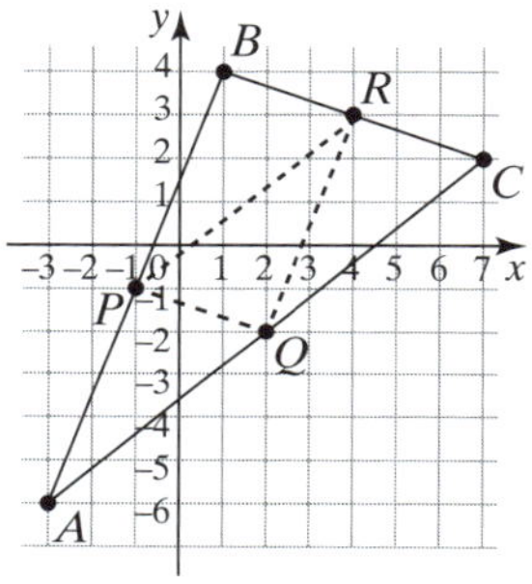

QR: $m = \frac{5}{2}$ ✓

PR: $m = \frac{4}{5}$ ✓

PQ: $m = \frac{-1}{3}$

$= -\frac{1}{3}$ ✓ (3 marks)

d The gradient of the line joining the midpoints of two sides is equal to the gradient of the other side of the triangle. The line joining the midpoints is parallel to the other side. ✓ (1 mark)

(Total: 23 marks)

MIDPOINT AND GRADIENT
ADVANCED TEST PAGE 48

1 a If the gradient of a line is zero the line is horizontal. ✓ (1 mark)

b The gradient of a vertical line is undefined. The line has infinite steepness. ✓ (1 mark)

2

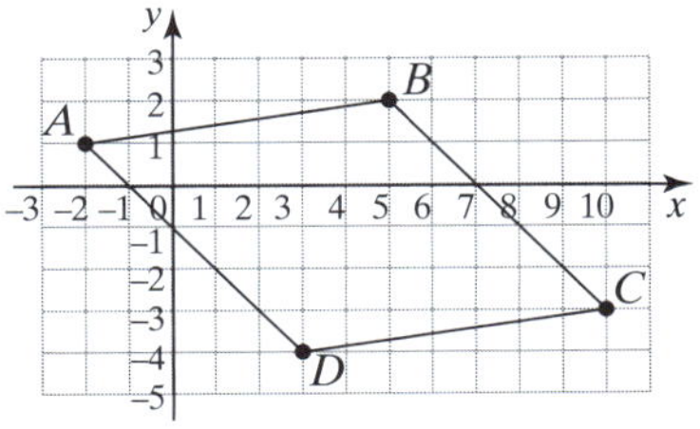

a $A(-2, 1), B(5, 2)$

$$m = \frac{y_2 - y_1}{x_2 - x_1} = \frac{2-1}{5-(-2)} = \frac{1}{7}$$ ✓ (1 mark)

b $D(3, -4), C(10, -3)$

$$m = \frac{y_2 - y_1}{x_2 - x_1} = \frac{-3-(-4)}{10-3} = \frac{1}{7}$$ ✓

$m_{AB} = m_{DC}$

$\therefore$ AB is parallel to DC ✓ (2 marks)

c $B(5, 2),\ C(10, -3)$

$$m = \frac{y_2 - y_1}{x_2 - x_1} = \frac{-3-2}{10-5} = -1$$ ✓ (1 mark)

d $A(-2, 1), D(3, -4)$

$$m = \frac{y_2 - y_1}{x_2 - x_1} = \frac{-4-1}{3-(-2)} = -1$$ ✓

Yes, AD is parallel to BC because the gradients are equal. ✓ (2 marks)

e Both pairs of opposite sides are parallel so it is a parallelogram. ✓ (1 mark)

f $A(-2, 1), C(10, -3)$

$$M(x,y) = \left(\frac{x_1+x_2}{2}, \frac{y_1+y_2}{2}\right) = \left(\frac{-2+10}{2}, \frac{1+(-3)}{2}\right) = (4, -1)$$ ✓ (1 mark)

g $B(5, 2), D(3, -4)$

$$M(x, y) = \left(\frac{x_1+x_2}{2}, \frac{y_1+y_2}{2}\right) = \left(\frac{5+3}{2}, \frac{2+(-4)}{2}\right) = (4, -1)$$ ✓ (1 mark)

h The diagonals of a parallelogram bisect each other (because both diagonals have the same midpoint). ✓ (1 mark)

3 $P(3, -7), Q(x, y), M(5, 2)$

M is the midpoint of PQ.

So $\frac{x+3}{2} = 5$

$x + 3 = 10$

$x = 7$ ✓

and $\frac{y+(-7)}{2} = 2$

$y - 7 = 4$

$y = 11$

Q is the point $(7, 11)$ ✓ (2 marks)

4 H lies on the y-axis so at H, $x = 0$. ✓

$G(-6, 5), H(0, y), m = -\frac{2}{3}$

$$m = \frac{y_2 - y_1}{x_2 - x_1}$$

$$-\frac{2}{3} = \frac{y-5}{0-(-6)}$$ ✓

$y - 5 = -\frac{2}{3} \times 6$

$y - 5 = -4$

$y = 1$

H is the point $(0, 1)$. ✓ (3 marks)

5 **a** $B(7, -4), D(-6, 6)$

$$m = \frac{y_2 - y_1}{x_2 - x_1} = \frac{6-(-4)}{-6-7} = -\frac{10}{13}$$ ✓ (1 mark)

b $A(3, 5), C(-2, -3)$

$$M(x,y) = \left(\frac{x_1+x_2}{2}, \frac{y_1+y_2}{2}\right) = \left(\frac{3+(-2)}{2}, \frac{5+(-3)}{2}\right) = \left(\frac{1}{2}, 1\right)$$ ✓ (1 mark)

c $B(7, -4), M(\frac{1}{2}, 1)$

$$m = \frac{y_2 - y_1}{x_2 - x_1} = \frac{1-(-4)}{\frac{1}{2}-7} = -\frac{10}{13}$$ ✓ (1 mark)

d M lies on the diagonal BD. ✓ The gradient of BM equals the gradient of BD and B is a common point so BM and BD are the same line. ✓ (2 marks)

6 **a** $A(4, 1), C(p, q), m = -1$

$$m = \frac{y_2 - y_1}{x_2 - x_1}$$

$$-1 = \frac{q-1}{p-4}$$ ✓

$-1(p - 4) = q - 1$

$-p + 4 = q - 1$

$-p + 5 = q$

So $q = -p + 5$ ✓ (2 marks)

b $B(-2, -5), C(p, q)$

$$m = \frac{y_2 - y_1}{x_2 - x_1}$$

$$3 = \frac{q-(-5)}{p-(-2)}$$

$3(p + 2) = q + 5$ ✓

$3p + 6 = q + 5$

$3p + 1 = q$

So $q = 3p + 1$ ✓ (2 marks)

c $-p + 5$ and $3p + 1$ are both equal to q. So they must be equal. ✓ (1 mark)

d $3p + 1 = -p + 5$

$4p + 1 = 5$

$4p = 4$

$p = 1$ ✓ (1 mark)

e $q = 3p + 1$

When $p = 1$,

$q = 3 \times 1 + 1$

$= 4$

So C is the point $(1, 4)$. ✓✓ (2 marks)

(Total: 30 marks)

WORKED SOLUTIONS

CHECK YOUR SOLUTIONS

SKETCHING LINEAR GRAPHS
SKILLS CHECK PAGE 50

1 **a** $y = x + 1$

x	0	1	2
y	1	2	3

b $y = 4x$

x	1	2	3
y	4	8	12

c $y = 6 - x$

x	–1	0	1
y	7	6	5

d $y = 2x + 5$

x	2	3	4
y	9	11	13

e $y = 7$

x	0	1	2
y	7	7	7

f $x = 5$

x	5	5	5
y	1	2	3

2 **a** $x + y = 2$
[Or $y = 2 - x$ or $y = -x + 2$]

b $y = x - 1$ [Or $x - y = 1$]

c $y = -2x + 2$ [Or $2x + y = 2$]

d $y = 2$

3 **a**

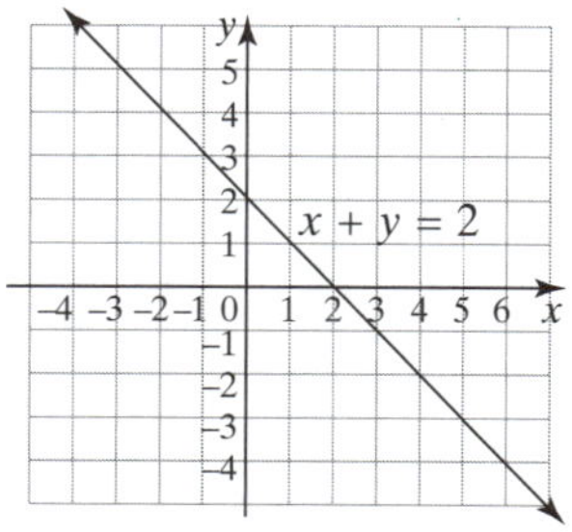

b

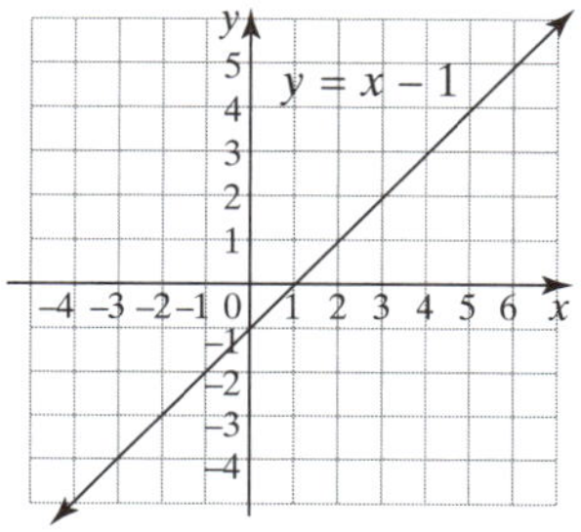

c

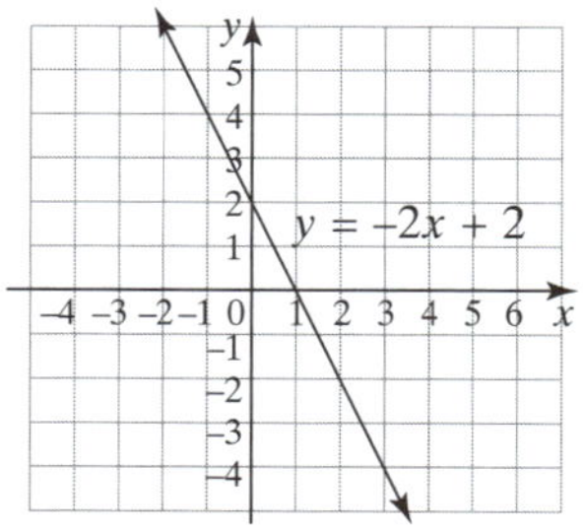

d

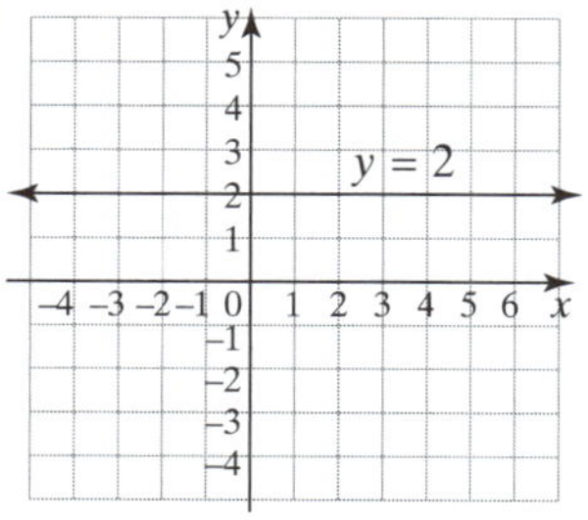

4 **a** gradient $= \frac{\text{rise}}{\text{run}} = \frac{-1}{1} = -1$

b gradient $= \frac{\text{rise}}{\text{run}} = \frac{1}{1} = 1$

c gradient $= \frac{\text{rise}}{\text{run}} = \frac{-2}{1} = -2$

d gradient $= \frac{\text{rise}}{\text{run}} = 0$

[There is no rise, so the gradient is zero.]

5 **a** 2 **b** –1
c 2 **d** 2

6 **a** x-intercept is 2 [The line cuts the x-axis at (2, 0).]
y-intercept is –2 [The line cuts the y-axis at (0, –2).]

b x-intercept is 5 [The line cuts the x-axis at (5, 0).]
y-intercept is 2 [The line cuts the y-axis at (0, 2).]

7 **a**

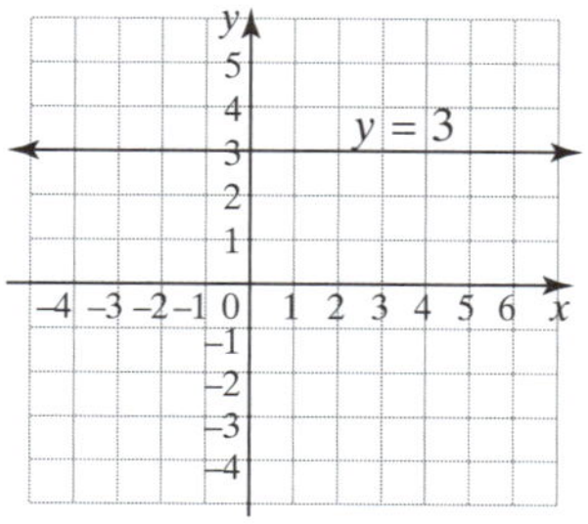

b

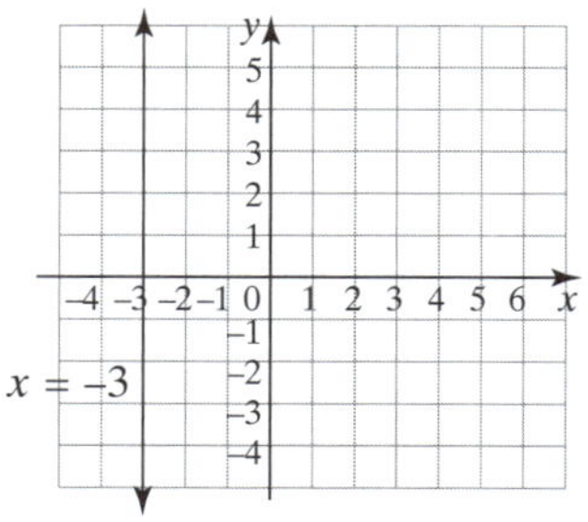

8 **a**

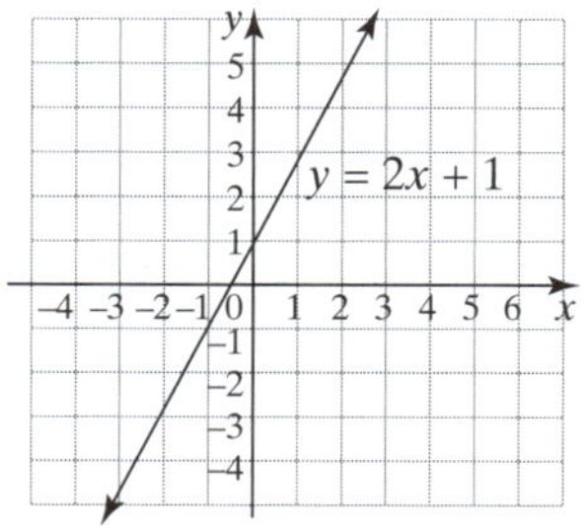

b

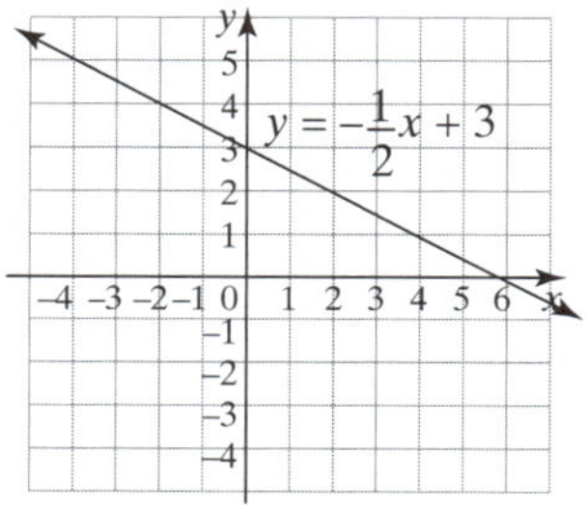

SKETCHING LINEAR GRAPHS
INTERMEDIATE TEST PAGE 51

1 Try each option.
$y = 4$
When $x = 0$, $y = 4$
$y = 2x - 2$
When $x = 0$, $y = -2$
$y = -x$
When $x = 0$, $y = 0$
$y = -3x + 3$
When $x = 0$, $y = 3$
Only $y = -x$ passes through (0, 0).
∴ **[C]** ✓ (1 mark)

2 $y = 2 - x$
$= -x + 2$
Gradient = –1,
y-intercept 2

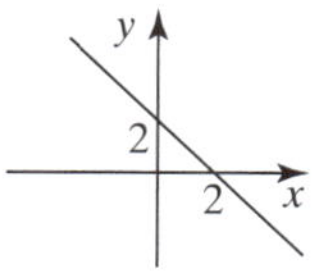

∴ **[B]** ✓ (1 mark)

3 When $x = 0$, $y = 3$
So only options B and C are possible.
When $x = 1$, $y = 1$ so the equation can only be $y = 3 - 2x$.

x	–1	0	1	2	3
y	5	3	1	–1	–3

∴ [C] ✓ (1 mark)

4 **a** $y = 2x - 3$

x	1	2	3	4
y	–1	1	3	5

✓✓ (2 marks)

b

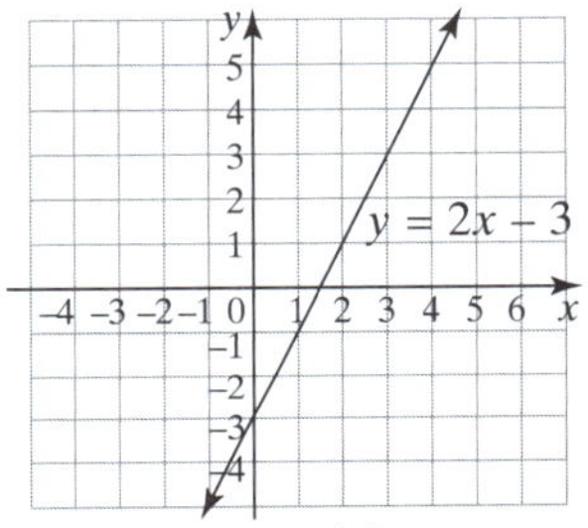

✓✓ (2 marks)

c y-intercept $= -3$ ✓ (1 mark)

5 **a** $y = \frac{1}{2}x - 2$

x	–1	0	1
y	$-2\frac{1}{2}$	–2	$-1\frac{1}{2}$

✓✓ (2 marks)

b

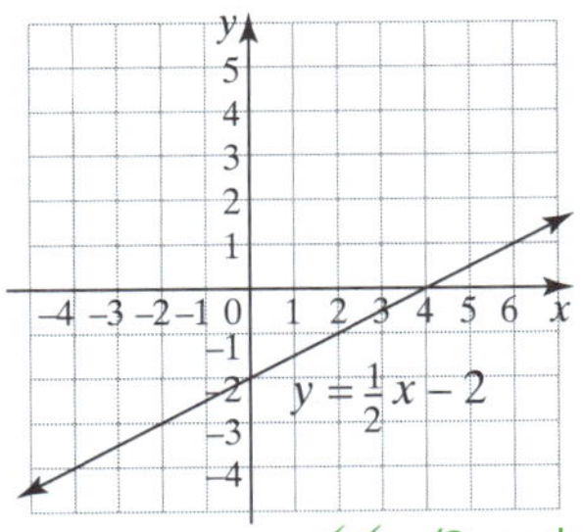

✓✓ (2 marks)

c x-intercept $= 4$ ✓ (1 mark)

d gradient $= \frac{\text{rise}}{\text{run}}$

$= \frac{1}{2}$ ✓ (1 mark)

6 **a** $x + y = 5$
When $x = 2$,
$2 + y = 5$
$y = 3$ ✓ (1 mark)

b $-2 + y = 5$
$y = 5 + 2$
$= 7$ ✓ (1 mark)

c

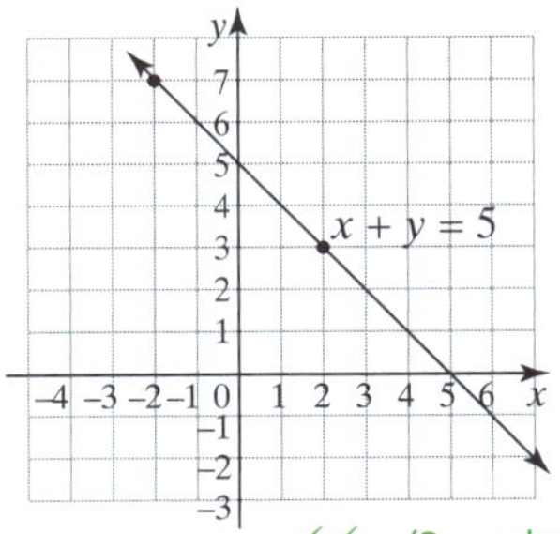

✓✓ (2 marks)

7 **a**

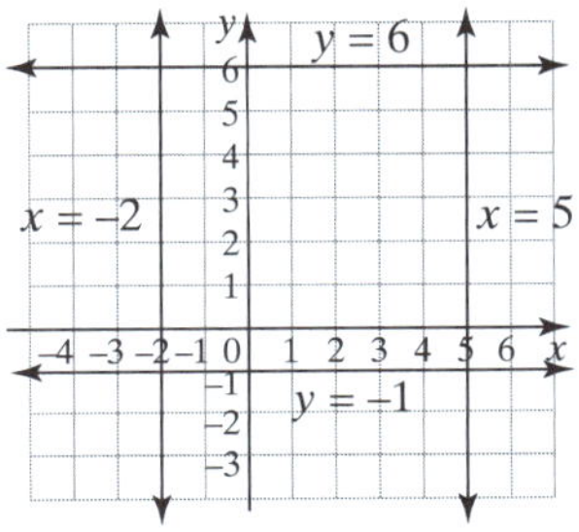

✓✓ (2 marks)

b Counting units:
length = 7 units;
breadth = 7 units ✓
$A = lb$
$= 7 \times 7$
$= 49$ units^2 ✓ (2 marks)

8 **a** $y = 3x - 2$
When $x = 0$, $y = -2$
When $x = 1$, $y = 3 \times 1 - 2$
$= 1$
$y = 4 - x$
When $x = 0$, $y = 4$
When $x = 1$, $y = 4 - 1$
$= 3$

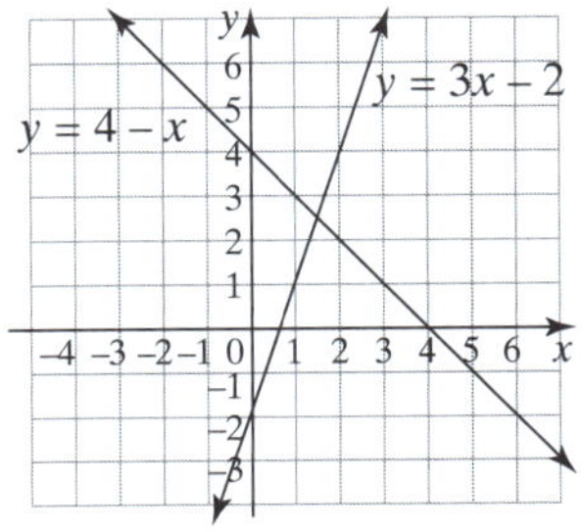

✓✓ (2 marks)

b Point of intersection is (1½, 2½). ✓ (1 mark)

(Total: 25 marks)

SKETCHING LINEAR GRAPHS
ADVANCED TEST PAGE 52

1 **a** $y = -3x + 8$
straight line
true ✓ (1 mark)

b $y = 5$
straight line
true ✓ (1 mark)

c $2x + 9y - 7 = 0$
straight line
true ✓ (1 mark)

d $5x = 2y + 1$
$5x - 2y - 1 = 0$
straight line
true ✓ (1 mark)

e $y = \frac{x}{4}$
$y = \frac{1}{4}x$
straight line
true ✓ (1 mark)

f $x = 0$
straight line
true ✓ (1 mark)

2 **a**

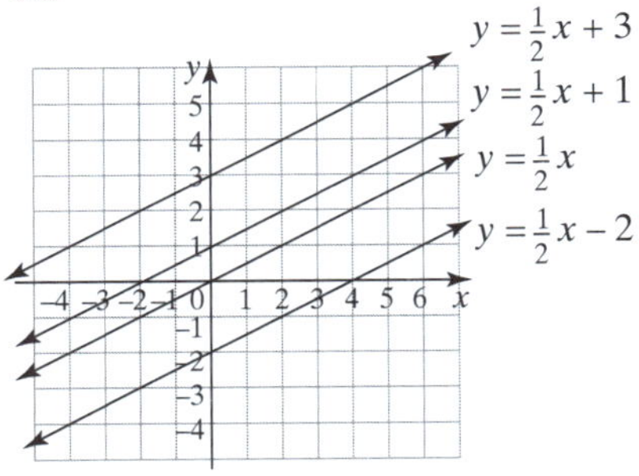

✓✓✓✓ (4 marks)

b The four lines are all parallel. They have the same gradient. ✓ (1 mark)

c The line moves up or down. A number added to the equation $y = ax$ will be the y-intercept. ✓ (1 mark)

3 **a** $3x + 2y - 6 = 0$
$2y = -3x + 6$
$y = -\frac{3}{2}x + 3$ ✓
(1 mark)

b $m = -\frac{3}{2}$ ✓

$b = 3$ ✓ (2 marks)

c

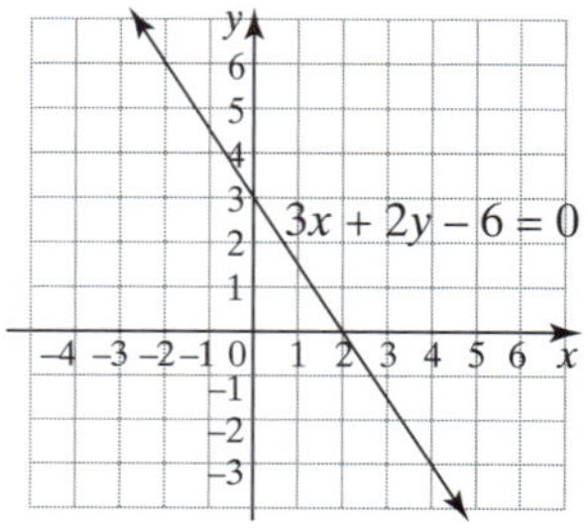

✓✓ (2 marks)

[Begin at 3, the y-intercept. Move down 3 units and across 2 units to find another point on the line.]

d Q is the point (2, 0) ✓ (1 mark)

4 $3x - 2y + 12 = 0$

a When $y = 0$,

$3x + 12 = 0$

$3x = -12$

$x = -4$ ✓ (1 mark)

b When $x = 0$,

$-2y + 12 = 0$

$2y = 12$

$y = 6$ ✓ (1 mark)

c

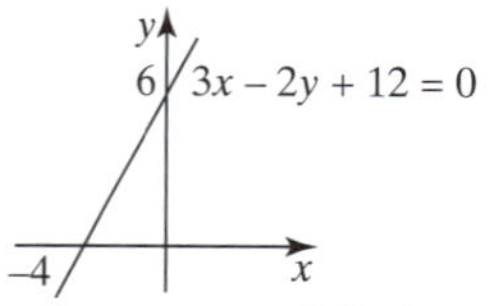

✓✓ (2 marks)

5 **a**

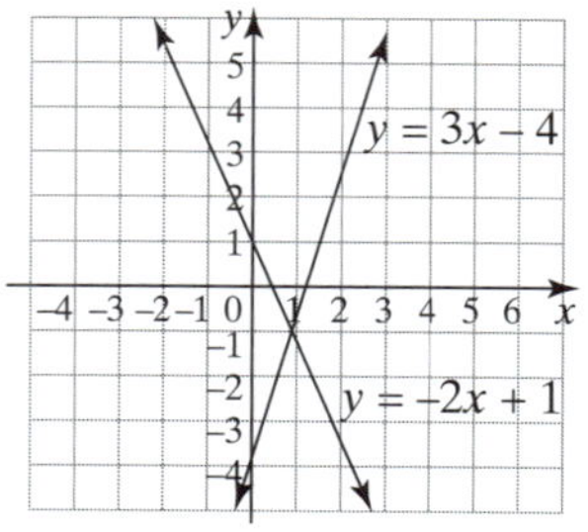

✓✓ (2 marks)

b A is the point (1, –1). ✓ (1 mark)

c

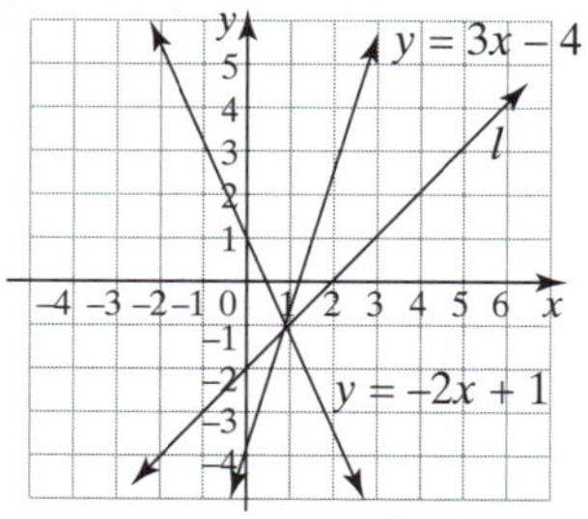

✓ (1 mark)

d $m = \frac{\text{rise}}{\text{run}}$

$= \frac{4}{4}$

$= 1$ ✓ (1 mark)

6 **a**

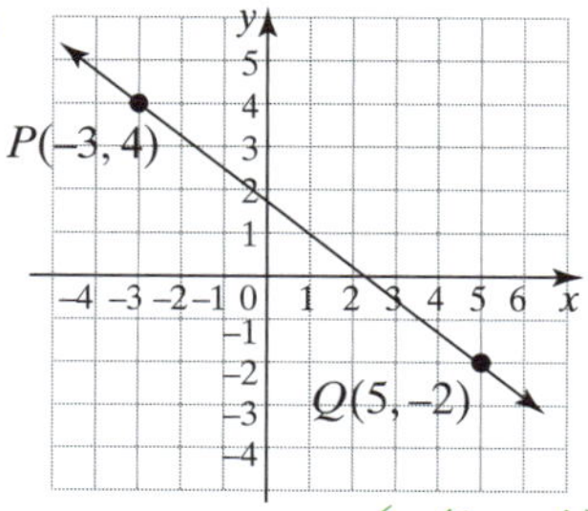

✓ (1 mark)

b

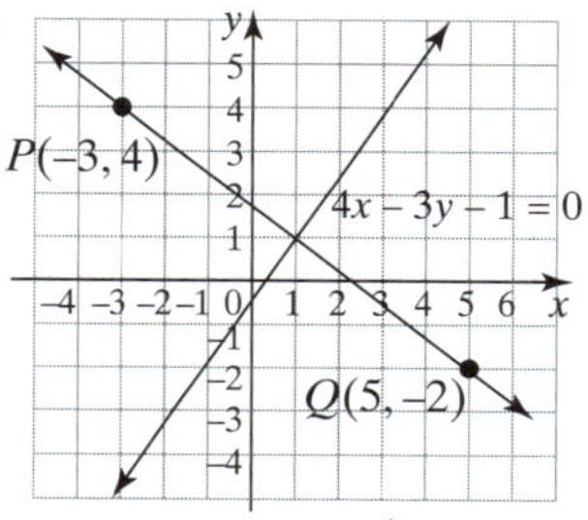

✓ (1 mark)

c The line $4x - 3y - 1 = 0$ is the perpendicular bisector of PQ. It passes through the midpoint of PQ and is at right-angles to the line joining P to Q. ✓ (1 mark)

(Total: 30 marks)

THE EQUATION OF A LINE
SKILLS CHECK PAGE 54

1 **a** $y = 2x + 5$; $m = 2$, $b = 5$

b $y = 3x - 7$; $m = 3$, $b = -7$

c $y = -4x + 1$; $m = -4$, $b = 1$

d $y = x - 3$; $m = 1$, $b = -3$

e $y = -x$; $m = -1$, $b = 0$

f $y = 4 - 3x$

$y = -3x + 4$; $m = -3$, $b = 4$

g $y = -\frac{3}{4}x - \frac{1}{2}$; $m = -\frac{3}{4}$, $b = -\frac{1}{2}$

h $y = \frac{x}{2} + 2$; $m = \frac{1}{2}$, $b = 2$

2 **a** $m = 5$, $b = 1$; $y = 5x + 1$

b $m = -2$, $b = 0$; $y = -2x$

c $m = \frac{2}{3}$, $b = -\frac{1}{3}$; $y = \frac{2}{3}x - \frac{1}{3}$

3 **i** **a** $m = \frac{\text{rise}}{\text{run}} = \frac{2}{3}$

b $b = 2$

c $y = \frac{2}{3}x + 2$

ii **a** $m = -\frac{1}{5}$

b $b = -1$

c $y = -\frac{1}{5}x - 1$

4 **a** $x + y = 3$

$y = -x + 3$

[subtracting x from both sides]

b $x - y = 1$

$-y = -x + 1$

$y = x - 1$

c $x + y + 4 = 0$

$y = -x - 4$

d $3x - y + 5 = 0$

$3x + 5 = y$

$y = 3x + 5$

e $4x - 2y = 1$

$-2y = -4x + 1$

$y = 2x - \frac{1}{2}$

f $x + 4y + 2 = 0$

$4y = -x - 2$

$y = -\frac{1}{4}x - \frac{1}{2}$

5 **a** When $x = 0$ and $y = 5$,

$2x + y = 2 \times 0 + 5$

$= 0 + 5$

$= 5$

True

b When $x = -2$ and $y = 1$,

$2x + y = 2 \times -2 + 1$

$= -4 + 1$

$= -3$ $(\neq 5)$

False

c When $x = 3$ and $y = -1$,
$2x + y = 2 \times 3 - 1$
$= 6 - 1$
$= 5$
True

6 **a** $y = 4x + 1$
$4x - y + 1 = 0$

b $y = -3x + 5$
$3x + y - 5 = 0$

c $y = \frac{1}{2}x - 3$
$2y = x - 6$
$x - 2y - 6 = 0$

d $y = \frac{2}{3}x + 4$
$3y = 2x + 12$
$2x - 3y + 12 = 0$

e $y = -\frac{4}{3}x + 2$
$3y = -4x + 6$
$4x + 3y - 6 = 0$

f $y = -\frac{x}{5} - 8$
$5y = -x - 40$
$x + 5y + 40 = 0$

7 **a** $x = 1$, $y = 5$ and $m = 3$
$y = mx + b$
$5 = 3 \times 1 + b$
$5 = 3 + b$
$b = 2$

b $x = 6$, $y = -7$ and $m = -\frac{1}{2}$
$y = mx + b$
$-7 = -\frac{1}{2} \times 6 + b$
$-7 = -3 + b$
$b = -4$

THE EQUATION OF A LINE
INTERMEDIATE TEST PAGE 55

1 $y = -3x + 6$
When $y = 0$,
$-3x + 6 = 0$
$-3x = -6$
$x = 2$
$\therefore$ [C] ✓ (1 mark)

2 $y = \frac{2}{3}x - \frac{1}{2}$
$6y = 4x - 3$
[multiplying both sides by 6]
$4x - 6y - 3 = 0$
$\therefore$ [A] ✓ (1 mark)

3 $2x + y - 3 = 0$
$y = -2x + 3$
Gradient is –2.
$\therefore$ [D] ✓ (1 mark)

4 $y = -3x + 1$

a (3, –10)
$-10 = -3 \times 3 + 1$
$-10 = -8$ False
(3, –10) does not lie on the line. ✓ (1 mark)

b (–2, 7)
$7 = -3 \times -2 + 1$
$7 = 7$ True
(–2, 7) does lie on the line. ✓ (1 mark)

c (5, –14)
$-14 = -3 \times 5 + 1$
$-14 = -14$ True
(5, –14) does lie on the line. ✓ (1 mark)

5 (–2, 5)

a $y = 2x + 1$
$5 = 2 \times -2 + 1$
$5 = -3$ False
(–2, 5) does not lie on $y = 2x + 1$ ✓ (1 mark)

b $y = 3x - 1$
$5 = 3 \times -2 - 1$
$5 = -7$ False
(–2, 5) does not lie on $y = 3x - 1$ ✓ (1 mark)

c $y = -x - 7$
$5 = -(-2) - 7$
$5 = -5$ False
(–2, 5) does not lie on $y = -x - 7$ ✓ (1 mark)

6

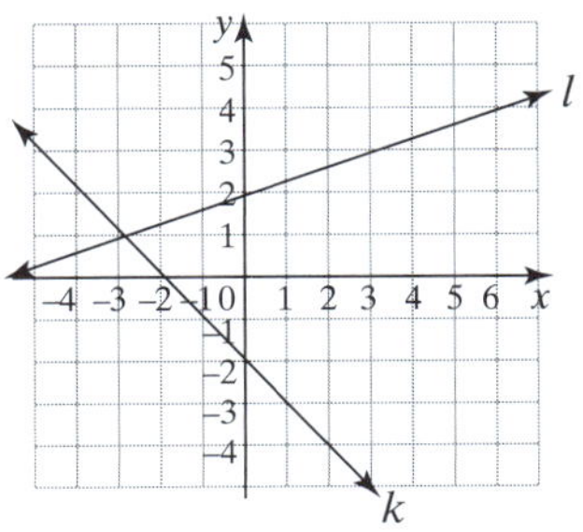

a The line k has gradient –1
[From any point move down one unit and across one unit to find another point. So the gradient is $\frac{-1}{1} = -1$.]

The y-intercept is –2 ✓
The equation is
$y = -1x - 2$
$y = -x - 2$ ✓ (2 marks)

b The line l has gradient $m = \frac{1}{3}$.
The y-intercept is $b = 2$ ✓
The equation is
$y = mx + b$
$y = \frac{1}{3}x + 2$ ✓ (2 marks)

7 **a** $A(-2, 5)$, $B(4, -1)$

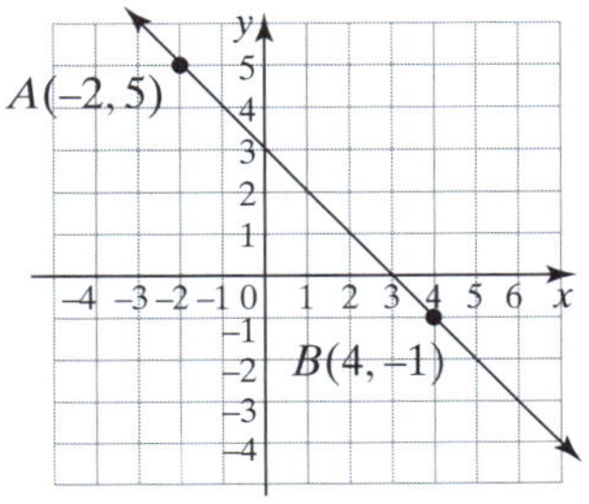

✓✓ (2 marks)

b $m = \frac{\text{rise}}{\text{run}}$
$= \frac{-6}{6}$
$= -1$ ✓ (1 mark)

c The line cuts the y-axis at (0, 3). [or when $y = 3$] ✓ (1 mark)

d Gradient is $m = -1$
y-intercept is $b = 3$
Equation is $y = -1x + 3$
$y = -x + 3$ ✓ (1 mark)

8 **a** (4, 3); gradient –2
$y = mx + b$
But $m = -2$,
$y = -2x + b$
When $x = 4$, $y = 3$
$3 = -2 \times 4 + b$
$3 = -8 + b$
$b = 11$
$y = -2x + 11$ ✓ (1 mark)

b (–1, –2); gradient $\frac{2}{3}$
$y - y_1 = m(x - x_1)$
$y - (-2) = \frac{2}{3}(x - (-1))$
$y + 2 = \frac{2}{3}(x + 1)$
$3y + 6 = 2x + 2$
$2x - 3y - 4 = 0$ ✓ (1 mark)
(Total: 20 marks)

THE EQUATION OF A LINE
ADVANCED TEST PAGE 56

1 **a** $2x + y - 8 = 0$
When $x = 0$,
$y - 8 = 0$
$y = 8$ ✓ (1 mark)

b $2x + y - 8 = 0$
When $y = 0$,
$2x - 8 = 0$
$2x = 8$
$x = 4$ ✓ (1 mark)

c

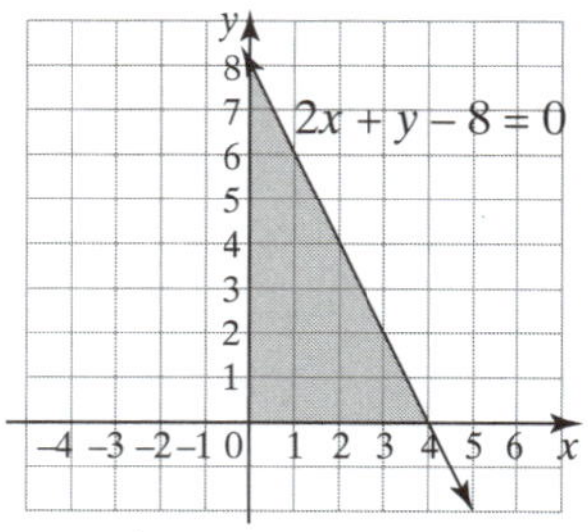

$A = \frac{1}{2}bh$
$= \frac{1}{2} \times 4 \times 8$
$= 16$ units^2 ✓ (1 mark)

2 **a** $m = -2, b = 5$
$y = -2x + 5$ ✓ (1 mark)

b When $x = 3$,
$y = -2 \times 3 + 5$
$= -1$
The point of intersection is $(3, -1)$. ✓ (1 mark)

c When $y = -7$,
$-7 = -2x + 5$
$2x = 12$
$x = 6$
The point of intersection is $(6, -7)$. ✓ (1 mark)

3 $(-2, 7), (5, -14)$
$m = \frac{y_2 - y_1}{x_2 - x_1}$
$= \frac{-14 - 7}{5 - (-2)}$
$= -3$ ✓
$y = -3x + b$
When $x = -2, y = 7$
$7 = -3 \times -2 + b$
$7 = 6 + b$
$b = 1$
$y = -3x + 1$ ✓ (2 marks)

4 **a** $(3, -1)$; gradient $\frac{2}{3}$
$y - y_1 = m(x - x_1)$
$y - (-1) = \frac{2}{3}(x - 3)$ ✓
$y + 1 = \frac{2}{3}(x - 3)$
$3y + 3 = 2x - 6$
$2x - 3y - 9 = 0$ ✓ (2 marks)

b When $x = -4$ and $y = -5$,
$2x - 3y - 9$
$= 2 \times -4 - 3 \times -5 - 9$
$= -2 \quad (\neq 0)$ ✓
So the point does not lie on the line. ✓ (2 marks)

5

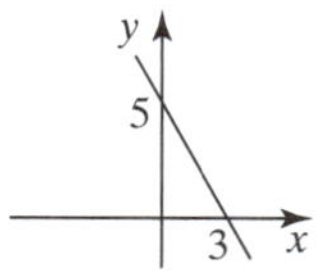

$b = 5$,
$m = -\frac{5}{3}$

$y = mx + b$
$y = -\frac{5}{3}x + 5$ ✓
$3y = -5x + 15$
$5x + 3y - 15 = 0$ ✓ (2 marks)

6 **a** $5x - 2y + 3 = 0$
$5x + 3 = 2y$
$2y = 5x + 3$
$y = \frac{5}{2}x + \frac{3}{2}$
✓ (1 mark)

b For the above line $m = \frac{5}{2}$. Parallel lines have the same gradient, so the gradient of any parallel line is $\frac{5}{2}$. ✓ (1 mark)

c $y = \frac{5}{2}x + b$
When $x = -1, y = 5$
$5 = \frac{5}{2} \times -1 + b$
$5 = -\frac{5}{2} + b$
$b = \frac{15}{2}$ ✓
$y = \frac{5}{2}x + \frac{15}{2}$
$2y = 5x + 15$
$5x - 2y + 15 = 0$ ✓ (2 marks)

d $y = \frac{5}{2}x + b$
When $x = 7, y = 1$
$1 = \frac{5}{2} \times 7 + b$
$1 = \frac{35}{2} + b$
$b = -\frac{33}{2}$ ✓
$y = \frac{5}{2}x - \frac{33}{2}$
$2y = 5x - 33$
$5x - 2y - 33 = 0$ ✓ (2 marks)

e The coefficients of x and the coefficients of y are the same in all the lines. The only difference in the equation is the constant term. ✓ (1 mark)

7 **a** $(1, 5)$ and $(-3, -1)$
$\frac{y - y_1}{x - x_1} = \frac{y_2 - y_1}{x_2 - x_1}$
$\frac{y - 5}{x - 1} = \frac{-1 - 5}{-3 - 1}$
$\frac{y - 5}{x - 1} = \frac{3}{2}$ ✓
$2(y - 5) = 3(x - 1)$
$2y - 10 = 3x - 3$
$3x - 2y + 7 = 0$ ✓ (2 marks)

b $(-1, 2)$ and $(5, 11)$
$\frac{y - y_1}{x - x_1} = \frac{y_2 - y_1}{x_2 - x_1}$
$\frac{y - 2}{x - (-1)} = \frac{11 - 2}{5 - (-1)}$
$\frac{y - 2}{x + 1} = \frac{3}{2}$ ✓
$2(y - 2) = 3(x + 1)$
$2y - 4 = 3x + 3$
$3x - 2y + 7 = 0$ ✓ (2 marks)

c The points all lie on the same straight line. They are collinear. ✓ (1 mark)

8

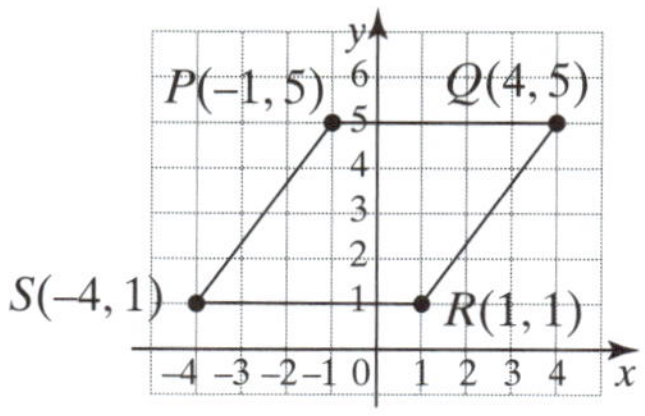

a $P(-1, 5), R(1, 1)$
$m = \frac{-4}{2}$
$= -2$
$y = -2x + b$
When $x = 1, y = 1$
$1 = -2 + b$
$b = 3$
$y = -2x + 3$ ✓ (1 mark)

b $Q(4, 5), S(-4, 1)$

$m = \frac{4}{8} = \frac{1}{2}$

$y = \frac{1}{2}x + b$

When $x = 4, y = 5$

$5 = \frac{1}{2} \times 4 + b$

$5 = 2 + b$

$b = 3$

$y = \frac{1}{2}x + 3$ ✓ (1 mark)

c Both lines have y-intercept 3. ✓ The diagonals intersect on the y-axis. Also, because the diagonals of a rhombus bisect each other this means that the midpoint of the diagonals is (0, 3). ✓

(2 marks)

(Total: 30 marks)

NON-LINEAR GRAPHS
SKILLS CHECK PAGE 58

1 a parabola

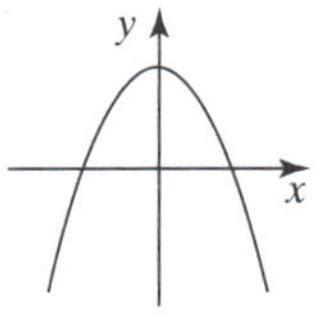

b circle

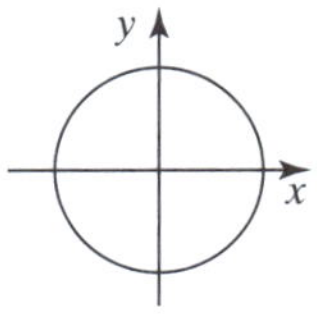

c straight line

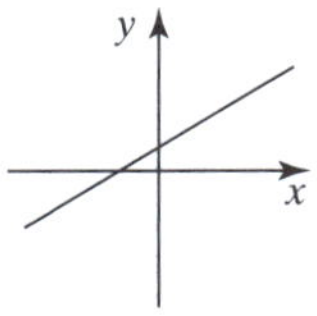

d parabola

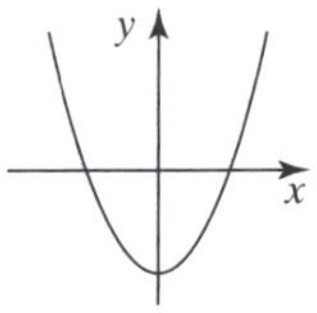

e exponential curve

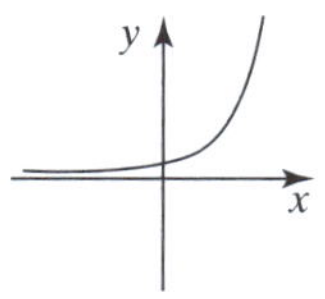

2 a $y = x^2 + 3$

x	–3	–2.5	–2	–1.5	–1	–0.5	0
y	12	9.25	7	5.25	4	3.25	3

x	0.5	1	1.5	2	2.5	3
y	3.25	4	5.25	7	9.25	12

b

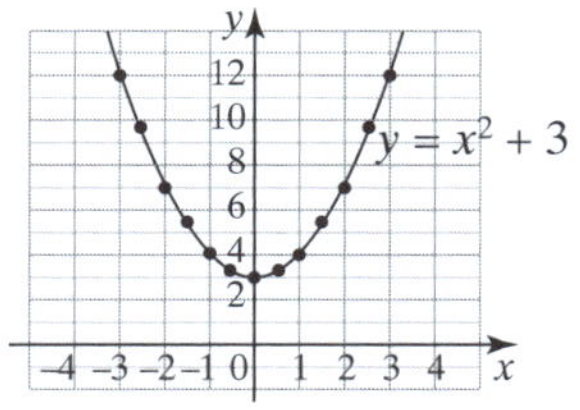

3 a $y = 2 - x$ straight line
[$y = -x + 2$, gradient –1, y-intercept 2]

b $x^2 + y^2 = 9$ circle
[centre (0,0), radius 3]

c $y = 2^x$ exponential curve

d $y = x^2$ parabola

e $y = \frac{x}{2}$ straight line

[$y = \frac{1}{2}x$, gradient $\frac{1}{2}$, y-intercept 0]

f $y = -x^2 + 2$ parabola

g $x + y = 9$ straight line
[$y = -x + 9$]

h $x^2 + y^2 = 2$ circle
[centre (0, 0), radius $\sqrt{2}$]

4 a $y = 4 - x^2$

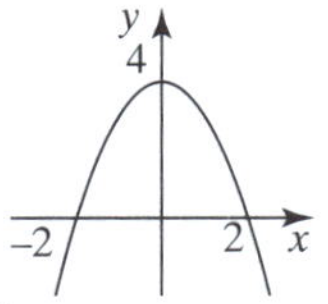

[Of form: $y = -ax^2 + c, c = 4$]

b $y = 4x^2$

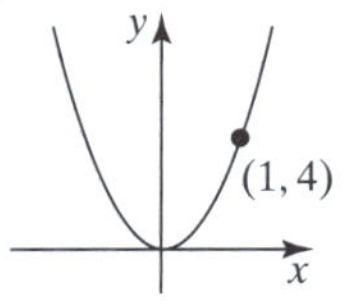

[Of form: $y = ax^2$
Passes through (1, 4)]

c $x^2 + y^2 = 8$

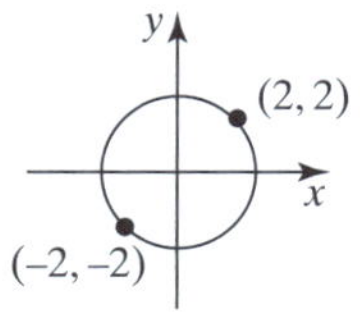

[Passes through (2, 2),
$2^2 + 2^2 = 4 + 4 = 8$]

d $y = 4 - x$

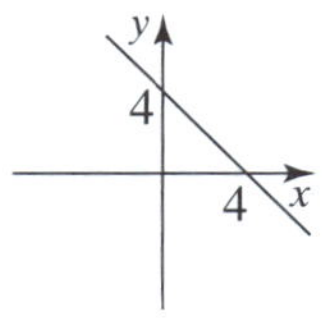

[y-intercept = 4,
gradient = –1]

e $y = \frac{x}{4}$

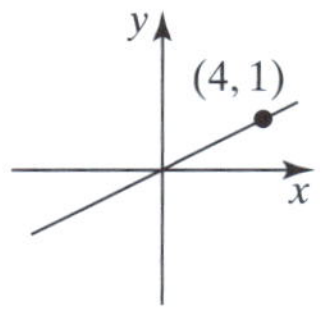

[y-intercept = 0, gradient = $\frac{1}{4}$]

f $x^2 + y^2 = 16$

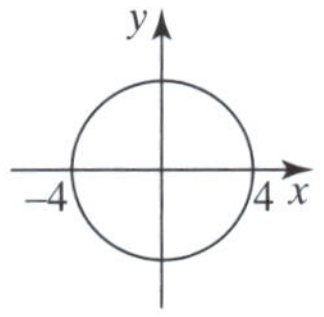

[Passes through (4, 0),
$4^2 + 0^2 = 16$]

g $y = x^2 - 4$

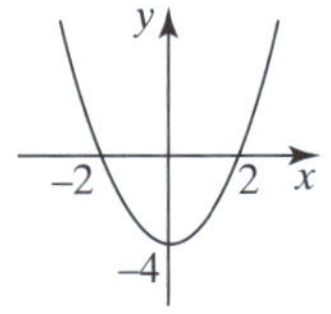

[Form: $y = ax^2 + c, c = -4$]

h $y = 4x$

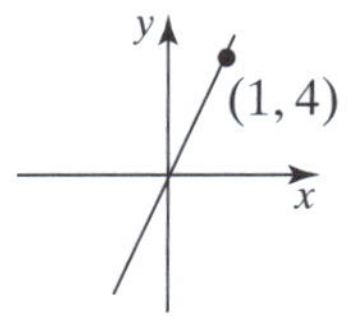

[y-intercept = 0
gradient = 4]

5 **a**

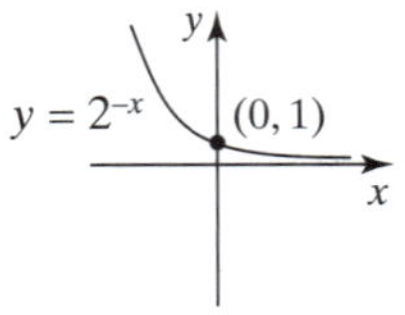

b

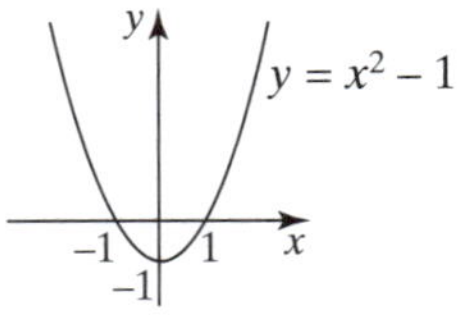

c

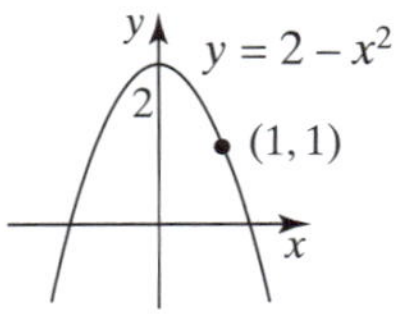

d

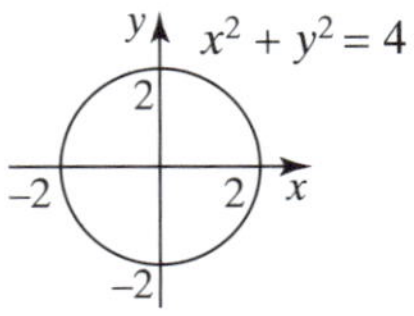

NON-LINEAR GRAPHS
INTERMEDIATE TEST PAGE 59

1 (3, 4) [Try each option.]
A: When $x = 3$ and $y = 4$,
$x^2 + y^2 = 3^2 + 4^2$
$= 9 + 16$
$= 25$
B: $y = 3x - 5$
When $x = 3$,
$y = 3 \times 3 - 5$
$= 4$
C: $y = (x - 1)^2$
When $x = 3$,
$y = (3 - 1)^2$
$= 4$
D: $y = 2^x$
When $x = 3$,
$y = 2^3$
$= 8$ $(\neq 4)$
$\therefore$ **[D]** ✓ (1 mark)

2 $y = -2^x$
When $x = 0$, $y = -2^0$
$= -1$
The curve goes through (0, –1).
When $x = 2$, $y = -2^2$
$= -4$
The curve goes through (2, –4).

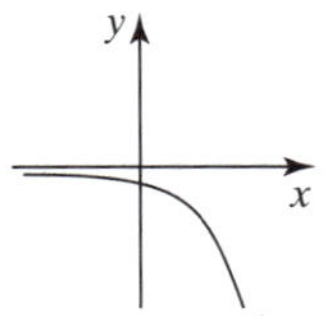

The only possible curve is C.
$\therefore$ **[C]** ✓ (1 mark)

3 $y = 2x^2 - 5$ [Try each option.]
When $x = 0$, $y = 2 \times 0^2 - 5$
$= -5$ (not –3)
So (0, –3) does not lie on the parabola.
When $x = -1$, $y = 2 \times (-1)^2 - 5$
$= -3$ (not –7)
So (–1, –7) does not lie on the parabola.
When $x = 2$, $y = 2 \times 2^2 - 5$
$= 3$ (not 11)
So (2, 11) does not lie on the parabola.
When $x = 5$, $y = 2 \times 5^2 - 5$
$= 45$
So (5, 45) does lie on the parabola.
$\therefore$ **[D]** ✓ (1 mark)

4 $x^2 + y^2 = 25$ [Try each option.]
(0, –5): $0^2 + (-5)^2 = 25$
(–3, –4): $(-3)^2 + (-4)^2 = 9 + 16$
$= 25$
(5, 0): $5^2 + 0^2 = 25$
(– 2, 3): $(-2)^2 + 3^2 = 4 + 9$
$= 13$ (not 25)
The point (–2, 3) does not lie on the circle.
$\therefore$ **[D]** ✓ (1 mark)

5 **a**

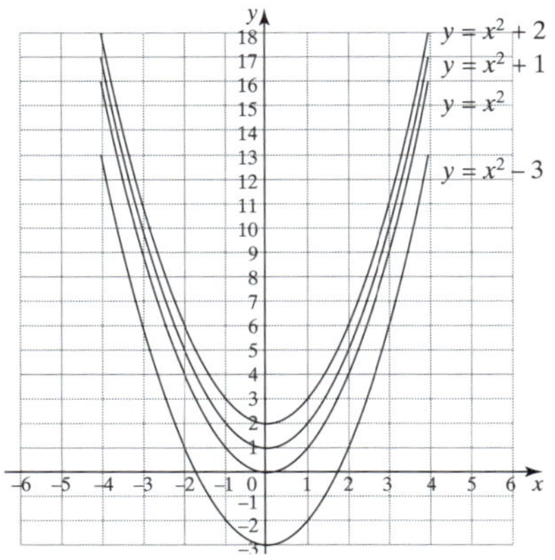

✓✓✓✓ (4 marks)

b The addition of the constant, c, moves the graph up or down the y-axis. ✓
The vertex of the parabola will be at $(0, c)$. ✓(2 marks)

6 **a** $c = 1$ ✓ (1 mark)

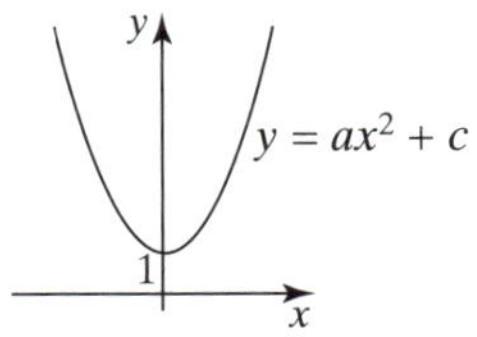

b $y = ax^2 + 1$
When $x = 2$, $y = 3$
$3 = a \times 2^2 + 1$ ✓
$3 = 4a + 1$
$2 = 4a$
$a = \frac{1}{2}$ ✓ (2 marks)

c $y = \frac{1}{2}x^2 + 1$
When $x = -4$,
$y = \frac{1}{2} \times (-4)^2 + 1$ ✓
$= 9$ ✓ (2 marks)

7 **a** $y = 9 - x^2$
When $y = 0$,
$0 = 9 - x^2$
$x^2 = 9$
$x = 3$ or $x = -3$ ✓✓
(2 marks)

b The largest possible value of y is 9. ✓
If $x = 0$, $y = 9$
For all other values of x,
$x^2 > 0$ so $9 - x^2 < 9$. ✓
(2 marks)

c

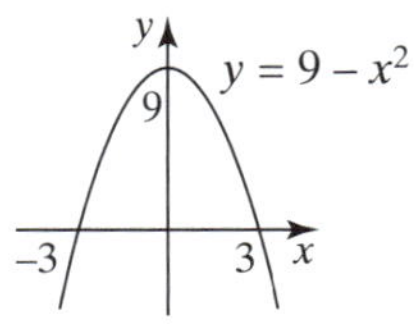

✓✓✓ (3 marks)

8 **a** Let OP be d.

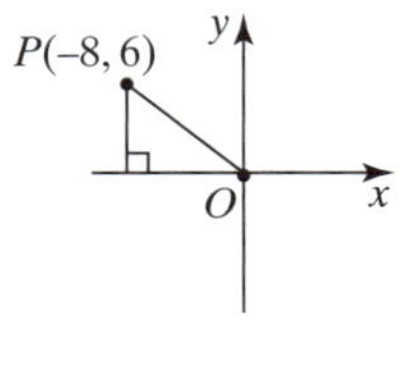

$d^2 = 6^2 + 8^2$
$= 36 + 64$
$= 100$

$d = \sqrt{100} \quad (d > 0)$
$= 10$

$\therefore$ The length of the interval joining O and P is 10 units. ✓ (1 mark)

b A circle with centre at O and passing through P will have a radius of 10 units. ✓
The equation of the circle is
$x^2 + y^2 = 10^2$
$\therefore x^2 + y^2 = 100$ ✓ (2 marks)

(Total: 25 marks)

NON-LINEAR GRAPHS
ADVANCED TEST PAGE 60

1 $y = 2x^2 - 3$

a y-intercept $= -3$ ✓ (1 mark)

b When $x = -2$,
$y = 2 \times (-2)^2 - 3$
$= 5$ ✓ (1 mark)

c When $y = 15$,
$2x^2 - 3 = 15$
$2x^2 = 18$
$x^2 = 9$
$x = 3$ or $x = -3$ ✓✓ (2 marks)

d $y = 2x^2 - 3$ will be a narrower curve than $y = x^2 - 3$. They both have the same y-intercept but $y = x^2 - 3$ is wider. ✓ (1 mark)

2 The curve $x^2 + y^2 = 12\frac{1}{4}$ is a circle, centre the origin. ✓
The radius is $3\frac{1}{2}$. ✓ (2 marks)

3 $y = 3^x$

a When $x = 0$, $y = 3^0$
$= 1$ ✓ (1 mark)

b As x gets large, the y-values get very large. ✓ (1 mark)

c

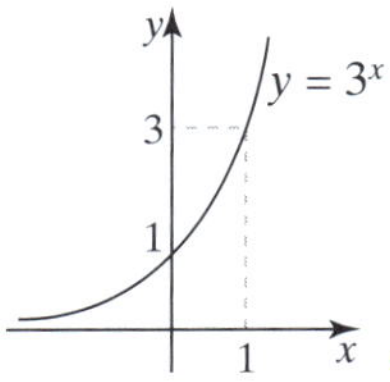

✓

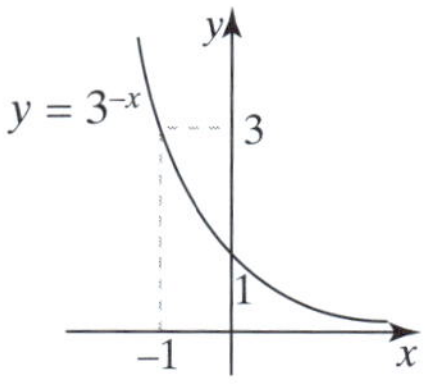

✓

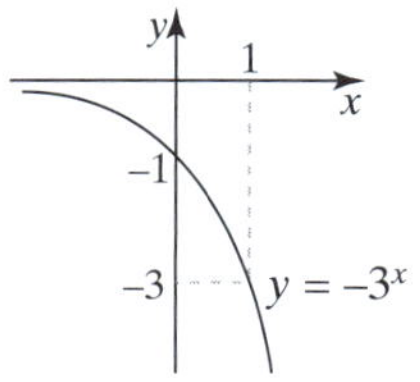

✓

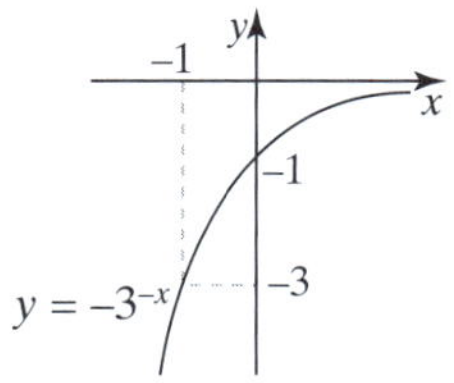

✓ (4 marks)

d Possible answers.
All of the curves are exponential curves and have the same basic shape, although facing different ways. All have y-intercepts 1 unit from the origin, and never cut the x-axis.
The curves $y = 3^x$ and $y = 3^{-x}$ are always positive, while the curves $y = -3^x$ and $y = -3^{-x}$ are always negative.
The curves $y = 3^x$ and $y = -3^x$ are close to the y-axis when $x < 0$, while $y = 3^{-x}$ and $y = -3^{-x}$ are close to the y-axis when $x > 0$. ✓ (1 mark)

4 a $r^2 = 3^2 + (-5)^2$
$= 34$ ✓
The equation is
$x^2 + y^2 = 34$ ✓ (2 marks)

b Q is 6 units from the origin along the x-axis. ✓
Now $6^2 = 36$ and $36 > 34$.
So the distance OQ is greater than the radius.
Q will lie outside the circle. ✓ (2 marks)

5 a The equation is of the form $y = ax^2 + c$.
The y-intercept is 8
So $c = 8$ ✓
$y = ax^2 + 8$
When $x = 2$, $y = 0$
So $a \times 2^2 + 8 = 0$ ✓
$4a + 8 = 0$
$4a = -8$
$a = -2$
The equation is
$y = -2x^2 + 8$. ✓ (3 marks)

b

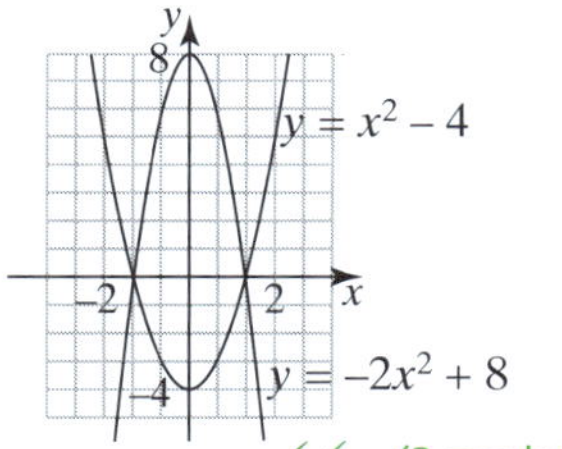

✓✓ (2 marks)

c Some possible answers.
Both curves are parabolas and both have the same x-intercepts. ✓
$y = x^2 - 4$ open upwards while $y = -2x^2 + 8$ opens downwards and is narrower than $y = x^2 - 4$. ✓ (2 marks)

6 a The parabola $y = x^2 + 5$ opens upwards. ✓ (1 mark)

b $x^2 + 5 = 2$
$x^2 = -3$
No solution. ✓ (1 mark)

c The curve $y = x^2 + 5$ is always greater than 2. ✓ (1 mark)

7 Adding a constant moves the curve up or down. Adding a positive constant means the curve will move up and adding a negative constant means the curve will move down. ✓✓ (2 marks)

(Total: 30 marks)

AREA OF COMPOSITE SHAPES
SKILLS CHECK PAGE 62

1 a $A = s^2$
$= 7^2$
$= 49$
$\therefore$ the area is 49 cm^2

b $A = \frac{1}{2}bh$

$= \frac{1}{2} \times 16 \times 7$

$= 56$

$\therefore$ the area is 56 cm^2

c $A = \frac{1}{2}bh$

$= \frac{1}{2} \times 28 \times 11$

$= 154$

$\therefore$ the area is 154 m^2

d $A = lb$

$= 42 \times 11$

$= 462$

Area = 462 m^2

e $A = bh$

$= 22 \times 12$

$= 264$

Area = 264 m^2

[h is the perpendicular height, so in this case it is 12 m.]

f Radius = 34 cm

$A = \pi r^2$

$= \pi \times 34^2$

$= 3631.681\,108\ldots$

Area = 3631.7 cm^2 (1 d.p.)

g $A = \frac{1}{2}xy$

$= \frac{1}{2} \times 17 \times 26$

$= 221$

Area = 221 m^2

h $A = \frac{1}{2}h(a + b)$

$= \frac{1}{2} \times 9 \times (26 + 41)$

$= 301.5$

Area = 301.5 cm^2

2 **a** $A = \frac{1}{2}bh$

$= \frac{1}{2} \times 12 \times 5$

$= 30$

The area is 30 m^2.

b $A = \frac{1}{2}h(a + b)$

$= \frac{1}{2} \times 4 \times (11 + 14)$

$= 50$

The area is 50 m^2.

c Diameter = 32 mm

Radius = 16 mm

$A = \frac{1}{2}\pi r^2$

$= \frac{1}{2} \times \pi \times 16^2$

$= 402.123\,8597\ldots$

Area is 402.1 mm^2 (1 d.p.).

3 **a**

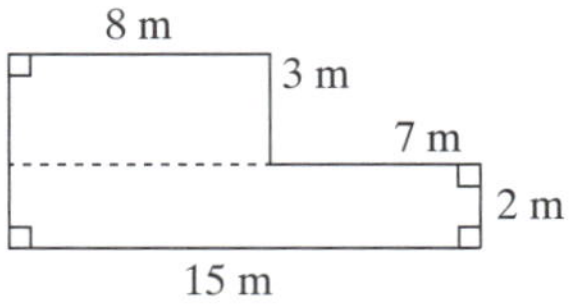

$A = 8 \times 3 + 15 \times 2$

$= 54$

The area is 54 m^2.

b

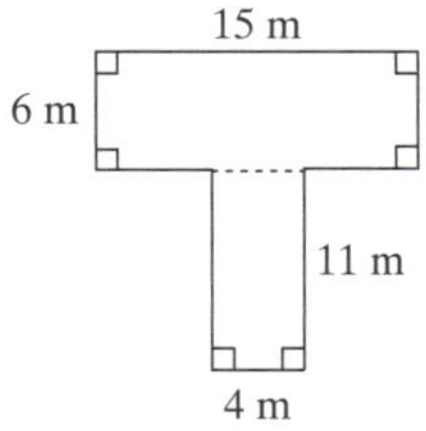

$A = 6 \times 15 + 11 \times 4$

$= 134$

The area is 134 m^2.

c Perpendicular height of triangle = 2.8 m – 1.1 m

= 1.7 m

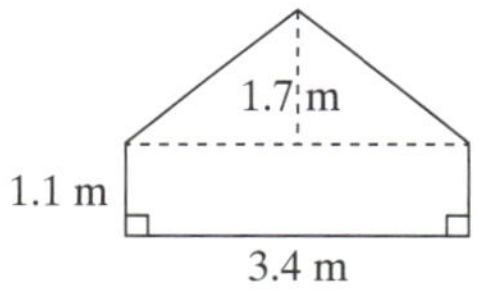

$A = 3.4 \times 1.1 + \frac{1}{2} \times 3.4 \times 1.7$

$= 3.74 + 2.89$

$= 6.63$

The area is 6.63 m^2.

4 **a** Shaded area

= area of outer rectangle

– area of inner rectangle

$A = 19 \times 9 - 11 \times 5$

$= 171 - 55$

$= 116$

The shaded area is 116 m^2.

b Shaded area

= area of square

– area of circle

Radius of circle = 44 mm

$A = 88^2 - \pi \times 44^2$

$= 7744 - 6082.123\,377\ldots$

$= 1661.876\,623\ldots$

$= 1661.9$ (1 d.p.)

Shaded area = 1661.9 mm^2 (1 d.p.)

c Shaded area

= area of trapezium

– area of triangle

$A = \frac{1}{2} \times 16 \times (15 + 19)$

$- \frac{1}{2} \times 19 \times 16$

$= 272 - 152$

$= 120$

Shaded area is 120 cm^2.

5 **a** Shaded part $= \frac{80}{360}$

$= \frac{2}{9}$

b $A = \frac{2}{9}\pi r^2$

$= \frac{2}{9} \times \pi \times 9^2$

$= 56.548\,66\ldots$

$= 56.5$ (1 d.p.)

$\therefore$ the area is 56.5 cm^2 to one decimal place.

AREA OF COMPOSITE SHAPES
INTERMEDIATE TEST PAGE 63

1 $A = 76 \times 23 - 18 \times 11$

$= 1748 - 198$

$= 1550$

The area is 1550 cm^2.

$\therefore$ **[C]** ✓ (1 mark)

2

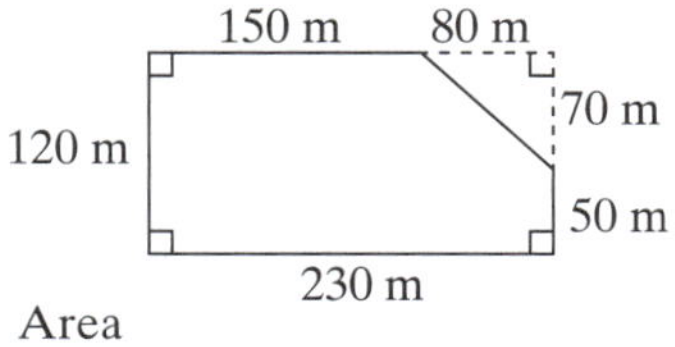

Area

= area of rectangle

– area of triangle

$= (230 \times 120 - \frac{1}{2} \times 80 \times 70)$ m^2

$= 24\,800$ m^2

$= 2.48$ ha

$\therefore$ **[A]** ✓ (1 mark)

3 $A = \frac{60}{360}\pi r^2$

$33.5 = \frac{1}{6} \times \pi \times r^2$

$r^2 = 33.5 \div 0.52359\ldots$

$= 63.9802\ldots$

≈ 64

So $r \approx 8$

The radius is 8 cm.

$\therefore$ **[D]** ✓ (1 mark)

4 Area of pool:

$A = lb$

$= 8 \times 3$

$= 24$

The area of the pool is 24 m². ✓

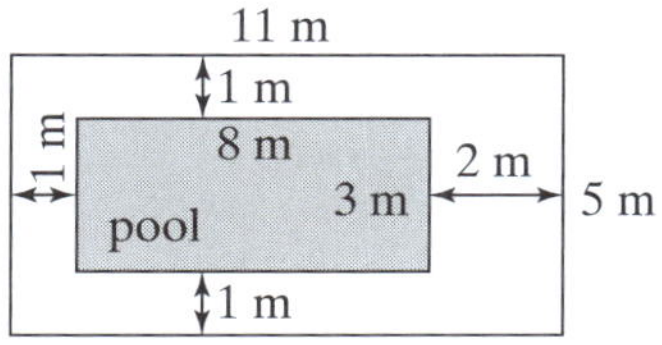

Area enclosed by fence:

$A = lb$

$= 11 \times 5$

$= 55$

The area enclosed by the fence is 55 m². ✓

Area to be tiled = 55 m² − 24 m²

= 31 m² ✓

Cost of tiling = 31 × \$23.00

= \$713.00 ✓

(4 marks)

5 Quadrant:

$A = \frac{1}{4}\pi r^2$

$= \frac{1}{4} \times \pi \times 6^2$

$= 28.274\,333\,88\ldots$ ✓

The area of the quadrant is 28 m² to the nearest square metre.

Triangle:

base = (19 − 6) m = 13 m

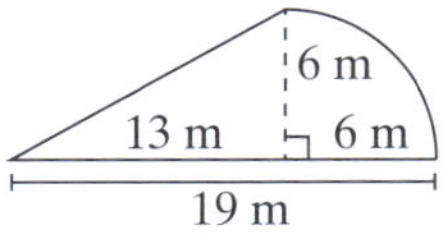

$A = \frac{1}{2}bh$

$= \frac{1}{2} \times 13 \times 6$

$= 39$

The area of the triangle is 39 m². ✓

Total area = (28 + 39) m²

= 67 m² ✓ (3 marks)

6

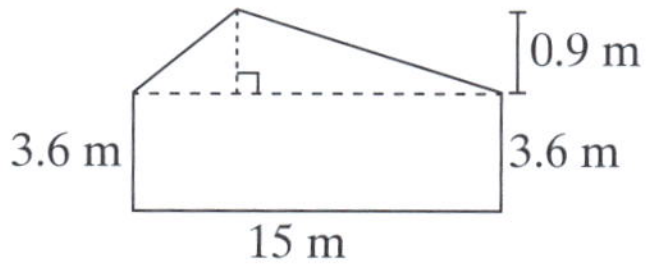

Area of rectangle:

Area = lb

= 15 m × 3.6 m

= 54 m² ✓

Area of triangle:

Area = $\frac{1}{2}bh$

$= \frac{1}{2} \times 15\text{ m} \times 0.9\text{ m}$

= 6.75 m² ✓

Total area = 54 m² + 6.75 m²

= 60.75 m² ✓

(3 marks)

7 Larger circle has radius $R = 3.15$

Smaller circle has radius $r = 2.35$ ✓

$A = \pi R^2 - \pi r^2$

$= \pi \times 3.15^2 - \pi \times 2.35^2$

$= 13.823\,007\ldots$ ✓

$= 13.82$ (2 d.p.)

The area of the annulus is 13.82 cm² to two decimal places. ✓ (3 marks)

8 **a** The area can be found by subtracting the area of the triangle from the area of the rectangle. ✓

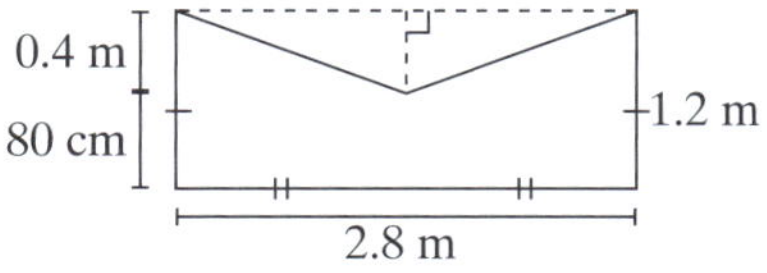

Base of triangle = 2.8 m

Height of triangle

= 1.2 m − 80 cm

= (1.2 − 0.8) m

= 0.4 m

$\therefore A = 2.8 \times 1.2 - \frac{1}{2} \times 2.8 \times 0.4$ ✓

(2 marks)

b The shape is symmetrical. The area is twice the area of the trapezium that makes up half the shape. ✓

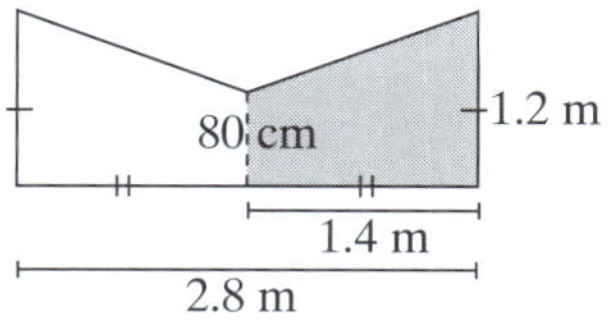

Trapezium:

Area = $\frac{1}{2}h(a + b)$

$\therefore A = 2 \times \frac{1}{2} \times 1.4 \times (1.2 + 0.8)$ ✓

(2 marks)

(Total: 20 marks)

AREA OF COMPOSITE SHAPES
ADVANCED TEST PAGE 64

1 The kite can be broken into 2 congruent right-angled triangles.

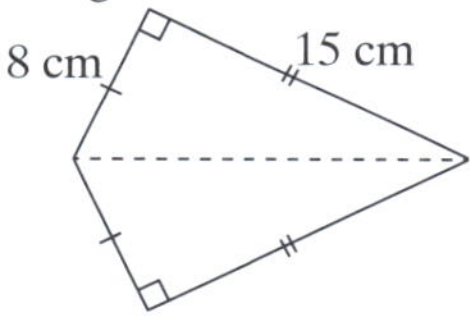

$A = 2 \times \frac{1}{2}bh$ ✓

$= 8 \times 15$

$= 120$

The area is 120 cm². ✓ (2 marks)

2

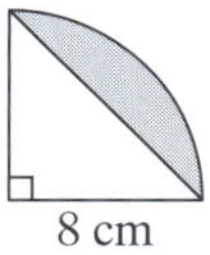

Area of quadrant:

$A = \frac{1}{4}\pi r^2$

$= \frac{1}{4} \times \pi \times 8^2$

$= 50.265\ldots$ cm² ✓

Area of triangle:

$A = \frac{1}{2}bh$

$= \frac{1}{2} \times 8 \times 8$

= 32 cm²

Shaded area

= (50.265 … − 32) cm²

= 18.265 … cm² ✓

Percentage that is shaded

$= \frac{18.265\ldots}{50.265\ldots} \times 100\%$

= 36.338 … %

Approximately 36% of the quadrant is shaded. ✓ (3 marks)

3 **a** $a = b = c = 6$

$s = \dfrac{a+b+c}{2}$

$= \dfrac{6+6+6}{2}$

$= 9$

$A = \sqrt{s(s-a)(s-b)(s-c)}$

$= \sqrt{9(9-6)(9-6)(9-6)}$

$= \sqrt{243}$ cm^2 ✓ (1 mark)

b A regular hexagon is made up of 6 equilateral triangles.

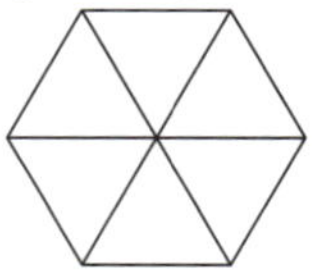

$A = 6 \times \sqrt{243}$

$= 93.5307\ldots$

$= 94$ cm^2

(nearest cm^2) ✓ (1 mark)

4

Radius of larger semicircle $= 7$ cm

Area of larger semicircle:

$A = \dfrac{1}{2}\pi r^2$

$= \dfrac{1}{2} \times \pi \times 7^2$

$= 76.969\ldots$ cm^2 ✓

The radius of smaller semicircles is half that of the larger one. So radius = 3.5 cm.

$A = 2 \times \dfrac{1}{2}\pi r^2$

$= \pi \times 3.5^2$

$= 38.484\ldots$ cm^2 ✓

Total area

$= 76.969\ldots + 38.484\ldots$

$= 115.45353\ldots$

$= 115.5$ cm^2 (1 d.p.) ✓

(3 marks)

5

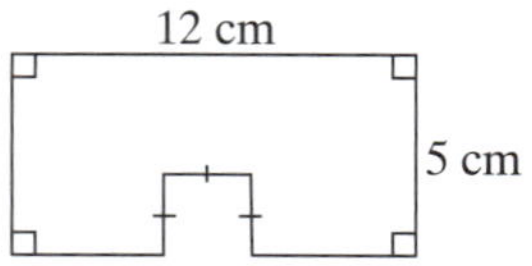

Perimeter = 37 cm

Let the unknown equal sides be x cm in length.

So $2 \times 12 + 2 \times 5 + 2x = 37$

[The whole bottom side, including the missing part of length x cm, will be 12 cm in length.]

$34 + 2x = 37$

$2x = 3$

$x = 1.5$ ✓

Area $= 12 \times 5 - 1.5^2$

$= 57.75$ cm^2 ✓ (2 marks)

6 Area of square:

$A = s^2$

$= 1.2^2$

$= 1.44$

The area of the square is 1.44 m^2. ✓

Semicircle: radius = 0.6 m

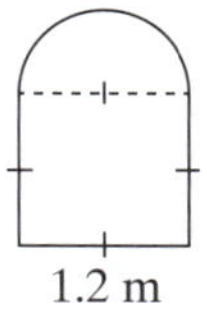

$A = \dfrac{1}{2}\pi r^2$

$= \dfrac{1}{2} \times \pi \times 0.6^2$

$= 0.565486677\ldots$

$= 0.57$ (2 d.p.)

The area of the semicircle is ≈ 0.57 m^2 ✓

Total area ≈ 1.44 m^2 $+ 0.57$ m^2

$= 2.01$ m^2

The area of the window is 2.01 m^2 correct to two decimal places. ✓ (3 marks)

7 $A = 134.3$ cm^2; $R = 7$ cm

$A = \pi(R^2 - r^2)$

$134.3 = \pi \times (7^2 - r^2)$ ✓

$\dfrac{134.3}{\pi} = 49 - r^2$

$r^2 = 49 - \dfrac{134.3}{\pi}$

$= 6.25098\ldots$

$r = \sqrt{6.25098\ldots}$ $(r > 0)$

$= 2.50019\ldots$

The radius of the smaller circle is 2.5 cm. ✓ (2 marks)

8

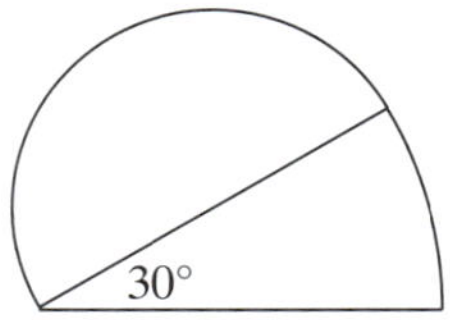

Radius of semicircle = 6 cm

Area of semicircle:

$A = \dfrac{1}{2}\pi r^2$

$= \dfrac{1}{2} \times \pi \times 6^2$

$= 56.5486\ldots$ cm^2 ✓

Sector has radius 12 cm.

[The radius of the sector is equal to the diameter of the semicircle.]

Area of sector:

$A = \dfrac{30}{360}\pi r^2$

$= \dfrac{30}{360} \times \pi \times 12^2$

$= 37.6991\ldots$ cm^2 ✓

Total area

$= 56.5486\ldots + 37.6991\ldots$

$= 94.24777\ldots$

$= 94.2$ cm^2 (1 d.p.) ✓

(3 marks)

9 The shape can be divided into a rectangle and trapezium.

Let x cm be the length of the rectangle.

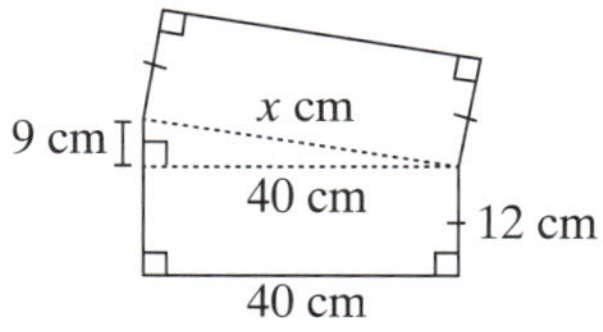

By Pythagoras' theorem:

$x^2 = 9^2 + 40^2$

$= 1681$

$x = \sqrt{1681}$ $(x > 0)$

$= 41$ ✓

Area of rectangle:

$A = lb$

$= 41 \times 12$

$= 492$ cm^2

Area of trapezium:

$A = \dfrac{1}{2}h(a+b)$

$= \dfrac{1}{2} \times 40 \times (21 + 12)$

$= 660$ cm^2

Total area $= (492 + 660)$ cm^2

$= 1152$ cm^2 ✓

(2 marks)

10 The house including the verandah is (14 + 2 × 2.4) m or 18.8 m long and (7.2 + 2 × 2.4) m or 12 m wide.
Area of verandah
$= (18.8 \times 12 - 14 \times 7.2)\ \text{m}^2$
$= 124.8\ \text{m}^2$ ✓

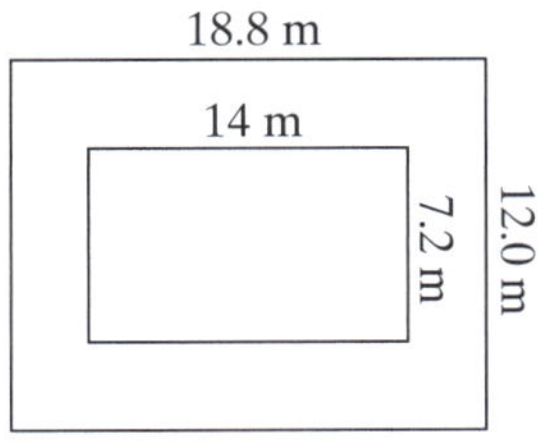

Now 96 mm = 0.096 m
Area of each board
$= (4.6 \times 0.096)\ \text{m}^2$
$= 0.4416\ \text{m}^2$ ✓
Number of boards
$= 124.8 \div 0.4416$
$= 282.6086\ldots$
So the number of boards needed = 283
[Although in practice a few more would be allowed for wastage.] ✓ (3 marks)
(Total: 25 marks)

SURFACE AREA OF PRISMS
SKILLS CHECK PAGE 66

1 a A cube has 6 faces.
b A rectangular prism has 6 faces.
c A triangular prism has 5 faces.

2 a i

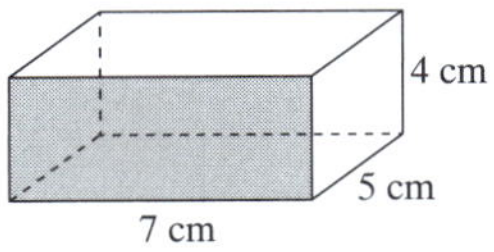

$A = lb$
$= 7 \times 4$
Area = 28 cm²

ii

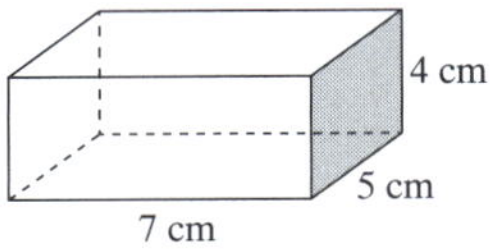

$A = lb$
$= 5 \times 4$
Area = 20 cm²

iii

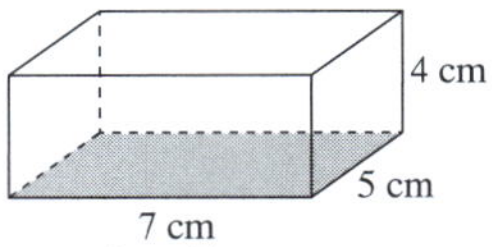

$A = lb$
$= 7 \times 5$
Area = 35 cm²

b [This is the rectangular prism from part **a**.]
$A = 2(28 + 20 + 35)$
$= 2 \times 83$
$= 166$
Surface area = 166 cm²
[Or $A = 2(lb + lh + bh)$
$= 2(7 \times 5 + 7 \times 4 + 5 \times 4)$
$= 2 \times 83$
$= 166$
Surface area = 166 cm²]

3 a cube [Or rectangular prism]
b triangular prism
c triangular prism

4 Rectangle 1:
Area = (4 × 5) cm²
= 20 cm²
Rectangle 2:
Area = (3 × 5) cm²
= 15 cm²
Rectangle 3:
Area = (3 × 4) cm²
= 12 cm²

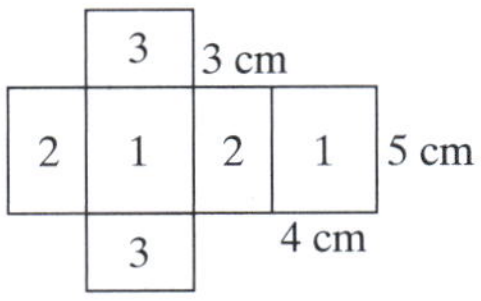

Total surface area
= 2 × (20 + 15 + 12) cm²
= 94 cm²

5 $A = 6s^2$
$= 6 \times 3^2$
$= 54$
Surface area = 54 m²

6 a

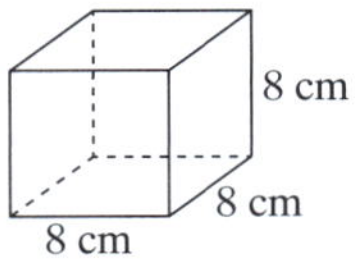

$A = 6s^2$
$= 6 \times 8^2$
$= 384$
Surface area = 384 cm²

b

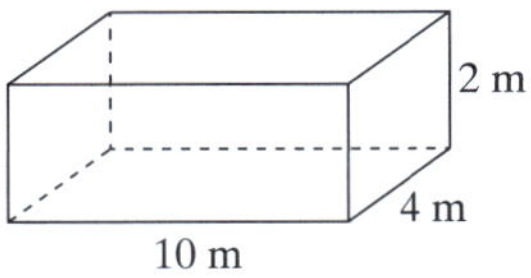

$A = 2(lb + lh + bh)$
$= 2(10 \times 4 + 10 \times 2 + 4 \times 2)$
$= 136$
Surface area = 136 m²

c

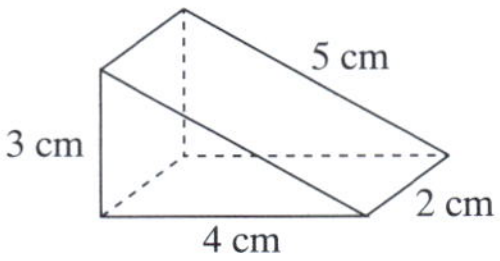

Triangular face:
$A = \frac{1}{2}bh$
$= \frac{1}{2} \times 4 \times 3$
$= 6$
Area = 6 cm²
[The surface area includes the two triangular faces plus three rectangles.]
$A = 2 \times 6 + 5 \times 2 + 3 \times 2 + 4 \times 2$
$= 36$
Surface area = 36 cm²

7 a

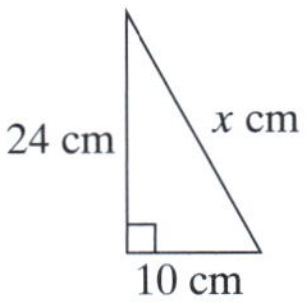

$x^2 = 10^2 + 24^2$
$= 100 + 576$
$= 676$
$x = \sqrt{676}$
$= 26$

b

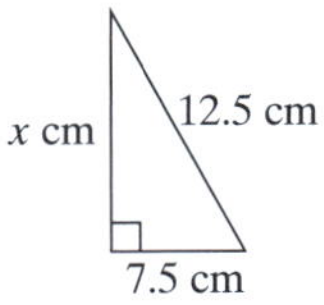

$x^2 = 12.5^2 - 7.5^2$
$= 156.25 - 56.25$
$= 100$
$x = 10$

8

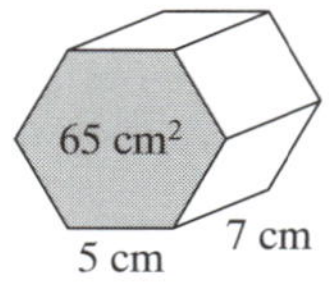

a There are 2 hexagonal faces.

b There are 6 rectangular faces.

c Each rectangular face is 5 cm by 7 cm.
Area = 5 cm × 7 cm
$= 35 \text{ cm}^2$

d Each hexagonal face has area 65 cm^2.
Total area
$= (2 \times 65 + 6 \times 35) \text{ cm}^2$
$= 340 \text{ cm}^2$

SURFACE AREA OF PRISMS
INTERMEDIATE TEST PAGE 67

1 $l = 7.5, \quad b = 4.5, \quad h = 1.5$
$A = 2 \times (7.5 \times 4.5 + 7.5 \times 1.5 + 4.5 \times 1.5)$
$= 103.5$
The surface area is 103.5 m^2.
∴ **[D]** ✓ (1 mark)

2 Find the surface area for each option.
A: $A = 2 \times (7 \times 3 + 7 \times 5 + 3 \times 5)$
$= 142$
B: $A = 2 \times (6 \times 3 + 6 \times 6 + 3 \times 6)$
$= 144$
C: $A = 2 \times (5 \times 5 + 5 \times 5 + 5 \times 5)$
$= 150$
D: $A = 2 \times (3 \times 2 + 3 \times 12 + 2 \times 12)$
$= 132$
The largest surface area is 150 m^2.
∴ **C** ✓ (1 mark)

3 Area of triangle: $A = \frac{1}{2}bh$
$= \frac{1}{2} \times 12 \times 5$
$= 30$
Surface area
$= (2 \times 30 + 12 \times 4 + 5 \times 4 + 13 \times 4) \text{ cm}^2$
$= 180 \text{ cm}^2$
∴ **[B]** ✓ (1 mark)

4 Surface area of a cube: $A = 6s^2$
But $A = 384 \text{ cm}^2$
So $6s^2 = 384$ ✓
$s^2 = 64$
$s = 8 \quad (s > 0)$
The sides are 8 cm long. ✓
(2 marks)

5

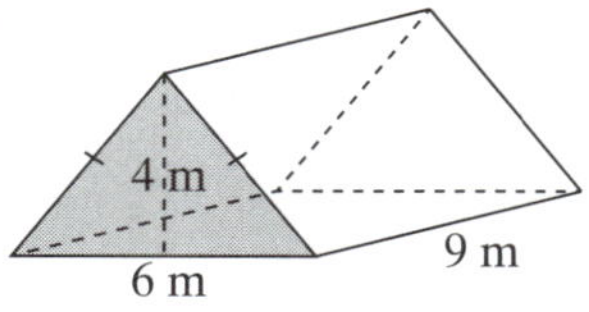

a

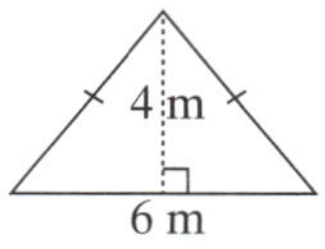

$A = \frac{1}{2}bh$
$= \frac{1}{2} \times 6 \times 4$
$= 12$
Area = 12 m^2 ✓ (1 mark)

b Let the length of the side be x m.

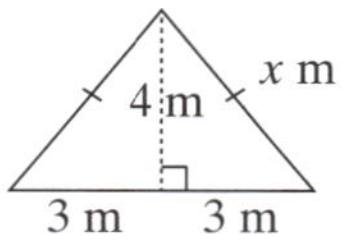

$x^2 = 3^2 + 4^2$ ✓
$= 9 + 16$
$= 25$
$x = \sqrt{25}$
$= 5$
The equal sides are both 5 m long. ✓ (2 marks)

c Area of 2 triangular ends
$= 2 \times 12 \text{ m}^2$
$= 24 \text{ m}^2$ ✓
Area of 2 sloping sides
$= 2 \times 5 \text{ m} \times 9 \text{ m}$
$= 90 \text{ m}^2$ ✓
Area of bottom face
$= 6 \text{ m} \times 9 \text{ m}$
$= 54 \text{ m}^2$
$A = 24 + 90 + 54$
$= 168$
The surface area is 168 m^2. ✓ (3 marks)

6 **a** $A = 2(lh + bh) + lb$ ✓
$= 2(5 \times 2.4 + 3 \times 2.4) + 5 \times 3$
$= 53.4$
The surface area is 53.4 m^2. ✓ (2 marks)

b Area of window
$= 1.5 \text{ m} \times 1 \text{ m}$
$= 1.5 \text{ m}^2$ ✓
Area of door
$= 0.85 \text{ m} \times 2 \text{ m}$
$= 1.7 \text{ m}^2$ ✓
Area to be painted
$= (53.4 - 1.5 - 1.7) \text{ m}^2$
$= 50.2 \text{ m}^2$ ✓ (3 marks)

c Total area of paint for 2 coats $= 50.2 \text{ m}^2 \times 2$
$= 100.4 \text{ m}^2$ ✓
Amount of paint
$= 100.4 \div 15$
$= 6.69333\ldots$
$= 6.7 \quad (1 \text{ d.p.})$
Approximately 6.7 litres of paint will be needed. ✓
(2 marks)

7 **a** Trapezium

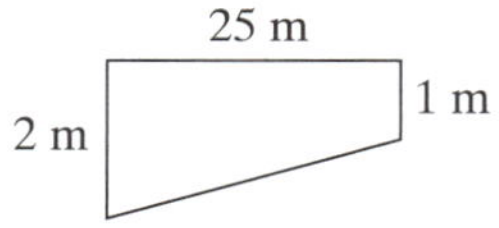

$A = \frac{1}{2}h(a + b)$
$= \frac{1}{2} \times 25 \times (2 + 1)$ ✓
$= 37.5$
The area of the shaded face is 37.5 m^2. ✓ (2 marks)

b Wall at deep end
$= 10 \text{ m} \times 2 \text{ m}$
$= 20 \text{ m}^2$
Wall at shallow end
$= 10 \text{ m} \times 1 \text{m}$
$= 10 \text{ m}^2$ ✓
Total area of walls
$= (20 + 10 + 2 \times 37.5) \text{ m}^2$
$= 105 \text{ m}^2$ ✓ (2 marks)

8 **a**

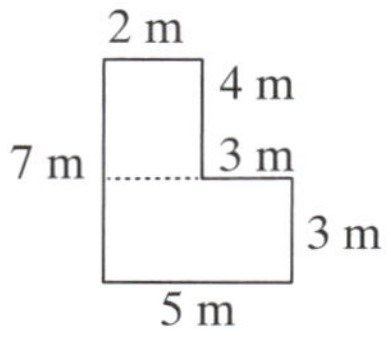

$A = 5 \times 3 + 2 \times 4$
$= 23$
The area of the hexagonal face is 23 m^2. ✓ (1 mark)

b Perimeter
$= (7 + 5 + 3 + 3 + 4 + 2)$ m
$= 24$ m ✓
Area of rectangular faces
$= (24 \times 12)\ m^2$
$= 288\ m^2$
Total surface area
$= (288 + 2 \times 23)\ m^2$
$= 334\ m^2$ ✓ (2 marks)
(Total: 25 marks)

SURFACE AREA OF PRISMS
ADVANCED TEST PAGE 68

1 **a** $A = 6s^2$
$= 6 \times 5^2$
$= 150\ cm^2$ ✓ (1 mark)

b If area doubles it will be $300\ cm^2$.
$6s^2 = 300$
$s^2 = 50$
$s = \sqrt{50}\quad (s > 0)$
$= 7.071\,06\ldots$ cm
Each side will be just over 2 cm longer. ✓ (1 mark)

2 **a**

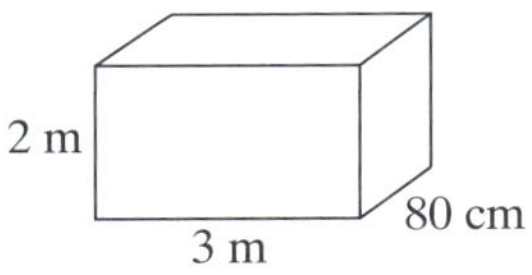

$A = 2(lb + lh + bh)$
$= 2 \times (3 \times 2 + 3 \times 0.8 + 2 \times 0.8)$ ✓
$= 20\ m^2$ ✓ (2 marks)

b

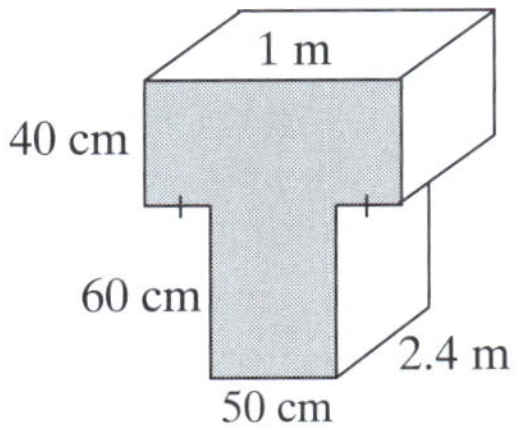

Area of shaded face
$= (1 \times 0.4 + 0.5 \times 0.6)\ m^2$
$= 0.7\ m^2$ ✓
Perimeter of shaded face
$= 4$ m
Surface area
$= (2 \times 0.7 + 4 \times 2.4)\ m^2$
$= 11\ m^2$ ✓ (2 marks)

c

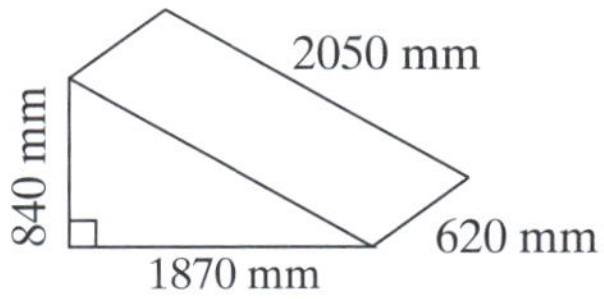

Area of triangle:
$A = \frac{1}{2}bh$
$= \frac{1}{2} \times 1.87 \times 0.84$
$= 0.7854\ m^2$ ✓
Surface area:
$A = 2 \times 0.7854 + (1.87 + 0.84 + 2.05) \times 0.62$
$= 4.522\ m^2$ ✓ (2 marks)

3 **a** $A = 2(lb + lh + bh)$
$= 2 \times (5 \times 4 + 5 \times 3 + 4 \times 3)$
$= 94\ cm^2$ ✓ (1 mark)

b $A = 2 \times (10 \times 8 + 10 \times 6 + 8 \times 6)$
$= 376\ cm^2$ ✓ (1 mark)

c If each dimension doubles the surface area of a rectangular prism will be 4 times larger. ✓ (1 mark)

d $A = 2 \times (15 \times 12 + 15 \times 9 + 12 \times 9)$
$= 846\ cm^2$ ✓ (1 mark)

e If each dimension is multiplied by 3 the surface area of a rectangular prism will be 9 times larger. ✓ (1 mark)

f The surface area will be increased by a factor equal to the square of the number by which each dimension is multiplied. ✓ (1 mark)

g Yes, this will be true for all right prisms. ✓ (1 mark)

4

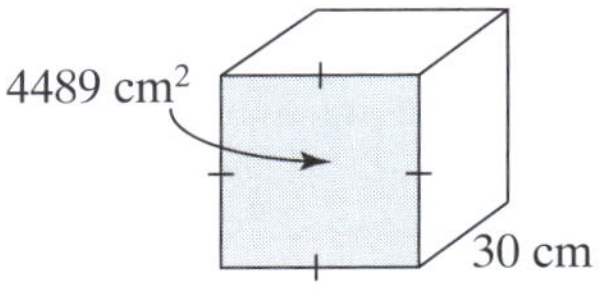

Shaded face is a square.
$s^2 = 4489$
$s = \sqrt{4489}\quad (s > 0)$
$= 67$ ✓

$A = 2 \times 4489 + 4 \times 67 \times 30$
$= 17\,018$
The surface area of the prism is $17\,018\ cm^2$. ✓ (2 marks)

5 **a** Rectangle:
$A = 7.48 \times 3.5$
$= 26.18\ m^2$
Triangle:
height $= (5.18 - 3.5)$ m
$= 1.68$ m ✓
$A = \frac{1}{2} \times 7.48 \times 1.68$
$= 6.2832\ m^2$
Area of pentagon
$= (26.18 + 6.2832)\ m^2$
$= 32.4632\ m^2$ ✓ (2 marks)

b Let x m be the length of the equal sides.

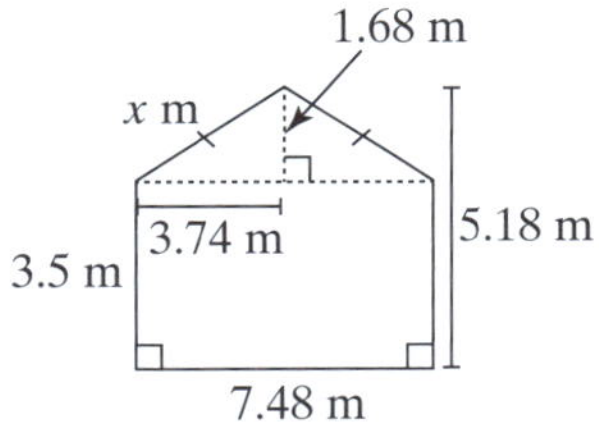

By Pythagoras' theorem:
$x^2 = 1.68^2 + 3.74^2$
$= 16.81$
$x = \sqrt{16.81}\quad (x > 0)$
$= 4.1$
The equal sides are both 4.1 m long. ✓ (1 mark)

c Area of rectangular faces
$= (2 \times 4.1 + 2 \times 3.5 + 7.48) \times 5.4$
$= 122.472\ m^2$ ✓
Total surface area
$= (2 \times 32.4632 + 122.472)\ m^2$
$= 187.3984\ m^2$ ✓ (2 marks)

6

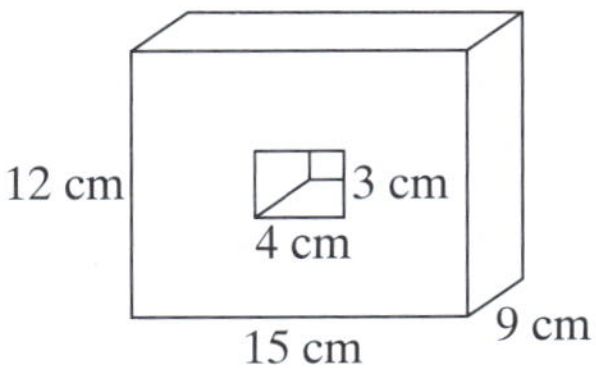

Area of front face
$= 15 \times 12 - 4 \times 3$
$= 168\ cm^2$ ✓
Area of outer rectangles
$= 2 \times 15 \times 9 + 2 \times 12 \times 9$
$= 486\ cm^2$

cont.

Area of inner rectangles
$= 2 \times 4 \times 9 + 2 \times 3 \times 9$
$= 126 \text{ cm}^2$ ✓

Total
$= 2 \times 168 + 486 + 126$
$= 948 \text{ cm}^2$ ✓ **(3 marks)**

(Total: 25 marks)

VOLUME OF PRISMS
SKILLS CHECK — PAGE 70

1 **a** shoe box: cm^3
b shipping container: m^3
c room: m^3
d book: cm^3

2 **a** $1 \text{ m} = 100 \text{ cm}$
$1 \text{ m}^3 = 100 \text{ cm} \times 100 \text{ cm} \times 100 \text{ cm}$
$= 1\,000\,000 \text{ cm}^3$

b $1 \text{ cm} = 10 \text{ mm}$
$1 \text{ cm}^3 = 10 \text{ mm} \times 10 \text{ mm} \times 10 \text{ mm}$
$= 1000 \text{ mm}^3$

3 **a** Parallelogram:
$A = bh$
$= 12 \times 5$
$= 60$
Area $= 60 \text{ m}^2$

b Trapezium:
$A = \frac{h}{2}(a + b)$
$= \frac{3}{2}(6 + 10)$
$= 24$
Area $= 24 \text{ m}^2$

c Rhombus:
$A = \frac{1}{2}xy$
$= \frac{1}{2} \times 11 \times 8$
$= 44$
Area $= 44 \text{ m}^2$

4 **a**

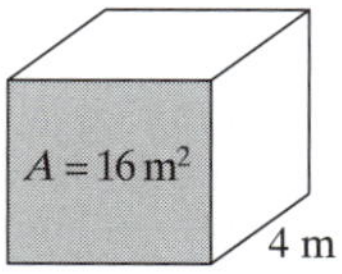

$V = Ah$
$= 16 \times 4$
Volume $= 64 \text{ m}^3$

b

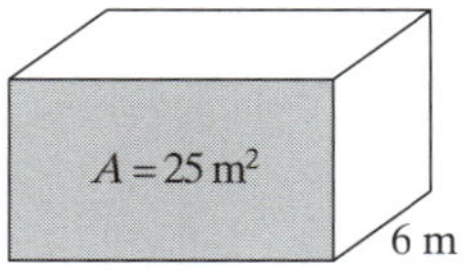

$V = Ah$
$= 25 \times 6$
Volume $= 150 \text{ m}^3$

c

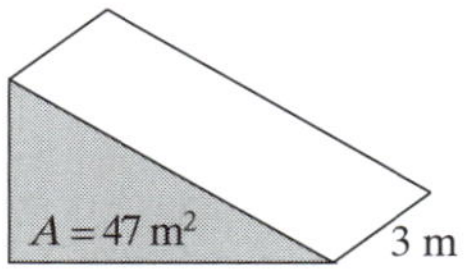

$V = Ah$
$= 47 \times 3$
Volume $= 141 \text{ m}^3$

d

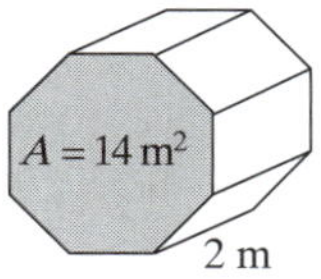

$V = Ah$
$= 14 \times 2$
Volume $= 28 \text{ m}^3$

e

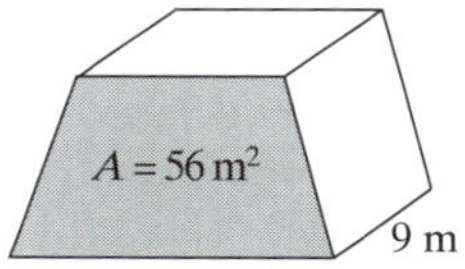

$V = Ah$
$= 56 \times 9$
$= 504$
Volume $= 504 \text{ m}^3$

f

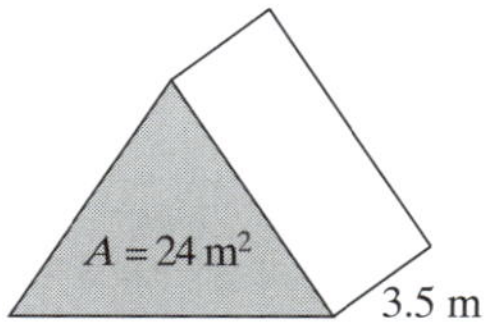

$V = Ah$
$= 24 \times 3.5$
$= 84$
Volume $= 84 \text{ m}^3$

5 By counting, there are 11 cubes in the front row. There are two rows.
volume $= 11 \times 2 \text{ cm}^3$
$= 22 \text{ cm}^3$

6 **a**

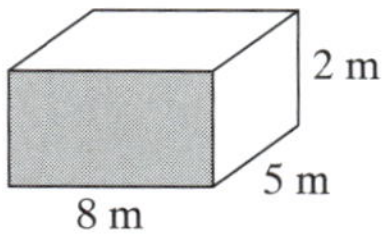

$V = lbh$
$= 8 \times 5 \times 2$
$= 80$
Volume $= 80 \text{ m}^3$

b

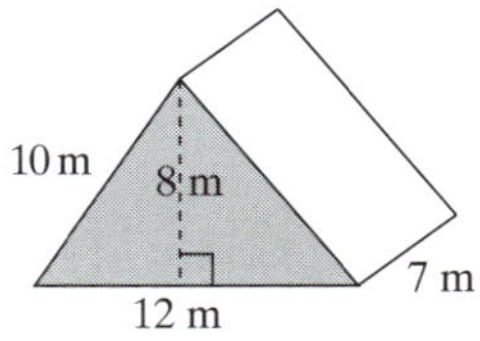

$V = \frac{1}{2}bhH$
$= \frac{1}{2} \times 12 \times 8 \times 7$
$= 336$
Volume $= 336 \text{ m}^3$

c $V = s^3$
$= 18^3$
$= 5832$
Volume $= 5832 \text{ cm}^3$

d

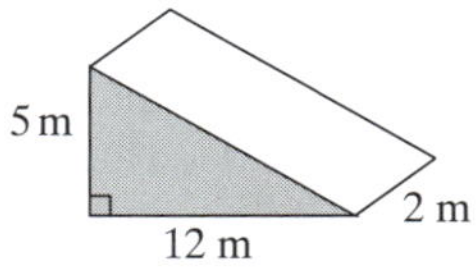

$V = \frac{1}{2}bhH$
$= \frac{1}{2} \times 12 \times 5 \times 2$
$= 60$
Volume $= 60 \text{ m}^3$

e

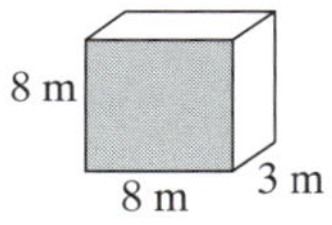

$V = lbh$
$= 8 \times 8 \times 3$
$= 192$
Volume $= 192 \text{ m}^3$

f

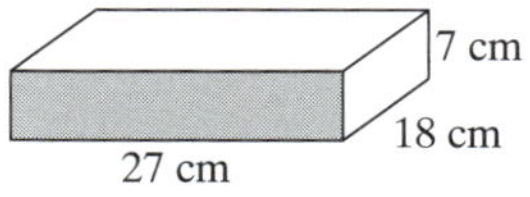

$V = lbh$
$= 27 \times 18 \times 7$
$= 3402$
Volume $= 3402 \text{ cm}^3$

7 **a** $1 \text{ m}^3 = 1000 \text{ L}$
$5.6 \text{ m}^3 = 5600 \text{ L}$

b $1 \text{ cm}^3 = 1 \text{ mL}$
$15 \text{ cm}^3 = 15 \text{ mL}$

VOLUME OF PRISMS
INTERMEDIATE TEST PAGE 71

1 $V = Ah$ only in option B
∴ [B] ✓ (1 mark)

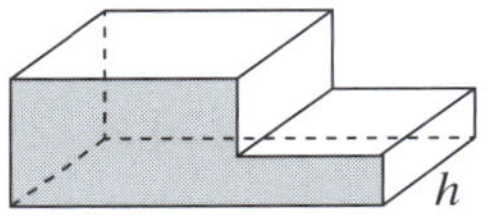

[In A, the solid is a cone, $V = \frac{1}{3}Ah$. In C, h is not the perpendicular height, it is not perpendicular to the base. In D, the shaded face is not the cross-section of the prism.]

2 length = 4.2 m; width = 2.6 m
depth = 30 cm = 0.3 m
$V = lbh$
$= 4.2 \times 2.6 \times 0.3$
$= 3.276$
The volume is closest to $3.3\ \text{m}^3$
∴ [A] ✓ (1 mark)

3 $V = \frac{1}{2} \times 15 \times 8 \times 6$
$= 360$
The volume is 360 cm^3.
∴ [A] ✓ (1 mark)

4 $V = 10 \times 10 \times 20$
$= 2000$
The volume is 2000 cm^3.
Capacity = 2000 mL
= 2 L
∴ [C] ✓ (1 mark)

5 $V = 90,\ h = 15$
$V = Ah$
$90 = A \times 15$ ✓
$= 15A$
$A = 90 \div 15$
$= 6$
The area of the base is 6 cm^2. ✓ (2 marks)

6

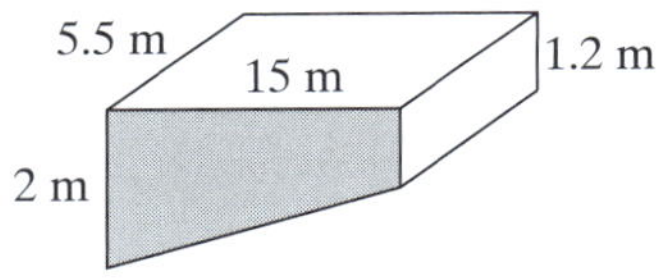

a The cross-section is a trapezium.
$A = \frac{1}{2}h(a + b)$
$= \frac{1}{2} \times 15 \times (2 + 1.2)$ ✓
$= 24$
Area of cross-section is $24\ \text{m}^2$. ✓
$V = AH$
$= 24 \times 5.5$
$= 132$
The volume of the pool is $132\ \text{m}^3$. ✓ (3 marks)
[The height of the trapezium is actually the length of the pool. The height of the prism is the width of the pool.]

b Volume of water
= 132 × 1000 litres
= 132 000 litres [or 132 kL]
✓ (1 mark)

7 **a**

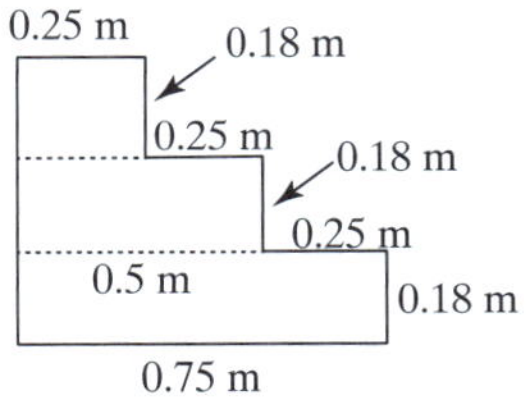

✓

$A = 0.75 \times 0.18 + 0.5 \times 0.18 + 0.25 \times 0.18$
$= 0.27$
The shaded area is $0.27\ \text{m}^2$. ✓ (2 marks)

b $V = Ah$
$= 0.27 \times 1.2$
$= 0.324$ ✓
$= 0.3$ (1 d.p.)
The volume of concrete needed is $0.3\ \text{m}^3$ to one decimal place. ✓ (2 marks)

8 **a** Let h m be the height of the triangle.

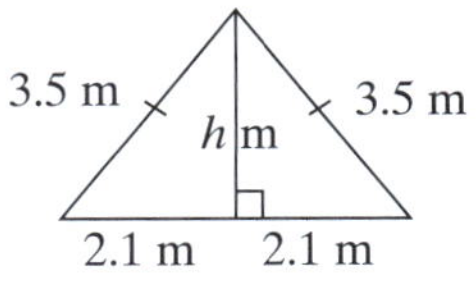

$h^2 = 3.5^2 - 2.1^2$
$= 12.25 - 4.41$
$= 7.84$
$h = \sqrt{7.84}$
$= 2.8$ ✓

$A = \frac{1}{2}bh$
$= \frac{1}{2} \times 4.2 \times 2.8$
$= 5.88$
The area of the shaded face is $5.88\ \text{m}^2$. ✓ (2 marks)

b

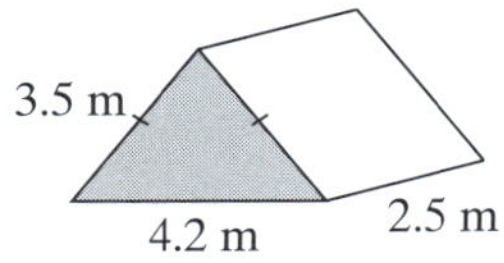

$V = AH$
$= 5.88 \times 2.5$
$= 14.7$
The volume of the prism is $14.7\ \text{m}^3$. ✓ (1 mark)

c

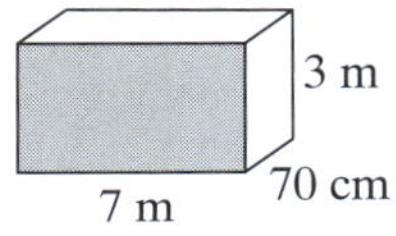

$V = lbh$
$= 7 \times 0.7 \times 3$
$= 14.7$
The volume of the rectangular prism is also $14.7\ \text{m}^3$. ✓ (1 mark)

d Surface area of triangular prism:
$A = 2 \times 5.88 + 2 \times 3.5 \times 2.5 + 4.2 \times 2.5$ ✓
$= 39.76$
The surface area of the triangular prism is $39.76\ \text{m}^2$. ✓
Surface area of rectangular prism:
$A = 2 \times (7 \times 0.7 + 7 \times 3 + 0.7 \times 3)$ ✓
$= 56$
The surface area of the rectangular prism is $56\ \text{m}^2$.
The rectangular prism has the greater surface area. ✓ (4 marks)

9

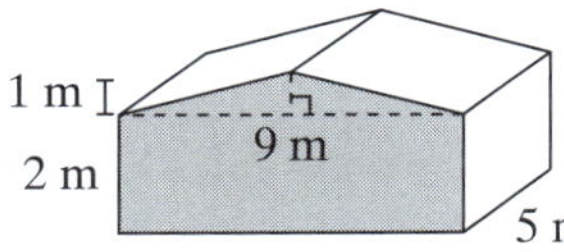

✓

$A = 9 \times 2 + \frac{1}{2} \times 9 \times 1$
$= 22.5$ ✓

cont.

$V = Ah$
$= 22.5 \times 5$
$= 112.5$
The volume is 112.5 m³. ✓
(3 marks)
(Total: 25 marks)

VOLUME OF PRISMS
ADVANCED TEST PAGE 72

1 **a**

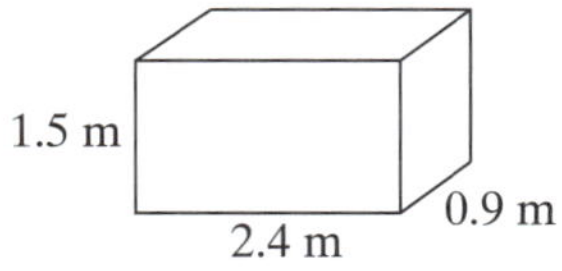

$V = lbh$
$= 2.4 \times 1.5 \times 0.9$ ✓
$= 3.24\ \text{m}^3$ ✓ (2 marks)

b

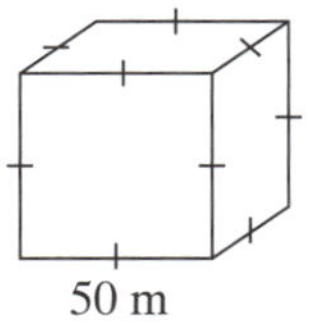

$0.05\ \text{km} = 50\ \text{m}$ ✓
$V = s^3$
$= 50^3$
$= 125\,000\ \text{m}^3$ ✓ (2 marks)

c

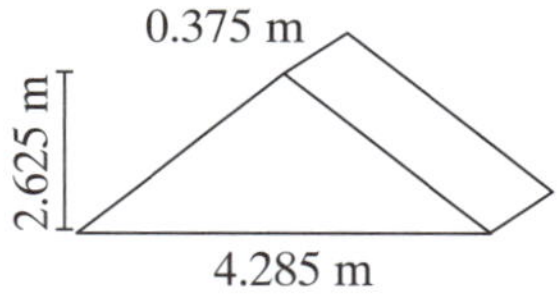

$V = Ah$
$= \frac{1}{2} \times 4.285 \times 2.625 \times 0.375$ ✓
$= 2.109\,023\ldots$
$= 2.1\ \text{m}^3$ (1 d.p.) ✓
(2 marks)

2 **a**

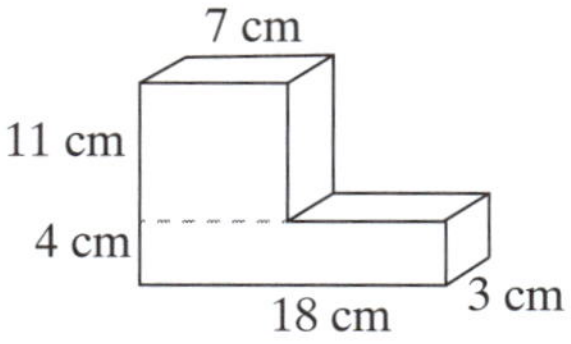

$V = Ah$
$= (18 \times 4 + 7 \times 11) \times 3$ ✓
$= 447\ \text{cm}^3$ ✓ (2 marks)

b $V = Ah$
$= \frac{1}{2} \times 6 \times (7 + 12) \times 9$ ✓
$= 513\ \text{cm}^3$ ✓ (2 marks)

c $A = 5.6 \times 2.4 + \frac{1}{2} \times 5.6 \times 1.6$
$= 17.92$ ✓

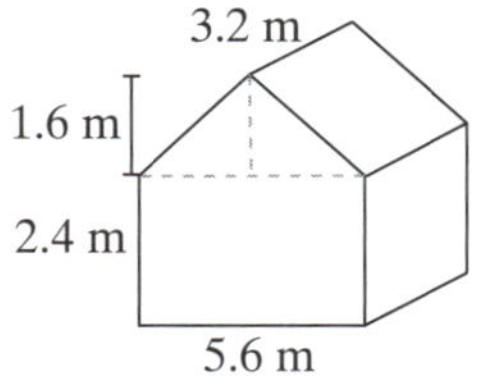

$V = Ah$
$= 17.92 \times 3.2$
$= 57.344\ \text{m}^3$ ✓ (2 marks)

3 Surface area = 433.5 cm²
$6s^2 = 433.5$
$s^2 = 72.25$
$s = \sqrt{72.25}$ $(s > 0)$
$= 8.5$ ✓
$V = s^3$
$= 8.5^3$
$= 614.125\ \text{cm}^3$ ✓ (2 marks)

4 **a** $600\ \text{L} = 0.6\ \text{m}^3$ ✓
$V = lbh$
$0.6 = 2.5 \times 0.6 \times h$
$= 1.5h$
$h = 0.4$
The height is 40 cm ✓
(2 marks)

b Extra height $= (55 - 40)$ cm
$= 15$ cm
$= 0.15$ m
Extra volume
$= 2.5 \times 0.6 \times 0.15$
$= 0.225\ \text{m}^3$ ✓
Extra capacity
$= 0.225 \times 1000$ L
$= 225$ L ✓ (2 marks)

5 The base is made up of 3 rectangular prisms.
Bottom prism:
height = 25 cm = 0.25 m
$V = lbh$
$= 2.5 \times 2 \times 0.25$
$= 1.25\ \text{m}^3$ ✓
Middle prism:
height = 40 cm = 0.4 m
$V = 1.5 \times 1 \times 0.4$
$= 0.6\ \text{m}^3$ ✓
Top prism:
height = 20 cm = 0.2 m
$V = 0.75 \times 0.3 \times 0.2$
$= 0.045\ \text{m}^3$
Total $= (1.25 + 0.6 + 0.045)\ \text{m}^3$
$= 1.895\ \text{m}^3$ ✓ (3 marks)

6 **a** $x^2 + x^2 = 8^2$
$2x^2 = 64$
$x^2 = 32$
$x = \sqrt{32}$ $(x > 0)$
$= 5.656\,854\ldots$
$= 5.657$ (3 d.p.) ✓
(1 mark)

b $A = \frac{1}{2}h(a + b)$
$= \frac{1}{2} \times 5.657 \times (8 + (8 + 2 \times 5.657))$
$= 77.257\,649$
$= 77.26\ \text{cm}^2$ (2 d.p.) ✓
(1 mark)

c A regular octagon can be divided into two trapeziums and a rectangle.

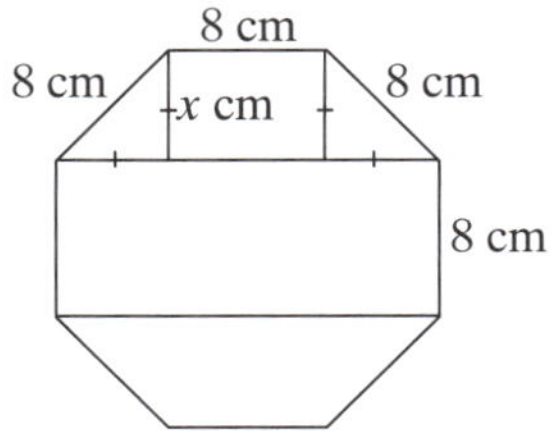

$A = 2 \times 77.26 + (8 + 2 \times 5.657) \times 8$
$= 309.032$
$= 309\ \text{cm}^2$
(nearest cm²) ✓ (1 mark)

d $V = Ah$
$= 309 \times 8$
$= 2472\ \text{cm}^3$ ✓ (1 mark)
(Total: 25 marks)

SURFACE AREA AND VOLUME OF CYLINDERS
SKILLS CHECK PAGE 74

1 $r = 7.6$
$d = 2r$
$= 2 \times 7.6$
$= 15.2$
∴ the length of the diameter is 15.2 cm.

2 $d = 856$
$r = d \div 2$
$= 856 \div 2$
$= 428$
∴ the length of the radius is 428 mm.

3 **a** $r = 8$
$C = 2\pi r$
$= 2 \times \pi \times 8$
$= 50.265\,482\,46\ldots$
$\therefore$ circumference = 50.3 cm (1 d.p.)

b $d = 1.5$
$C = \pi d$
$= \pi \times 1.5$
$= 4.712\,388\,98\ldots$
$\therefore$ circumference = 4.7 m (1 d.p.)

4 **a** $r = 9$
$A = \pi r^2$
$= \pi \times 9^2$
$= 254.469\,0049\ldots$
$\therefore$ area = 254.5 m^2 (1 d.p.)

b $d = 3.6$, $r = 1.8$
$A = \pi r^2$
$= \pi \times 1.8^2$
$= 10.178\,7602\ldots$
$\therefore$ area = 10.2 m^2 (1 d.p.)

5 **a** $d = 12$
$C = \pi d$
$= \pi \times 12$
$= 37.699\,111\,84\ldots$
$\therefore$ circumference = 37.7 cm (1 d.p.)

b $r = 6$
$A = \pi r^2$
$= \pi \times 6^2$
$= 113.097\,3355\ldots$
$\therefore$ area = 113.1 cm^2 (1 d.p.)

6 **a** $A = lb$
$= 9.6 \times 3.7$
$= 35.52$
$\therefore$ area = 35.5 cm^2 (1 d.p.)

b $r = 11$
$A = \pi r^2$
$= \pi \times 11^2$
$= 380.132\,7111\ldots$
$\therefore$ area = 380.1 m^2 (1 d.p.)

c $d = 35$, $r = 17.5$
$A = \pi r^2$
$= \pi \times 17.5^2$
$= 962.112\,7502\ldots$
$\therefore$ area = 962.1 cm^2 (1 d.p.)

d $d = 73$, $r = 36.5$
$A = \frac{1}{2}\pi r^2$
$= \frac{1}{2} \times \pi \times 36.5^2$
$= 2092.693\,406\ldots$
$\therefore$ area = 2092.7 cm^2 (1 d.p.)

e $A = \frac{1}{4}\pi r^2$
$= \frac{1}{4} \times \pi \times 28^2$
$= 615.752\,1601\ldots$
$\therefore$ area = 615.8 cm^2 (1 d.p.)

f $A = \frac{1}{2}\pi r^2$
$= \frac{1}{2} \times \pi \times 32^2$
$= 1608.495\,439\ldots$
$\therefore$ area = 1608.5 cm^2 (1 d.p.)

7 $C = \frac{1}{2}\pi d$
$= \frac{1}{2} \times \pi \times 2.1$
$= 3.298\,672\,286\ldots$
$= 3.3$ m (1 d.p.)
Perimeter = 3.3 m + 2.1 m
= 5.4 m (1 d.p.)

8

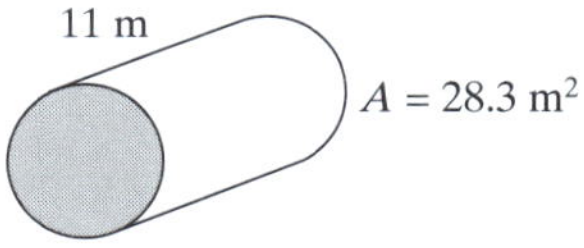

$V = Ah$
$= 28.3 \times 11$
$= 311.3$
Volume = 311.3 m^3

9 $V = \pi r^2 h$
$= \pi \times 9^2 \times 2$
$= 508.938\,0099\ldots$
Volume = 508.9 m^3 (1 d.p.)

10 **a** $S = 2\pi rh$
$= 2 \times \pi \times 4 \times 9$
$= 226.194\,6711\ldots$
Surface area = 226.2 cm^2 (1 d.p.)

b $A = \pi r^2$
$= \pi \times 4^2$
$= 50.265\,482\,46\ldots$
Area = 50.3 cm^2 (1 d.p.)

c $A = 226.194\,6711\ldots + 50.265\,482\,46\ldots$
$= 276.460\,1535\ldots$
Surface area = 276.5 cm^2 (1 d.p.)

d $A = 276.460\,1535\ldots + 50.265\,482\,46\ldots$
$= 326.725\,636\ldots$
Total surface area
= 326.7 cm^2 (1 d.p.)

SURFACE AREA AND VOLUME OF CYLINDERS
INTERMEDIATE TEST PAGE 75

1 $V = \pi r^2 h$
[Try each option.]
A: $V = \pi \times 1.2^2 \times 1.5$
$= 6.785\,84\ldots$
B: $V = \pi \times 1.6^2 \times 0.8$
$= 6.433\,98\ldots$
C: $V = \pi \times 0.9^2 \times 2.6$
$= 6.616\,19\ldots$
D: $V = \pi \times 1.05^2 \times 1.8$
$= 6.234\,49\ldots$
The largest volume is 6.785 84... m^3.
$\therefore$ [A] ✓ (1 mark)

2 Radius = 40 cm = 0.4 m
$A = 2\pi rh$
$= 2 \times \pi \times 0.4 \times 12$
$= 30.159\,289\ldots$
$= 30.16$ (2 d.p.)
The external surface area is 30.16 m^2 to two decimal places.
$\therefore$ [A] ✓ (1 mark)

3 $A = 2\pi r^2 + 2\pi rh$
Try each option.
A: $A = 2 \times \pi \times 5^2 + 2 \times \pi \times 5 \times 28$
$= 1036.7255\ldots$
B: $A = 2 \times \pi \times 6^2 + 2 \times \pi \times 6 \times 22$
$= 1055.5751\ldots$
C: $A = 2 \times \pi \times 7^2 + 2 \times \pi \times 7 \times 18$
$= 1099.5574\ldots$
D: $A = 2 \times \pi \times 8^2 + 2 \times \pi \times 8 \times 13$
$= 1055.5751\ldots$
The largest surface area is 1099.5574... m^2.
$\therefore$ [C] ✓ (1 mark)

4 Diameter = 60 cm = 0.6 m
radius = 0.3 m, height 0.9 m

a $V = \pi r^2 h$
$= \pi \times 0.3^2 \times 0.9$ ✓
$= 0.254\,469\,004\ldots$
$= 0.25$ (2 d.p.)
The volume of the drum is 0.25 m^3, to two decimal places. ✓ (2 marks)

b Number of drums
$= 1$ m$^3 \div 0.254\,469\ldots$ m^3
$= 3.929\,751\,681\ldots$
Approximately four drums would be needed to hold 1 m^3 of sand. ✓ (1 mark)

5 $A = \pi r^2 + 2\pi rh$
$= \pi \times 1.2^2 + 2 \times \pi \times 1.2 \times 0.2$ ✓
$= 6.031\,857\ldots$
$= 6.0$ (1 d.p.)
The amount of iron required is $6.0\ \text{m}^2$ to one decimal place. ✓
(2 marks)

6 **a** Capacity = 22 700 L
Volume = $22.7\ \text{m}^3$
[dividing by 1000] ✓
(1 mark)

b $V = 22.7,\ r = 1.7$
$V = \pi r^2 h$
$22.7 = \pi \times 1.7^2 \times h$ ✓
$= 9.079\,2027\ldots \times h$
$h = 22.7 \div 9.079\,2027\ldots$ ✓
$= 2.500\,2195\ldots$
$= 2.5$ (1 d.p.)
The height of the tank is 2.5 metres to one decimal place. ✓ (3 marks)

7 **a** $A = 2\pi r^2 + 2\pi rh$
$= 2 \times \pi \times 5^2 + 2 \times \pi \times 5 \times 18$ ✓
$= 722.566\,3103\ldots$
$= 722.6$ (1 d.p.)
The surface area is $722.6\ \text{m}^2$ to one decimal place. ✓
(2 marks)

b Area of four walls of rectangular prism:
$A = 2(lh + bh)$
$= 2 \times (18 \times 3 + 10 \times 3)$
$= 168$
Area of 4 walls is $168\ \text{m}^2$. ✓
Area of roof is half the area of a cylinder of radius 5 m and height 18 m.
$A = \frac{1}{2} \times 722.566\,31\ldots\ \text{m}^2$
[from part **a**]
$= 361.3\ \text{m}^2$ (1 d.p.) ✓
Total area
$= 168\ \text{m}^2 + 361.3\ \text{m}^2$
$= 529.3\ \text{m}^2$ ✓ (3 marks)

8 $V = \frac{1}{4}\pi r^2 h$
$= \frac{1}{4} \times \pi \times 20^2 \times 44$ ✓
$= 13\,823.007\,68\ldots$ ✓
$= 13\,823$ (nearest unit)
The volume is $13\,823\ \text{cm}^3$ to nearest cubic centimetre. ✓
(3 marks)
(Total: 20 marks)

SURFACE AREA AND VOLUME OF CYLINDERS
ADVANCED TEST PAGE 76

1 **a** Curved surface:
$A = 2\pi rh$
$226.2 = 2 \times \pi \times r \times 8$
$r = \frac{226.2}{16\pi}$
$= 4.500\,10\ldots$
The radius is 4.5 cm. ✓
$V = \pi r^2 h$
$= \pi \times 4.5^2 \times 8$
$= 508.938\ldots$
$= 508.9\ \text{cm}^3$ (1 d.p.) ✓
(2 marks)

b Closed cylinder:
$A = 2\pi rh + 2\pi r^2$
$= 226.2 + 2 \times \pi \times 4.5^2$
$= 353.4345\ldots$
$= 353.4\ \text{cm}^2$ (1 d.p.) ✓
(1 mark)

2 **a** diameter = 2.6 m
radius = 1.3 m
$V = \pi r^2 h$
$= \pi \times 1.3^2 \times 2$
$= 10.618\,583\ldots\ \text{m}^3$ ✓
Capacity
$= 10.618\,583\ldots \times 1000$ L
$= 10\,618.583\,17\ldots$ L
$= 10\,619$ L
(nearest litre) ✓ (2 marks)

b Difference = 619 L
% difference
$\approx \frac{619}{10\,619} \times 100\%$
$= 5.829\ldots\%$
$= 5.8\%$ (1 d.p.) ✓ (1 mark)

c Possible answers:
The tank cannot be filled to the very top.
The sides of the tank might be corrugated not flat. ✓
(1 mark)

3 $h = 0.825$ m
$V = \pi r^2 h$
$0.13 = \pi r^2 \times 0.825$ ✓
$r^2 = \frac{0.13}{\pi \times 0.825}$
$= 0.050\,15\ldots$
$r = \sqrt{0.050\,15\ldots}$ $(r > 0)$
$= 0.223\,959\ldots$ ✓
$A = 2\pi rh + \pi r^2$
$= 2 \times \pi \times 0.2239\ldots \times 0.825 + \pi \times (0.223\,959\ldots)^2$
$= 1.318\,499\ldots$
$= 1.3\ \text{m}^2$ (1 d.p.) ✓
(3 marks)

4 **a** $V = \pi r^2 h$
$= \pi \times 55^2 \times 115$
$= 1\,092\,881.54\ldots\ \text{cm}^3$
$= 1\,092\,900\ \text{cm}^3$
(nearest $100\ \text{cm}^3$) ✓
(1 mark)

b 0.6 t = 600 kg
= 600 000 g ✓
Density $= \frac{600\,000}{1\,092\,900}$
$\approx 0.55\ \text{g/cm}^3$ ✓
(2 marks)

c $A = 2\pi rh + 2\pi r^2$
$= 2 \times \pi \times 0.55 \times 1.15 + 2 \times \pi \times 0.55^2$
$= 5.874\,778\ldots$
$= 5.9\ \text{m}^2$ (1 d.p.) ✓
(1 mark)

d 5 layers:
$5 \times 5.9 = 29.5$
So each bale requires about $30\ \text{m}^2$ of wrap.
Area of wrap in packet
$= (1500 \times 0.75)\ \text{m}^2$
$= 1125\ \text{m}^2$ ✓
Number of bales per packet
$= 1125 \div 30$
$= 37.5$
So around 38 bales could be wrapped from each packet of wrap. ✓ (2 marks)

5 **a** Outer curved surface:
$A = 2\pi Rh$
$= 2 \times \pi \times 8 \times 19$
$= 955.044\,16\ldots\ \text{cm}^2$ ✓
Inner curved surface:
$A = 2\pi rh$
$= 2 \times \pi \times 3 \times 19$
$= 358.141\,56\ldots\ \text{cm}^2$ ✓

Top:
$A = \pi(R^2 - r^2)$
$= \pi(8^2 - 3^2)$
$= 172.7875\ldots \text{ cm}^2$
Total
$= 955.044\ldots + 358.141\ldots + 2 \times 172.787\ldots$
$= 1658.7609\ldots$
$= 1659 \text{ cm}^2$
(nearest cm^2) ✓ (3 marks)

b

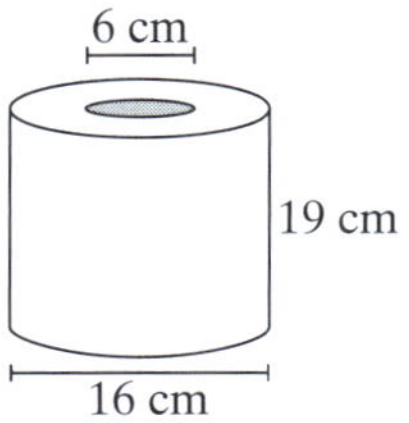

$V = Ah$
$= \pi \times (8^2 - 3^2) \times 19$ ✓
$= 3282.964\ldots$
$= 3283 \text{ cm}^3$
(nearest cm^3) ✓ (2 marks)

6 **a**

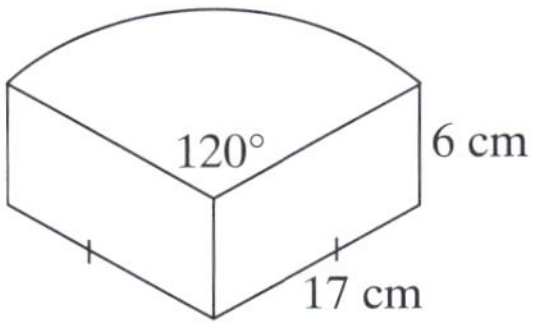

Area top:
$A = \frac{120}{360} \times \pi r^2$
$= \frac{1}{3} \times \pi \times 17^2$
$= 302.64009\ldots \text{ cm}^2$ ✓
Curved surface:
$A = \frac{120}{360} \times 2\pi rh$
$= \frac{1}{3} \times 2 \times \pi \times 17 \times 6$
$= 213.6283\ldots \text{ cm}^2$ ✓
Total surface area:
$A = 2 \times 302.64\ldots + 213.628\ldots + 2 \times 17 \times 6$
$= 1022.9084\ldots$
$= 1022.9 \text{ cm}^2$ (1 d.p.) ✓
(3 marks)

b

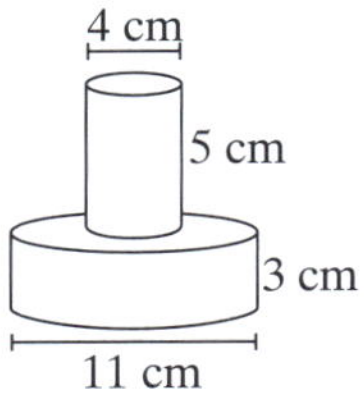

Bottom cylinder:
radius = 5.5 cm
[The circle at the top of top cylinder can be added to the bottom cylinder so that it is a complete closed cylinder.]
$A = 2\pi rh + 2\pi r^2$
$= 2 \times \pi \times 5.5 \times 3 + 2 \times \pi \times 5.5^2$
$= 293.7389\ldots \text{ cm}^2$ ✓
Top cylinder is now open.
radius = 2 cm
$A = 2\pi rh$
$= 2 \times \pi \times 2 \times 5$
$= 62.83185\ldots \text{ cm}^2$ ✓
Total surface area:
$= 293.7389\ldots + 62.83185\ldots$
$= 356.5707\ldots$
$= 356.6 \text{ cm}^2$ (1 d.p.) ✓
(3 marks)

7 radius = 25 cm
= 0.25 m
$V = Ah$
$= \frac{1}{2} \times \pi \times 0.25^2 \times 3.2$ ✓
$= 0.314159\ldots \text{ m}^3$ ✓
Capacity
$= 0.314159\ldots \times 1000 \text{ L}$
$= 314 \text{ L}$ (nearest litre) ✓
(3 marks)
(Total: 30 marks)

ENLARGEMENT AND SIMILARITY
SKILLS CHECK PAGE 78

1 **a** i $AB = 30$ mm
ii $BC = 25$ mm
iii $CD = 46$ mm
iv $DA = 22$ mm

b i $EF = 60$ mm
ii $FG = 50$ mm
iii $GH = 92$ mm
iv $HE = 44$ mm

c Each side of $EFGH$ is twice as long as the corresponding side of $ABCD$.

d i $\angle DAB = 100°$
ii $\angle ABC = 120°$
iii $\angle BCD = 60°$
iv $\angle CDA = 80°$

e i $\angle HEF = 100°$
ii $\angle EFG = 120°$
iii $\angle FGH = 60°$
iv $\angle GHE = 80°$

f The corresponding angles are equal.

2 **a** True. [All squares are exactly the same shape (but not necessarily the same size).]

b False. [A rectangle could be long and thin or short and wide. They are not exactly the same shape.]

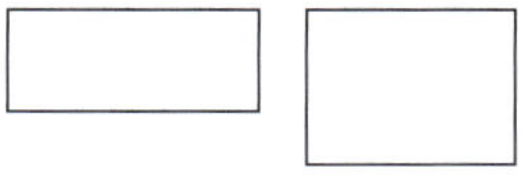

c True. [All circles are exactly the same shape (but not necessarily the same size).]

d False. [All regular hexagons are similar, but hexagons that are not regular are not necessarily the same shape.]

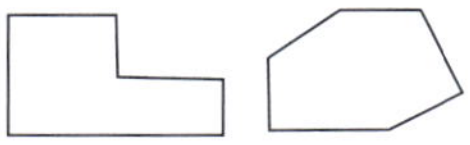

3

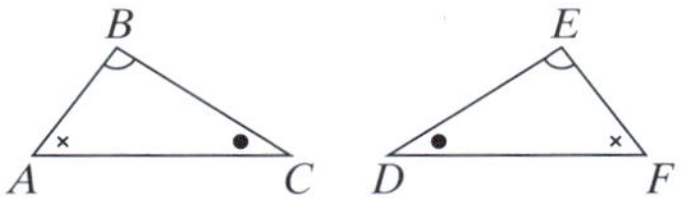

a AB corresponds to FE.
b BC corresponds to ED.
c AC corresponds to FD.

4 **a** $\Delta ABC \ ||| \ \Delta DFE$

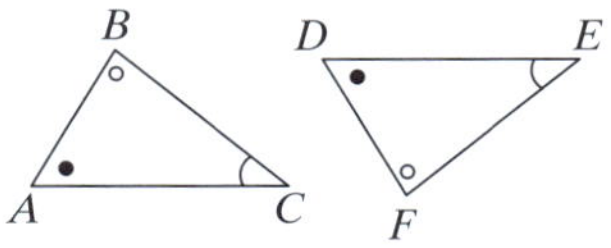

[$\angle A$ corresponds to $\angle D$, $\angle B$ corresponds to $\angle F$ and $\angle C$ corresponds to $\angle E$.]

b $\Delta PQR \ ||| \ \Delta TSR$

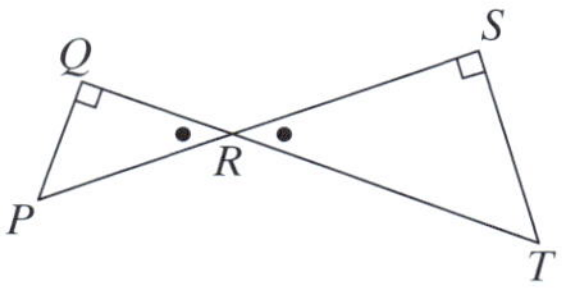

[$\angle P$ corresponds to $\angle T$, $\angle Q$ corresponds to $\angle S$ and $\angle PRQ$ corresponds to $\angle TRS$.]

c $\Delta JKL \parallel\!| \Delta LJM$

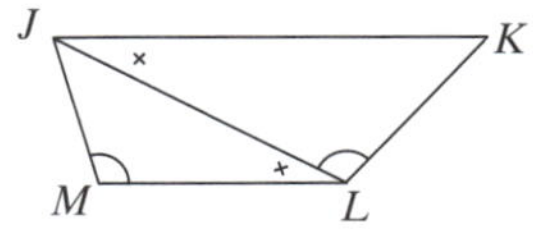

[$\angle KJL$ corresponds to $\angle JLM$, $\angle JKL$ corresponds to $\angle LJM$ and $\angle JLK$ corresponds to $\angle LMJ$.]

5 **a** True

$\angle ABC = 180° - 80° - 50°$ (angle sum of a triangle)

$= 50°$

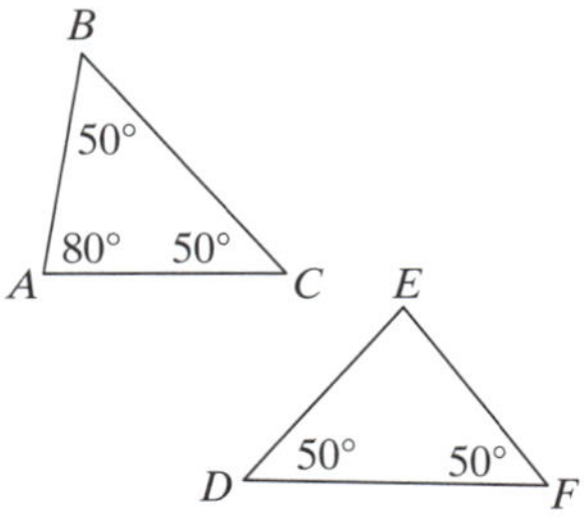

$\angle ABC = \angle EDF$

and $\angle ACB = \angle EFD$

$\therefore\ \Delta ABC \parallel\!| \Delta EDF$

b True

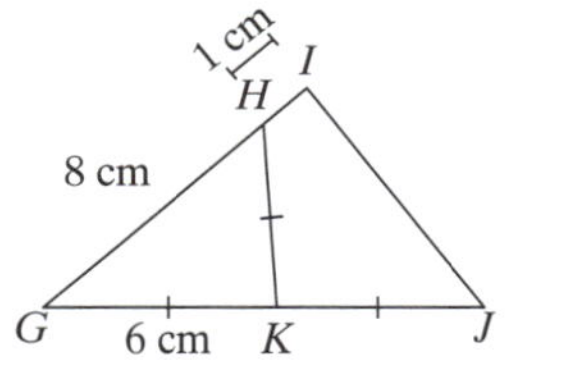

$\angle HGK = \angle JGI$ (common)

$\frac{GK}{GI} = \frac{6}{9} = \frac{2}{3}$

$\frac{GH}{GJ} = \frac{8}{12} = \frac{2}{3}$

$\therefore\ \Delta HGK \parallel\!| \Delta JGI$

c False

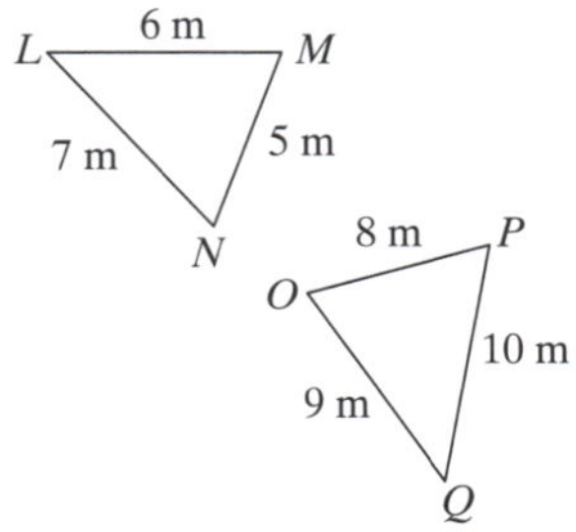

$\frac{MN}{OP} = \frac{5}{8} \neq \frac{LM}{QO} \neq \frac{LN}{QP}$

d True

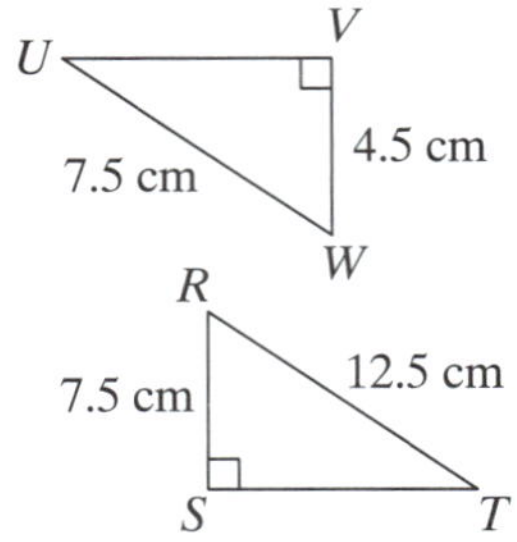

$\angle RST = \angle WVU$ (both 90°)

$\frac{RS}{WV} = \frac{7.5}{4.5} = \frac{5}{3}$

$\frac{RT}{WU} = \frac{12.5}{7.5} = \frac{5}{3}$

$\therefore\ \Delta RST \parallel\!| \Delta WVU$

ENLARGEMENT AND SIMILARITY
INTERMEDIATE TEST PAGE 79

1 $\angle ABC$ is the angle between the two smallest sides.

$\therefore \angle PSR$ corresponds to $\angle ABC$.

$\therefore$ **[C]** ✓ (1 mark)

2 All equilateral triangles are similar (equiangular).

$\therefore$ **[C]** ✓ (1 mark)

3 $\angle HEG = \angle DCB$ (matching vertices)

$\therefore \angle HEG = 95°$

$\therefore$ **[C]** ✓ (1 mark)

4 **a** Largest side of $\Delta STU = 50$ m

Largest side of $\Delta PQR = 20$ m

Times larger $= 50 \div 20$

$= 2.5$ ✓

(1 mark)

b Smallest side of $\Delta STU = 35$ m

Smallest side of $\Delta PQR = 14$ m

Times larger $= 35 \div 14$

$= 2.5$ ✓

(1 mark)

c Third side of $\Delta STU = 45$ m

Third side of $\Delta PQR = 18$ m

Times larger $= 45 \div 18$

$= 2.5$

$\therefore \Delta STU \parallel\!| \Delta RPQ$ (sides in proportion) ✓

$\angle STU = \angle RPQ$

(corresponding angles of similar triangles)

$\therefore \angle STU = 43°$ ✓ (2 marks)

5

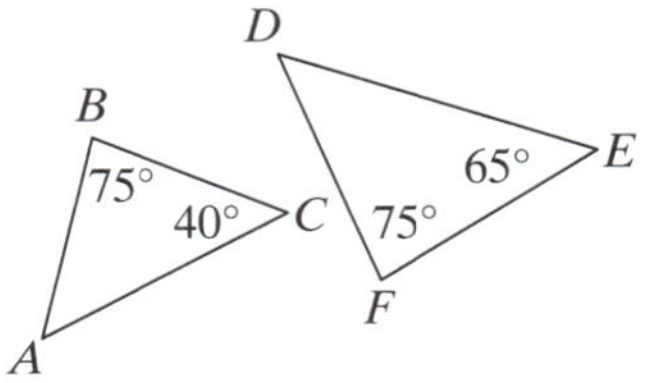

a $\angle FDE + 75° + 65° = 180°$ (angle sum of Δ)

$\angle FDE + 140° = 180°$

$\angle FDE = 40°$ ✓

(1 mark)

b Equiangular

$\angle FDE = \angle BCA = 40°$

$\angle EFD = \angle ABC = 75°$ ✓

(1 mark)

c $\Delta DEF \parallel\!| \Delta CAB$

ED corresponds to AC. ✓

(1 mark)

6

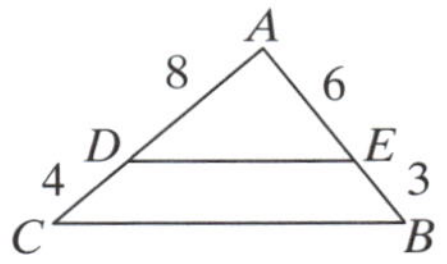

a In Δs ABC, AED

$\angle BAC = \angle EAD$

(common angle) ✓

$\frac{AE}{AB} = \frac{6}{9} = \frac{2}{3}$

$\frac{AD}{AC} = \frac{8}{12} = \frac{2}{3}$ ✓

$\therefore\ \frac{AE}{AB} = \frac{AD}{AC}$

$\therefore\ \Delta ABC \parallel\!| \Delta AED$

(2 sides in proportion, included angle) ✓ (3 marks)

b $\angle ADE = \angle ACB$

(corresponding angles, similar triangles) ✓

$\therefore\ DE \parallel CB$

(corresponding angles equal) ✓ (2 marks)

7

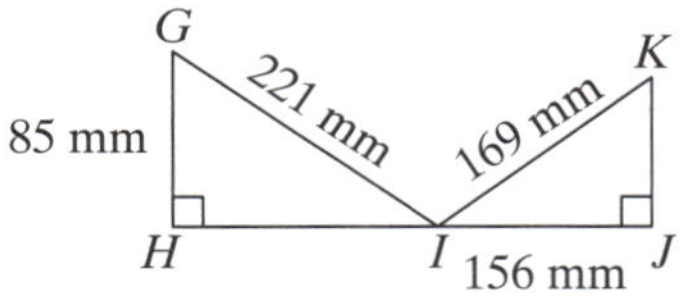

a Let the length of HI be x mm.

$x^2 = 221^2 - 85^2$

$= 41\,616$ ✓

$x = \sqrt{41\,616}$

$= 204$

The length of HI is 204 mm. ✓ (2 marks)

b In Δs GHI, KJI
$\angle GHI = \angle KJI$ (both 90°, given) ✓

$\frac{GI}{KI} = \frac{221}{169}$

$= \frac{17}{13}$

$\frac{HI}{JI} = \frac{204}{156}$

$= \frac{17}{13}$

$\therefore \frac{GI}{KI} = \frac{HI}{JI}$ ✓

$\therefore \Delta GHI \,|||\, \Delta KJI$ (right angle, hypotenuse and side in proportion) ✓ (3 marks)

(Total: 20 marks)

ENLARGEMENT AND SIMILARITY ADVANCED TEST PAGE 80

1 **a** ΔABC and ΔPQR

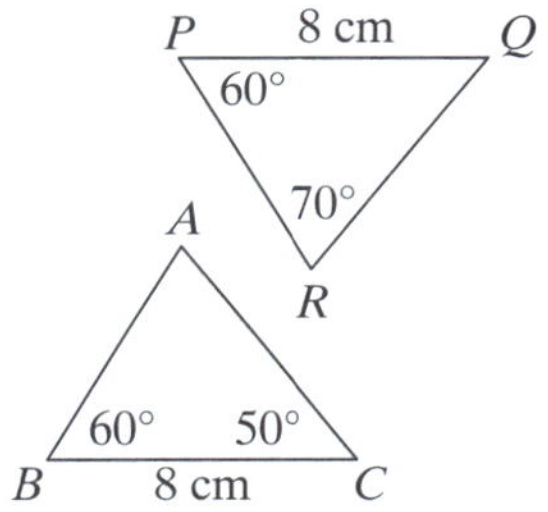

$\angle PQR = 50°$
(angle sum of triangle)
$\angle B = \angle P$; $BC = PQ$;
$\angle C = \angle Q$
Congruent. AAS ✓
$\Delta ABC \equiv \Delta RPQ$ ✓ (2 marks)

b ΔJKL and ΔLMN

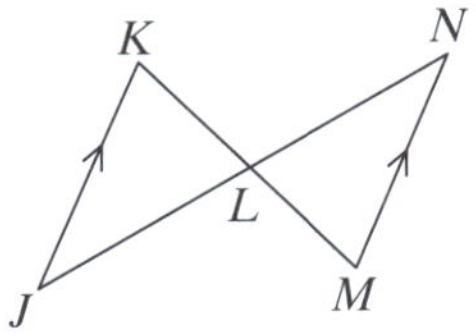

Similar, but not necessarily congruent. ✓
$\angle J = \angle N$ and $\angle K = \angle M$
(alternate, $JK \parallel MN$)
$\Delta JKL \,|||\, \Delta NML$
(equiangular) ✓ (2 marks)

c ΔUVW and ΔXYZ

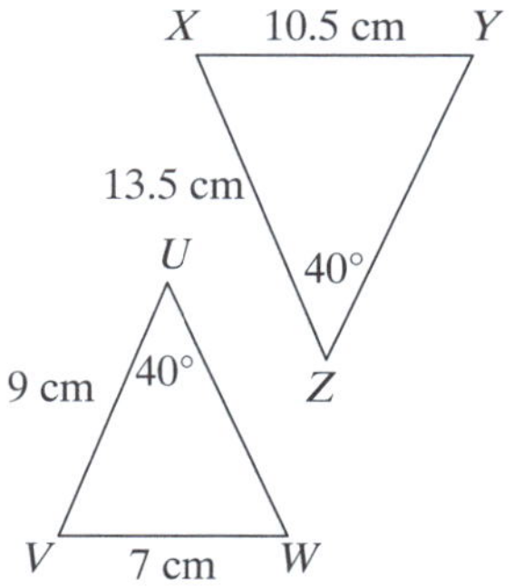

Not congruent ✓
Not necessarily similar.
[Two pairs of sides are in proportion but the equal angles are not between those sides.] ✓ (2 marks)

2

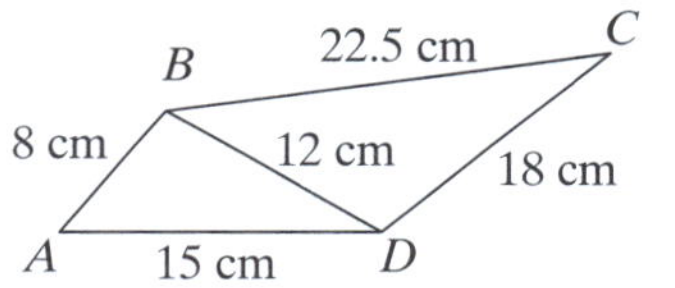

a In ΔABD, ΔCBD

$\frac{AB}{BD} = \frac{8}{12} = \frac{2}{3}$ ✓

$\frac{BD}{DC} = \frac{12}{18} = \frac{2}{3}$

$\frac{AD}{BC} = \frac{15}{22.5} = \frac{2}{3}$

$\therefore \Delta ABD \,|||\, \Delta BDC$
(sides in proportion) ✓
(2 marks)

b $\angle ABD = 95°$, $\angle ABC = 148°$
$\angle DBC = 148° - 95°$
$= 53°$
$\angle BAD = \angle DBC$
(corresponding angles of similar triangles)
$\therefore \angle BAD = 53°$ ✓
$\angle BDA + \angle BAD + \angle ABD$
$= 180°$
(angle sum of triangle)
$\angle BDA + 53° + 95° = 180°$
$\angle BDA = 32°$ ✓
$\angle BDC = \angle ABD$
(corresponding angles of similar triangles)
$\angle BDC = 95°$
$\angle ADC = 32° + 95°$
$= 127°$ ✓ (3 marks)

3 **a** Two triangles are similar if all the angles are equal. If two angles are equal, the remaining angles must also be equal because the angle sum of any triangle is 180°.
✓ (1 mark)

b

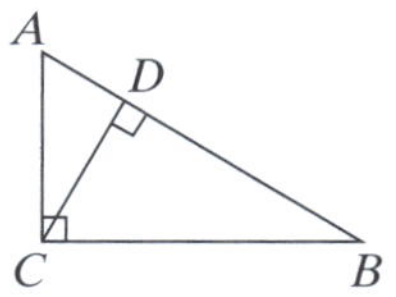

In Δs ADC, ACB
$\angle DAC = \angle CAB$
(common angle)
$\angle ADC = \angle ACB$
(both 90°, given) ✓
$\therefore \Delta ADC \,|||\, \Delta ACB$
(equiangular) ✓(2 marks)

4

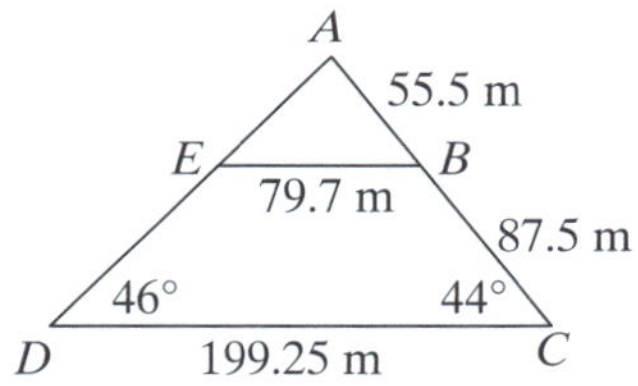

a $\angle DAC + 46° + 44° = 180°$
(angle sum of a triangle)
$\angle DAC = 90°$
ΔACD is a right-angled triangle ✓ (1 mark)

b $AC = (55.5 + 87.5)$ m
$= 143$ m
$AD^2 = 199.25^2 - 143^2$
$= 19\,251.5625$
$AD = \sqrt{19\,251.5625}$ $(AD > 0)$
$= 138.75$ m ✓ (1 mark)

c $\frac{EB}{DC} = \frac{79.7}{199.25}$

$= \frac{2}{5}$ ✓

$\frac{AB}{AC} = \frac{55.5}{143}$

$\neq \frac{2}{5}$

$\frac{AB}{AD} = \frac{55.5}{138.75}$

$= \frac{2}{5}$ ✓

So $\Delta ABE \,|||\, \Delta ADC$ (right angle, hypotenuse and side in proportion) ✓ (3 marks)

5

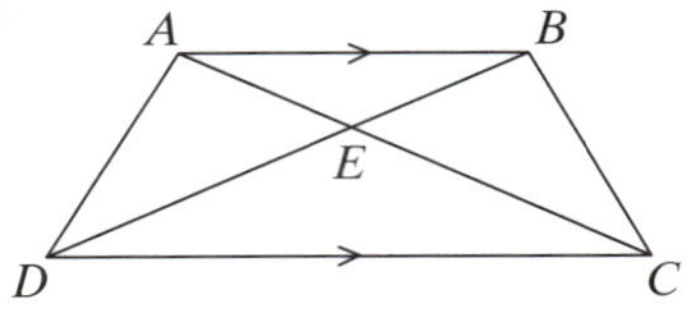

a In ΔABE, ΔCDE
$\angle BAE = \angle DCE$
(alternate, $AB \parallel DC$)
$\angle ABE = \angle CDE$
(alternate, $AB \parallel DC$) ✓
$\therefore \Delta ABE \,|||\, \Delta CDE$
(equiangular) ✓ (2 marks)

b Justin is correct that the vertically opposite angles are equal. It is also true that the corresponding sides of the triangles ABE and CDE are in proportion. ✓
But the corresponding sides of triangles ABE and CDE are not the corresponding sides of triangles BEC and AED. $\frac{BE}{DE} = \frac{AE}{CE}$ but it does not follow that $\frac{BE}{CE} = \frac{AE}{DE}$. ✓ (2 marks)

6 **a** **i** Side 4 cm
Perimeter $= 4 \times 4$ cm
$= 16$ cm
New side $= 1.5 \times 4$ cm
$= 6$ cm
New perimeter
$= 4 \times 6$ cm
$= 24$ cm ✓
Increase
$= (24 - 16)$ cm
$= 8$ cm
% increase
$= \frac{8}{16} \times 100\%$
$= 50\%$ ✓ (2 marks)

ii Area $= 4^2$
$= 16\text{ cm}^2$
New area $= 6^2$
$= 36\text{ cm}^2$ ✓
Increase $= (36 - 16)\text{ cm}^2$
$= 20\text{ cm}^2$
% increase
$= \frac{20}{16} \times 100\%$
$= 125\%$ ✓ (2 marks)

b Yes, the percentages will be the same for any square. ✓
Each new side will be 1.5 times as long as it was.
So the increase in each is 50%.
So the increase in the perimeter will also be 50%. ✓
The new area will be 1.5^2 times what it was.
Now $1.5^2 = 2.25$
So the increase is 1.25 what it was.
The increase will be 125%. ✓ (3 marks)
(Total: 30 marks)

RATIO AND SCALE FACTORS
SKILLS CHECK PAGE 82

1 **a** $8:10 = 4:5$
[dividing both parts by 2]

b $12:6 = 2:1$
[dividing both parts by 6]

c $25:35 = 5:7$
[dividing both parts by 5]

d $7:14 = 1:2$
[dividing both parts by 7]

e $16:20 = 4:5$
[dividing both parts by 4]

f $24:8 = 3:1$
[dividing both parts by 8]

2 **a** 9 months : 1 year
= 9 months : 12 months
$= 9:12$
$= 3:4$

b 25 cm : 1 m
= 25 cm : 100 cm
$= 25:100$
$= 1:4$

c 30 g : 2 kg
= 30 g : 2000 g
$= 30:2000$
$= 3:200$

d 1 cm : 1 m
= 1 cm : 100 cm
$= 1:100$

e 5 cm : 100 m
= 5 cm : 10 000 cm
$= 5:10\,000$
$= 1:2000$

f 1 cm : 1 km
= 1 cm : 1000 m
= 1 cm : 100 000 cm
$= 1:100\,000$

3 **a** $\frac{x}{8} = \frac{1}{2}$
$2x = 8$
$x = 4$

b $\frac{x}{6} = \frac{2}{3}$
$3x = 12$
$x = 4$

c $\frac{x}{3} = \frac{4}{5}$
$5x = 12$
$x = 2.4$

d $\frac{3}{4} = \frac{x}{8}$
$4x = 24$
$x = 6$

e $\frac{6}{72} = \frac{1}{x}$
$6x = 72$
$x = 12$

f $\frac{2}{x} = \frac{5}{7}$
$5x = 14$
$x = 2.8$

4

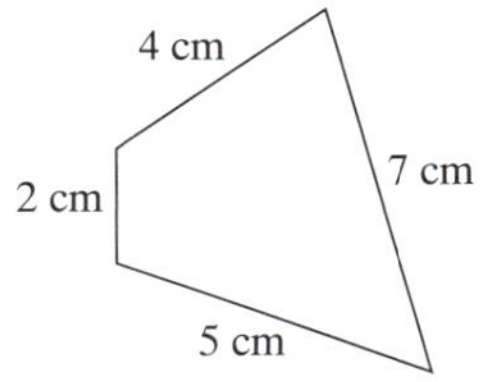

a On the original figure the shortest side is 2 cm.
New shortest side
$= 9 \times 2$ cm
$= 18$ cm

b On the original figure the longest side is 7 cm.
New longest side
$= 9 \times 7$ cm
$= 63$ cm

c On the original figure another side is 4 cm.
New side $= 9 \times 4$ cm
$= 36$ cm
The remaining side on the original figure is 5 cm.
New side $= 9 \times 5$ cm
$= 45$ cm
New perimeter
$= (18 + 63 + 36 + 45)$ cm
$= 162$ cm
[Or find the perimeter of the original figure, 18 cm, and multiply by 9.]

5 **a**

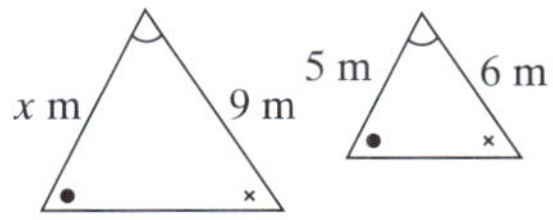

$\frac{x}{5} = \frac{9}{6}$
(corresponding sides of similar triangles)
$x = \frac{9 \times 5}{6}$
$= 7.5$

b

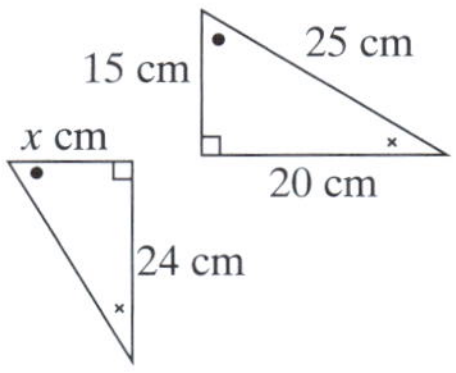

$\frac{x}{15} = \frac{24}{20}$
(corresponding sides of similar triangles)
$x = \frac{15 \times 24}{20}$
$= 18$

c $AB = (15 + 7.5)$ m $= 22.5$ m

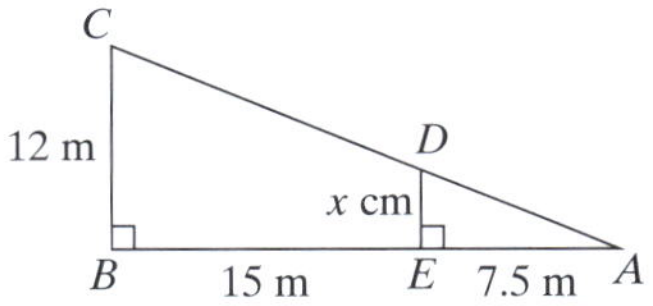

$\frac{x}{12} = \frac{7.5}{22.5}$
(corresponding sides of similar triangles)
$x = \frac{12 \times 7.5}{22.5}$
$= 4$

d $\Delta PQO \parallel\!| \Delta SRO$

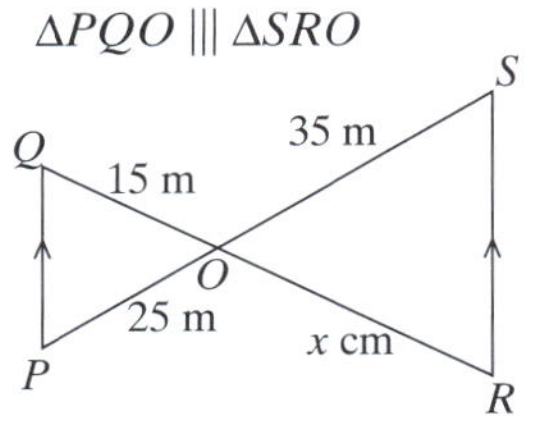

[$\angle S = \angle P$ and $\angle R = \angle Q$ (alternate) so RO corresponds to QO]
$\frac{x}{15} = \frac{35}{25}$
(corresponding sides of similar triangles)
$x = \frac{15 \times 35}{25}$
$= 21$

RATIO AND SCALE FACTORS
INTERMEDIATE TEST PAGE 83

1 3:4:5
Shortest side = 60 mm
Enlargement factor $= 60 \div 3$
$= 20$
Longest side $= 20 \times 5$ mm
$= 100$ mm
∴ **[D]** ✓ (1 mark)

2 Enlargement factor $= 90 \div 6$
$= 15$
New length $= 15 \times 10$ cm
$= 150$ cm
∴ **[C]** ✓ (1 mark)

3 Scale = 4 cm to 80 km
= 1 cm to 20 km
= 1 cm to 20 000 m
= 1 cm to 2 000 000 cm
= 1:2 000 000
∴ **[D]** ✓ (1 mark)

4 Enlargement factor $= 91 \div 39$
$= 2\frac{1}{3}$
New length $= 2\frac{1}{3} \times 69$ mm
$= 161$ mm ✓
(1 mark)

5 **a** Enlargement factor
$= 72$ m $\div 144$ mm
$= 72\,000$ mm $\div 144$ mm
$= 500$ ✓ (1 mark)
The northern boundary of the paddock is 500 times longer than the drawing.

b Western boundary on plan
$= 45$ m $\div 500$
$= 0.09$ m
$= 90$ mm ✓ (1 mark)

c Actual boundary
$= 170$ mm $\times 500$
$= 85\,000$ mm
$= 85$ m ✓ (1 mark)

6 **a**

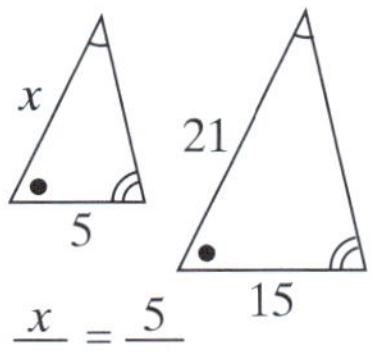

$\frac{x}{21} = \frac{5}{15}$
$x = \frac{5}{15} \times 21$
$= 7$ ✓ (1 mark)

b

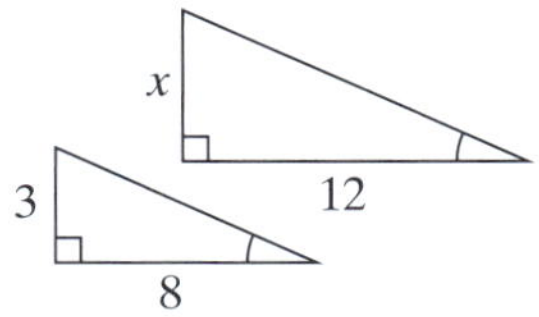

$\frac{x}{3} = \frac{12}{8}$
$x = \frac{12}{8} \times 3$
$= 4.5$ ✓ (1 mark)

c

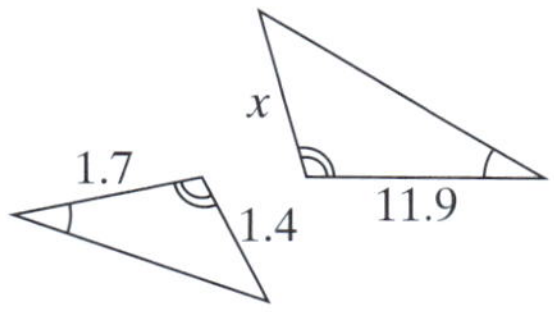

$\frac{x}{1.4} = \frac{11.9}{1.7}$ ✓
$x = \frac{11.9}{1.7} \times 1.4$
$= 9.8$ ✓ (2 marks)

7

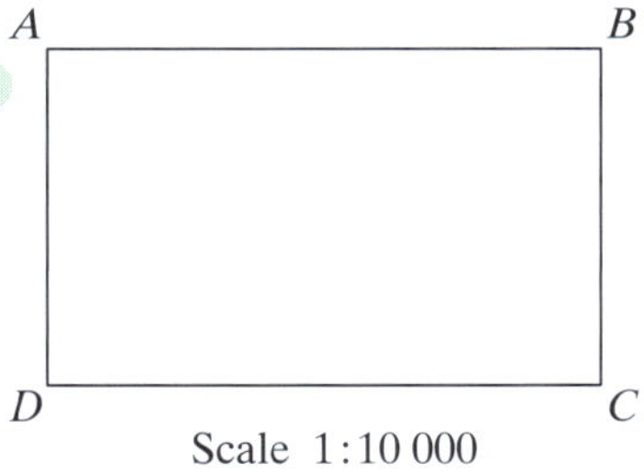

Scale 1:10 000

a 1 cm represents 10 000 cm
Now 10 000 cm = 100 m
So 1 cm represents 100 m. ✓
(1 mark)

b By measurement,
$AB = 4$ cm.
So the actual length is 400 m. ✓ (1 mark)

c By measurement,
$BC = 2.5$ cm
So the actual length is 250 m. ✓
Perimeter
$= 2 \times (400 + 250)$ m
$= 1300$ m ✓ (2 marks)

d Area $= (400 \times 250)$ m^2
$= 100\,000$ m^2 ✓
$= 10$ ha ✓ (2 marks)

8 **a** BC [the hypotenuse of the triangle] ✓ (1 mark)

b Let the length of AB be x cm.

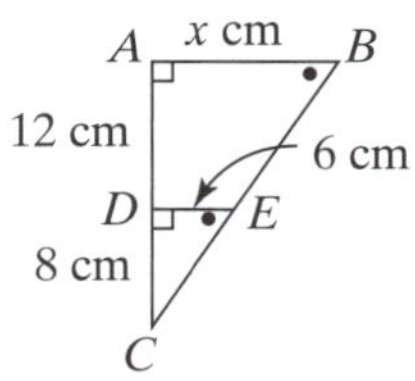

$\frac{AB}{DE} = \frac{AC}{DC}$

$\frac{x}{6} = \frac{20}{8}$ ✓

$x = 6 \times \frac{20}{8}$

$= 15$

The length of AB is 15 cm. ✓ (2 marks)

(Total: 20 marks)

RATIO AND SCALE FACTORS
ADVANCED TEST PAGE 84

1 1 cm represents 1 000 000 cm
Now 1 000 000 cm = 10 000 m
= 10 km
So 4.75 cm represents 47.5 km ✓ (1 mark)

2 **a** By measurement the three sides are 6 cm, 5.4 cm and 4.2 cm. ✓
The scale is 1:2000.
So 6 cm represents
6 × 2000 cm = 12 000 cm
= 120 m
5.4 cm represents
5.4 × 2000 cm = 10 800 cm
= 108 m
4.2 cm represents
4.2 cm × 2000 cm = 8400 cm
= 84 m ✓
Perimeter
= (120 + 108 + 84) m
= 312 m ✓ (3 marks)

b By measurement the perpendicular height is 3.7 cm. 3.7 cm represents
3.7 × 2000 cm
= 7400 cm
= 74 m ✓

$A = \frac{1}{2}bh$

$= \frac{1}{2} \times 120 \times 74$

$= 4440\text{ m}^2$ ✓ (2 marks)

3

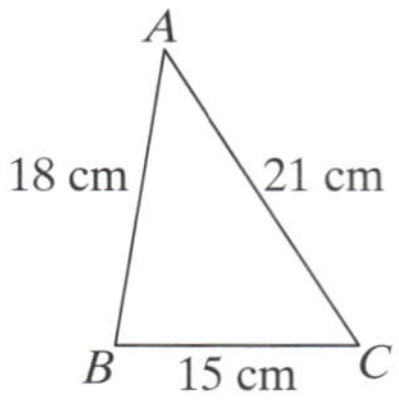

a $\angle DFE$ corresponds to $\angle BAC$.
[Matching the order of the vertices.] ✓ (1 mark)

b Side DE corresponds to BC.
Enlargement factor
$= \frac{48}{15}$
$= 3.2$ ✓ (1 mark)

c Side DF corresponds to BA
$DF = 3.2 \times 18$ cm
$= 57.6$ cm ✓ (1 mark)

4

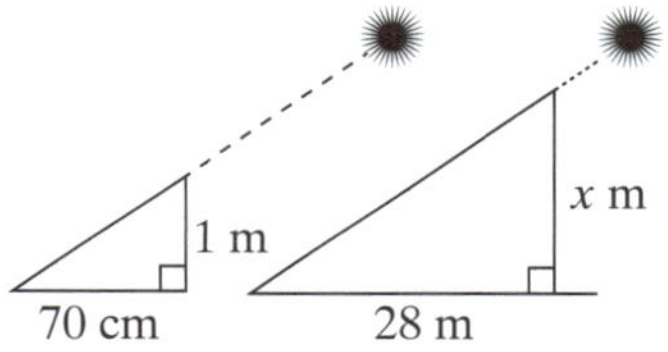

a The triangles are equiangular.
Both triangles have a right angle.
In both triangles the angle of the sun is the same. ✓ (1 mark)

b Let x m be the height of the tree.

$\frac{x}{1} = \frac{28}{0.7}$ ✓

$x = 40$

$\therefore$ the tree is 40 m tall. ✓ (2 marks)

5 **a** In Δs PQR, PRS
$\angle PQR = \angle PRS$ (given)
$\angle QPR = \angle RPS$ (given) ✓
$\therefore\ \Delta PQR \,|||\, \Delta PRS$
(equiangular) ✓ (2 marks)

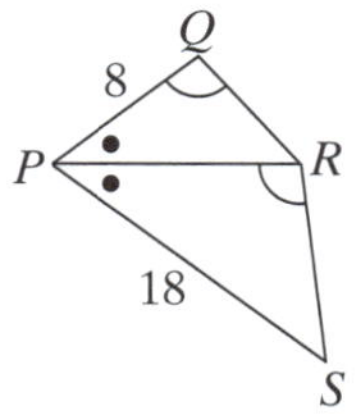

b Let the length of PR be x cm.

$\frac{PR}{PQ} = \frac{PS}{PR}$

(corresponding sides, similar Δs) ✓

$\frac{x}{8} = \frac{18}{x}$

$x^2 = 144$

$x = 12 \quad (x > 0)$

The length of PR is 12 cm. ✓ (2 marks)

6 All circles are similar.
2.5 m = 250 cm
= 5 × 50 cm
So the enlargement factor is 5. ✓
Radius ≈ 5 × 8 cm
= 40 cm ✓ (2 marks)

7 Original photo is 10 cm by 8 cm. ✓
[After subtracting 1 cm from each side]
Now 7.5 × 10 cm = 75 cm
and 7.5 × 8 cm = 60 cm
So the enlarged photo is 75 cm by 60 cm. ✓
New frame is 77 cm by 62 cm. ✓ (3 marks)

8 **a** [The diagram needs to be drawn to scale, the measurements to be 2.1 cm, 6.5 cm, 4.8 cm, 2.9 cm.] ✓✓ (2 marks)

b By measurement the remaining boundary is 4.5 cm ✓
Using the scale the actual boundary would be 450 m. ✓ (2 marks)

(Total: 25 marks)

TRIGONOMETRY OF RIGHT-ANGLED TRIANGLES
SKILLS CHECK PAGE 86

1 **a**

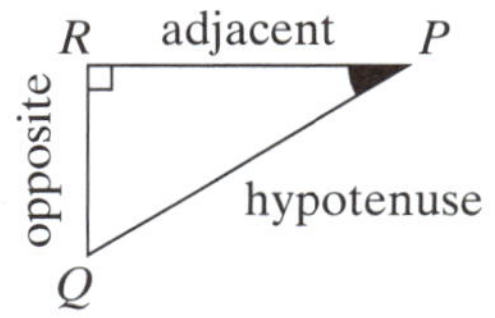

b

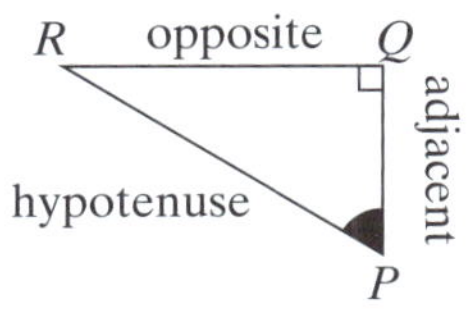

c

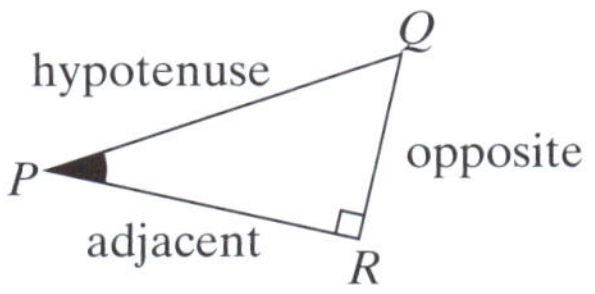

2 **a** The side opposite angle A is side $\boldsymbol{a}$.

b The angle opposite side c is angle $\boldsymbol{C}$.

c The angle opposite side b is angle $\boldsymbol{B}$.

d The side opposite angle Q is side $\boldsymbol{q}$.

3

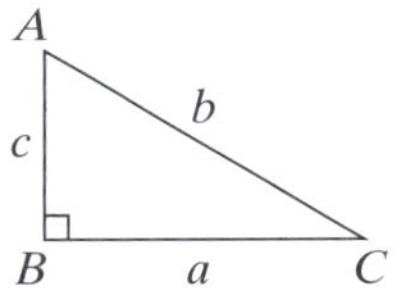

a $\tan A = \frac{a}{c}$ **b** $\sin A = \frac{a}{b}$

c $\cos A = \frac{c}{b}$ **d** $\tan C = \frac{c}{a}$

e $\sin C = \frac{c}{b}$ **f** $\cos C = \frac{a}{b}$

4 **a** $\sin 32° = 0.529\,919\,264\ldots$
$= 0.530$ (3 d.p.)

b $\tan 65° = 2.144\,506\,921\ldots$
$= 2.145$ (3 d.p.)

c $\cos 49° = 0.656\,059\,029\ldots$
$= 0.656$ (3 d.p.)

d $\tan 21° = 0.383\,864\,035\ldots$
$= 0.384$ (3 d.p.)

5 $\sin 23°48' = 0.403\,545\,296\ldots$
$= 0.4035$ (4 d.p.)

6 **a** $\tan\theta = 0.2375$
$\theta = 13.360\,218\,44\ldots°$
$= 13°$ (nearest degree)

b $\sin\theta = 0.7183$
$\theta = 45.914\,303\,07\ldots°$
$= 46°$ (nearest degree)

c $\cos\theta = 0.0795$
$\theta = 85.440\,173\,69\ldots°$
$= 85°$ (nearest degree)

d $\tan\theta = 2.4$
$\theta = 67.380\,135\,05\ldots°$
$= 67°$ (nearest degree)

7 $\cos A = 0.2785$
$A = 73.829\,299\,59\ldots°$
$= 73°50'$ (nearest minute)

8 **a** $\cos 58° = \frac{x}{8}$
$x = 8 \times \cos 58°$
$= 4.239\,354\,114\ldots$
$= 4.2$ (1 d.p.)

b $\tan 23° = \frac{x}{9}$
$x = 9 \times \tan 23°$
$= 3.820\,273\,346\ldots$
$= 3.8$ (1 d.p.)

c $\sin 33° = \frac{3}{x}$
$x = \frac{3}{\sin 33°}$
$= 5.508\,235\,376\ldots$
$= 5.5$ (1 d.p.)

9 **a** $\tan A = \frac{2}{9}$
$A = 12.528\,807\,71\ldots°$
$= 13°$ (nearest degree)

b $\cos A = \frac{5}{7}$
$A = 44.415\,3086\ldots°$
$= 44°$ (nearest degree)

c $\sin A = \frac{37}{61}$
$A = 37.340\,993\,02\ldots°$
$= 37°$ (nearest degree)

TRIGONOMETRY OF RIGHT-ANGLED TRIANGLES
INTERMEDIATE TEST PAGE 87

1 65°41′3.34″ = 66° (nearest degree)
∴ **[D]** ✓ (1 mark)

2

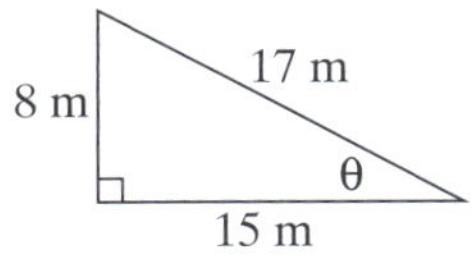

The correct statement is $\tan\theta = \frac{8}{15}$.
∴ **[C]** ✓ (1 mark)

3

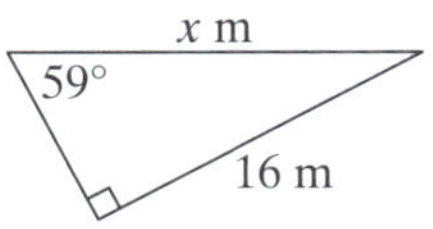

$\sin 59° = \frac{16}{x}$

So $x = \frac{16}{\sin 59°}$

∴ **[A]** ✓ (1 mark)

4 Let the length of the side be x m.

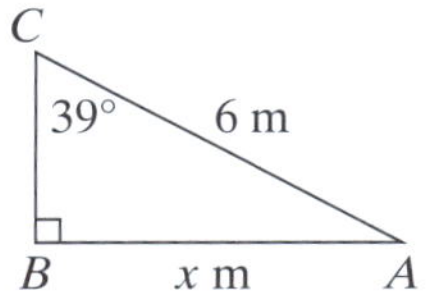

$\sin 39° = \frac{x}{6}$ ✓
$x = 6 \times \sin 39°$
$= 3.775\,922\,346\ldots$
$= 3.8$ (1 d.p.)

The length of the side is 3.8 m to one decimal place. ✓ (2 marks)

5

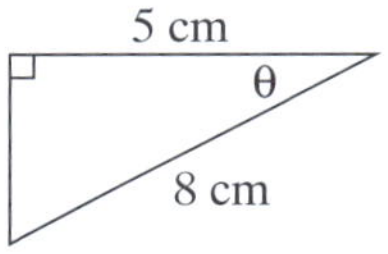

$\cos\theta = \frac{5}{8}$ ✓
$\theta = 51.317\,812\,55\ldots°$
$= 51°$ (nearest degree) ✓
(2 marks)

6 Let the length of the hypotenuse be x cm.

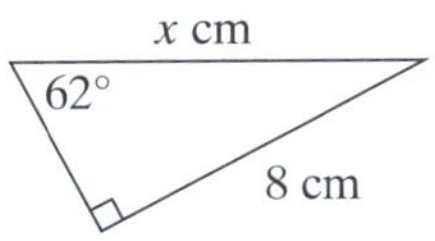

$\sin 62° = \frac{8}{x}$ ✓
$x = \frac{8}{\sin 62°}$
$= 9.060\,560\,405\ldots$
$= 9.1$ (1 d.p.)

The hypotenuse is 9.1 cm long to one decimal place. ✓
(2 marks)

7 **a**

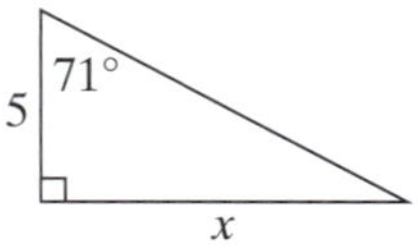

$\tan 71° = \frac{x}{5}$ ✓

$x = 5 \times \tan 71°$
$= 14.52105\ldots$
$= 14.5$ (1 d.p.) ✓

(2 marks)

b

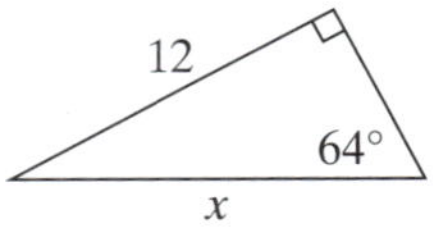

$\sin 64° = \frac{12}{x}$ ✓

$x = \frac{12}{\sin 64°}$
$= 13.35122\ldots$
$= 13.4$ (1 d.p.) ✓

(2 marks)

c

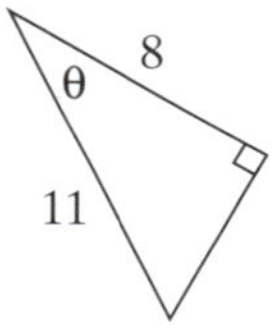

$\cos\theta = \frac{8}{11}$ ✓

$\theta = \cos^{-1}\left(\frac{8}{11}\right)$
$= 43.34175\ldots°$
$= 43°$ (nearest degree)

✓ (2 marks)

d

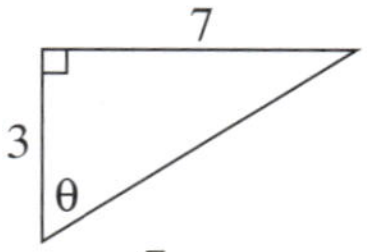

$\tan\theta = \frac{7}{3}$ ✓

$\theta = \tan^{-1}\left(\frac{7}{3}\right)$
$= 66.8014\ldots°$
$= 67°$ (nearest degree)

✓ (2 marks)

8 Let the width of the dam be x m.

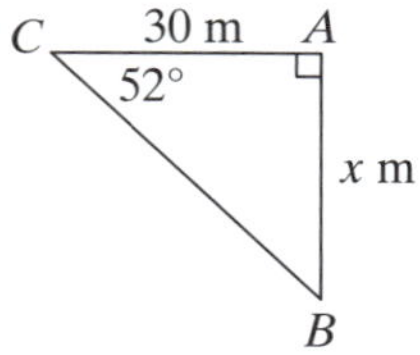

$\tan 52° = \frac{x}{30}$ ✓

$x = 30 \times \tan 52°$
$= 38.39824897\ldots$ ✓
$= 38$ (nearest whole number)

The width of the dam is 38 m to the nearest metre. ✓ (3 marks)

(Total: 20 marks)

TRIGONOMETRY OF RIGHT-ANGLED TRIANGLES
ADVANCED TEST PAGE 88

1 **a**

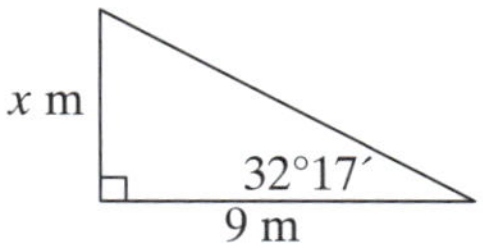

$\tan 32°17' = \frac{x}{9}$ ✓

$x = 9 \times \tan 32°17'$
$= 5.6859\ldots$
$= 5.7$ (1 d.p.) ✓

(2 marks)

b

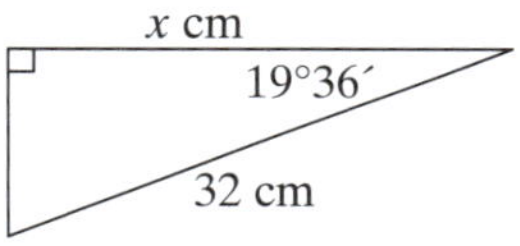

$\cos 19°36' = \frac{x}{32}$ ✓

$x = 32 \times \cos 19°36'$
$= 30.1458\ldots$
$= 30.1$ (1 d.p.) ✓

(2 marks)

c

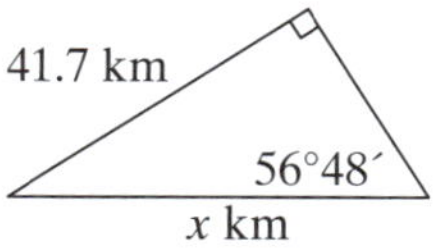

$\sin 56°48' = \frac{41.7}{x}$ ✓

$x = \frac{41.7}{\sin 56°48'}$
$= 49.8348\ldots$
$= 49.8$ (1 d.p.) ✓

(2 marks)

2 **a**

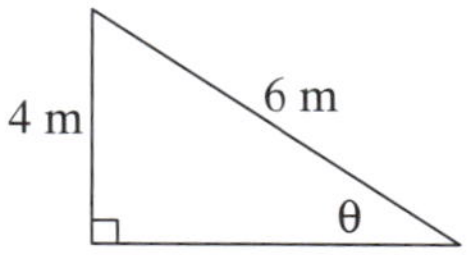

$\sin\theta = \frac{4}{6}$ ✓

$\theta = \sin^{-1}\left(\frac{4}{6}\right)$
$= 41.8103\ldots°$
$= 41°49'$
(nearest minute) ✓

(2 marks)

b

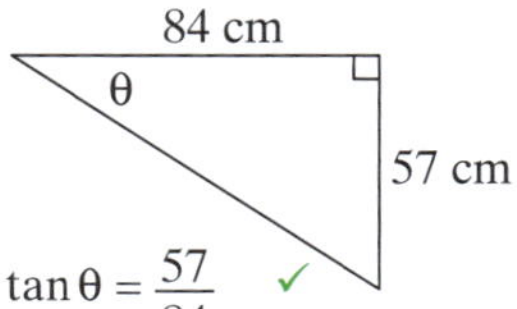

$\tan\theta = \frac{57}{84}$ ✓

$\theta = \tan^{-1}\left(\frac{57}{84}\right)$
$= 34.1596\ldots°$
$= 34°10'$
(nearest minute) ✓

(2 marks)

c

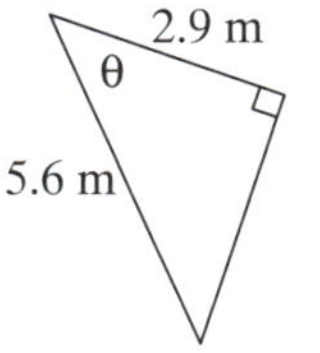

$\cos\theta = \frac{2.9}{5.6}$ ✓

$\theta = \cos^{-1}\left(\frac{2.9}{5.6}\right)$
$= 58.8113\ldots°$
$= 58°49'$
(nearest minute) ✓

(2 marks)

3 Let x m be the length of the brace.

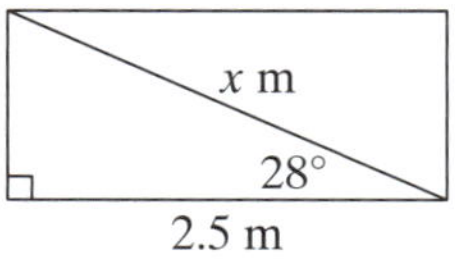

$\cos 28° = \frac{2.5}{x}$ ✓

$x = \frac{2.5}{\cos 28°}$
$= 2.831425127\ldots$ ✓
$= 2.8$ (1 d.p.)

The length of the brace is 2.8 m to one decimal place. ✓ (3 marks)

4 Let θ be the angle the roof makes with the ceiling.

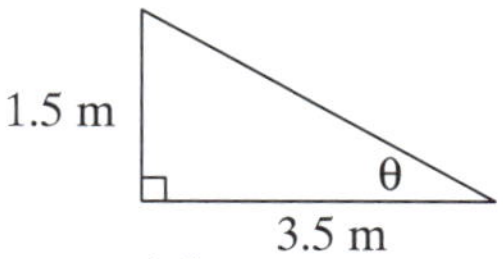

$\tan\theta = \frac{1.5}{3.5}$ ✓

$\theta = 23.19859051\ldots°$ ✓
$= 23°$ (nearest degree)

The angle the roof makes with

the ceiling is 23° to the nearest whole degree. ✓ (3 marks)

5

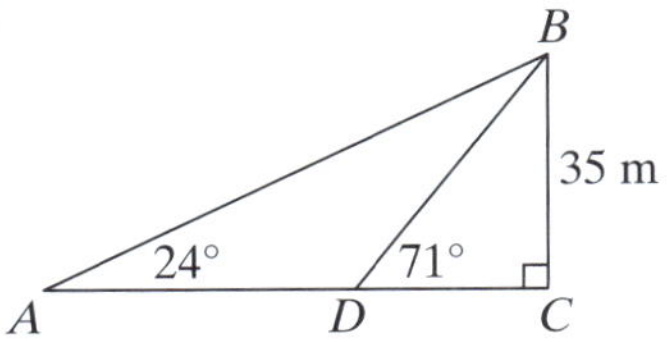

a In ΔBCD,

$\tan 71° = \frac{35}{DC}$ ✓

$DC = \frac{35}{\tan 71°}$

$= 12.051\,46\ldots$

$= 12.1$ m (1 d.p.) ✓

(2 marks)

b In ΔABC,

$\tan 24° = \frac{35}{AC}$ ✓

$AC = \frac{35}{\tan 24°}$

$= 78.611\,28\ldots$

$= 78.6$ m (1 d.p.) ✓

(2 marks)

c $AD = AC - DC$

$= 78.6112\ldots - 12.0514\ldots$

$= 66.559\ldots$

$= 67$ m

(nearest metre) ✓

(1 mark)

6

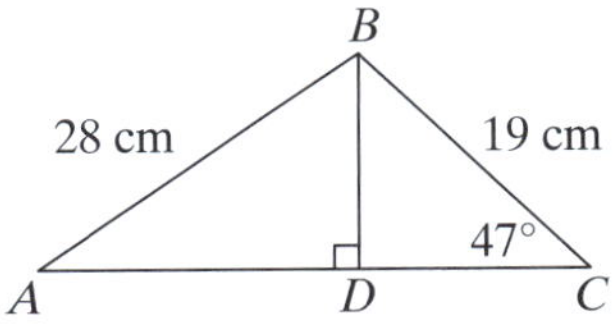

a In ΔBCD,

$\sin 47° = \frac{BD}{19}$ ✓

$BD = 19 \times \sin 47°$

$= 13.895\,72\ldots$

$= 13.9$ cm (1 d.p.) ✓

(2 marks)

b In ΔBAD,

$\sin A = \frac{BD}{28}$ ✓

$\angle A = 29.7539\ldots°$

$= 30°$

(nearest degree) ✓

(2 marks)

7

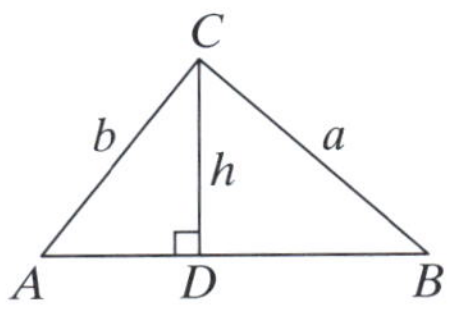

a In ΔACD,

$\sin A = \frac{h}{b}$

$\therefore h = b \sin A$ ✓ (1 mark)

b In ΔCBD,

$\sin B = \frac{h}{a}$

$\therefore h = a \sin B$ ✓

But $h = b \sin a$ from part **a**

$\therefore a \sin B = b \sin A$ ✓

(2 marks)

(Total: 30 marks)

COLLECTING AND EVALUATING DATA

SKILLS CHECK PAGE 90

1 **a** sample

b census

2

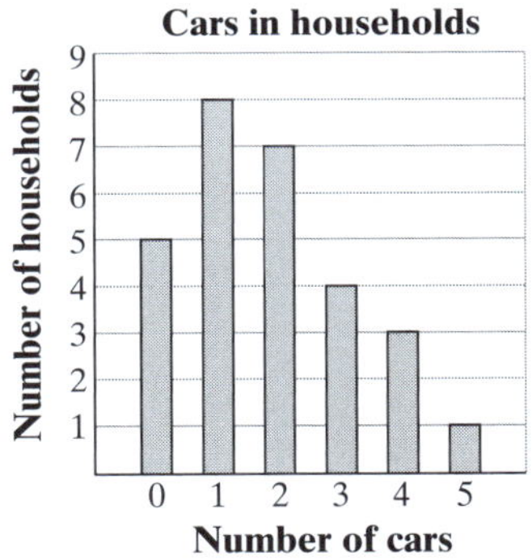

a 7 households had 2 cars.

b 4 households had 3 cars, 3 households had 4 cars and 1 household had 5 cars.
The number of households with more than 2 cars
$= 4 + 3 + 1 = 8$

c Number of households
$= 5 + 8 + 7 + 8 = 28$

d Number of cars
$= 8 \times 1 + 7 \times 2 + 4 \times 3 + 3 \times 4 + 5$
$= 51$

3

Candidates	Number of votes received
Annabel	18
Ethan	24
Henry	31
Laura	37
Pia	26

a Pia got 26 votes.

b Laura got the most votes, so Laura won the ballot.

c Henry got 31 votes while Ethan got 24.
Extra votes $= 31 - 24$
$= 7$
Henry got 7 more votes than Ethan.

d Votes
$= 18 + 24 + 31 + 37 + 26$
$= 136$
So 136 people voted.

4 **a**

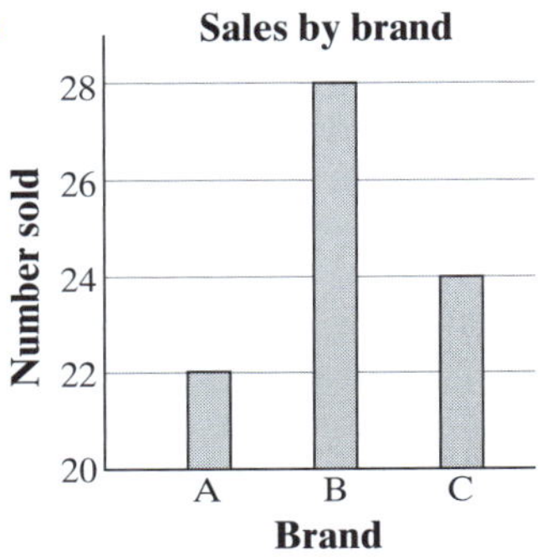

The scale on the vertical axis (number sold) does not begin at zero. This makes sales of brand B look larger than the others. If drawn correctly the difference would not be as great.

b

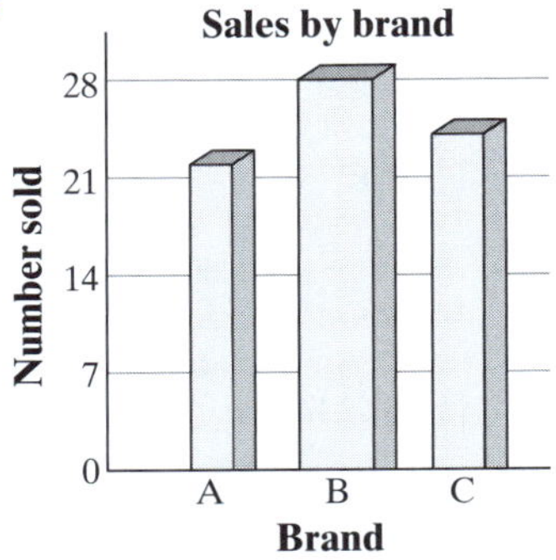

The use of 3D columns and the greater width of column B make it seem much larger than the others. This makes it seem as if the sales of brand B are much more than either brand A or brand C.

c

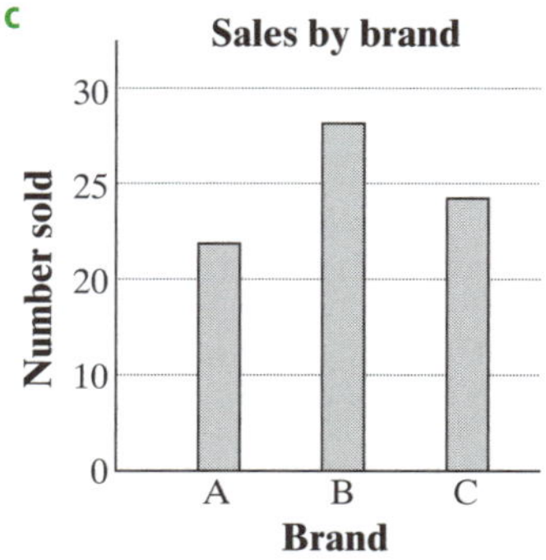

The scale on the vertical axis (number sold) is not even. It goes up by 10 at first and then by 5. This makes the differences between the brands appear larger than they really are.

COLLECTING AND EVALUATING DATA
INTERMEDIATE TEST PAGE 91

1 Unbiased data is the only one of the options that would be desirable in a survey.
∴ **[B]** ✓ (1 mark)

2 Positions in a race are categorical data.
∴ **[D]** ✓ (1 mark)

3 If everyone was questioned as they came in the door, then it would be a census not a random sample.
∴ **[C]** ✓ (1 mark)

4 **a** The highest average monthly rainfall was 204 mm. The month with the highest average monthly rainfall was February. ✓ (1 mark)

b Summer rainfall
= 115 + 113 + 204
= 432
Autumn rainfall
= 180 + 137 + 111
= 428
Winter rainfall
= 133 + 64 + 62
= 259
Spring rainfall
= 58 + 101 + 147
= 306
Winter has the least average rainfall. ✓ (1 mark)

c Difference between summer and winter rainfall
= 432 – 259
= 173
On average, 173 mm more rain falls in summer than in winter. ✓ (1 mark)

d Rainfall for April and May
= 137 + 111
= 248 ✓
Rainfall for October and November = 101 + 147
= 248
The months are October and November. ✓ (2 marks)

e Total rainfall
= 432 + 428 + 259 + 306
= 1425
The average annual rainfall is 1425 mm. ✓ (1 mark)

f Although summer had the highest rainfall, it had only 4 mm more than autumn. ✓ So, you would expect that in some years autumn will receive more rainfall than summer. ✓ (2 marks)

g September is the month with the least rainfall but it might not be the month with the least wet days. ✓ Ellen should also find out the average number of wet days per month. ✓ (2 marks)

5 **a** Total students
= 35 + 43 + 20 + 22 + 30
= 150 ✓ (1 mark)

b Number in favour
= 35 + 43
= 78 ✓ (1 mark)

c 30 students didn't know.
Percentage = $\frac{30}{150} \times 100\%$
= 20% ✓ (1 mark)

d 35 were strongly in favour and 22 strongly against.
Difference = 35 – 22
= 13 ✓ (1 mark)

e Some possible answers:
- There is bias in the question.
- The percentage in favour is $\frac{78}{150} \times 100\% = 52\%$
  This is only a small majority.
- The sample of students might be too small [e.g. if there are 1500 students at the school, the sample would only be 10% of those].
- The sample might not be representative.
  [All the students in the survey might be from year 7, for example, and year 12 might have a totally different idea.] ✓✓✓ (3 marks)

(Total: 20 marks)

COLLECTING AND EVALUATING DATA
ADVANCED TEST PAGE 92

1

No. of children under 10	No. of households
0	1512
1	823
2	764
3	189
4	75
5	21
6	3

a Total households
= 1512 + 823 + 764 + 189 + 75 + 21 + 3
= 3387 ✓ (1 mark)

b Total children
= $823 \times 1 + 764 \times 2 + 189 \times 3 + 75 \times 4 + 21 \times 5 + 6 \times 3$ ✓
= 3341 ✓ (2 marks)

c 1512 out of 3387 households had no children under 10.
Estimation for neighbouring town = $\frac{1512}{3387} \times 5300$ ✓
= 2365.9876...
= 2366
(nearest whole number)
Approx. 2366 households would have no children

under 10. ✓ (2 marks)

2

Make	Length	Width	Height	Ground clearance
Ozdrive	4485	1735	1590	200
Zippy	4610	1785	1710	170
Fortuna	4415	1825	1775	205
Prestige	4600	1815	1695	195

a The Fortuna was both the shortest and widest. ✓ (1 mark)

b The longest car is the Zippy. The ground clearance of the Zippy is 170 mm. ✓ (1 mark)

c $1815 - 1695 = 120$
So, the Prestige is the required car.
The length of the Prestige is 4600 mm. ✓ (1 mark)

d Ozdrive: $4485 - 1735 = 2750$
Zippy: $4610 - 1785 = 2825$
Fortuna: $4415 - 1825 = 2590$
Prestige: $4600 - 1815 = 2785$
The biggest difference is 2825 mm. ✓ (1 mark)

e Ozdrive. It has the lowest height by more than 100 mm but the second highest ground clearance, only 5 mm less than the highest. ✓ (1 mark)

3

Revenue	This year \$m	Last year \$m
Premiums	6672	6423
Dividends	138	112
Interest	425	396
Trusts	59	41
Other	218	178
Total		7150

a $6672 + 138 + 425 + 59 + 218$
$= 7512$
The total income this year
$= \$7\,512\,000\,000$. ✓ (1 mark)

b Last year, interest as a % of income
$= \frac{396}{7150} \times 100\%$
$= 5.538\,46\ldots\ \%$
$= 5.5\%$ (1 d.p.) ✓ (1 mark)

c This year, interest as a % of income
$= \frac{425}{7512} \times 100\%$
$= 5.657\,61\ldots\ \%$
$= 5.7\%$ (1 d.p.) ✓ (1 mark)

d Increase $= 425 - 396$
$= 29$
Increase in interest was \$29 000 000.
% increase
$= \frac{29}{396} \times 100\%$
$= 7.323\,23\ldots\ \%$
$= 7.3\%$ (1 d.p.) ✓ (1 mark)

4

Votes

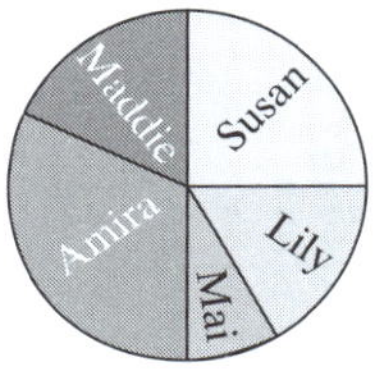

a Amira won ✓ (1 mark)

b Angle of sector for Mai is 30°.
Fraction of votes $= \frac{30}{360}$
$= \frac{1}{12}$ ✓ (1 mark)

c Angle of sector for Lily
$= 60°$ ✓
Now $\frac{60}{360} = \frac{1}{6}$
So 12 votes is $\frac{1}{6}$ of the total votes.
Total votes $= 6 \times 12$
$= 72$ ✓
Susan received a quarter of the votes.
Now $72 \div 4 = 18$
So Susan received 18 votes. ✓ (3 marks)

d 60° represents 12 votes.
So 5° would represent 1 vote. ✓
Angle for Maddie = 65°
Maddie received 13 votes.
So Maddie got 1 more vote than Lily. ✓ (2 marks)

5 **a** Possible answers:
It might not have been a random sample. There is no information about the sample; where the survey was taken or who commissioned it. ✓
There is no information about the sample size. There may have only been a small number who voted. ✓
There may have been bias in the wording of the question or the report upon which the survey was based.
✓ (3 marks)

b The question immediately implies that the survey is not worthwhile. It is prejudicial. ✓ (1 mark)
(Total: 25 marks)

STEM-AND-LEAF PLOTS, HISTOGRAMS AND DESCRIBING DATA
SKILLS CHECK PAGE 94

1 **a** mode = 8

b mode = 6 [There are three 6s which is more than any other number.]

c modes are 1 and 4

d modes are 23 and 37

2

5	3 8
6	0 4 8
7	1 2 3 3 5 7 9
8	0 1 4 6 9
9	2

a There are 18 students in the class. [counting the number of scores]

b The lowest score is 53

c The highest score is 92.

d The mode is 73 [There are two scores of 73 and only one of all the others.]

e 8

3

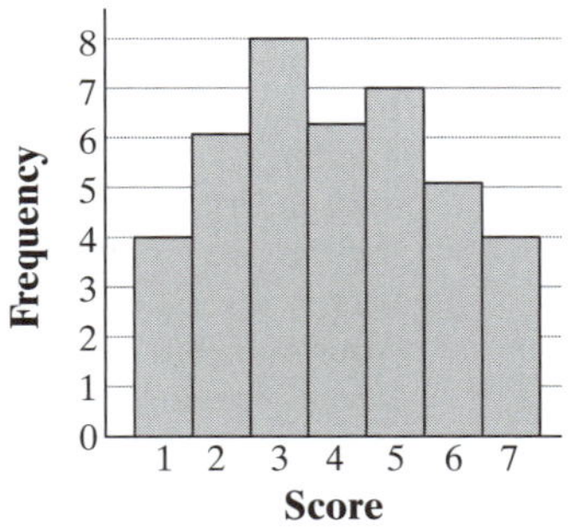

a 7 students scored 5.

b The lowest score was 1.

c The highest score was 7.

d 5 students scored 6.

e The mode was 3.

f Total frequency
$= 4 + 6 + 8 + 6 + 7 + 5 + 4$
$= 40$
So 40 people took part in the quiz.

4

Girls		Boys
7 4	13	
9 6 5 2 0	14	
8 7 3 3 1	15	3 4 5 6 8 9
6 4 2	16	1 2 4 5 5 7 9
	17	0 3 6
	18	1

a Total boys = 17

b The tallest boy is 181 cm tall.

c Shortest girl = 134 cm
Shortest boy = 153 cm
Difference = 153 – 134
= 19
The difference in height is 19 cm.

d mode for girls = 153 cm

e mode for boys = 165 cm

f The tallest girl is 166 cm tall. 6 boys are taller than 166 cm.

5 a

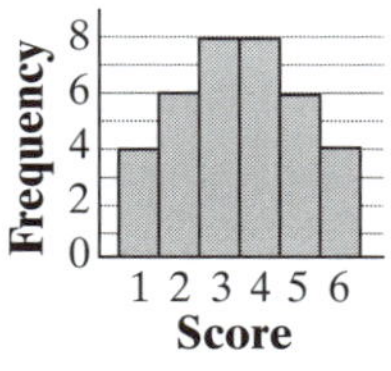

symmetric

b

1	2 4 6 6 7 9
2	1 3 3 5 5 8 9
3	0 2 6 7
4	3 4 8
5	1 2

positively skewed

c

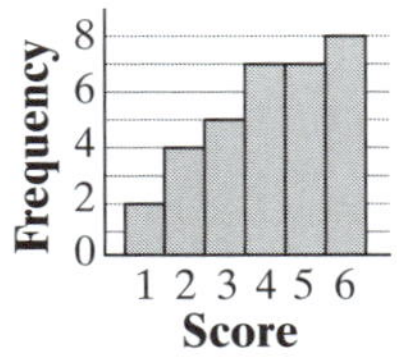

negatively skewed

d

1	4 7 8
2	2 3 6 7 9
3	0 1 4 8 8 9
4	0 2 2 2 7
5	3 5 6

symmetric

e

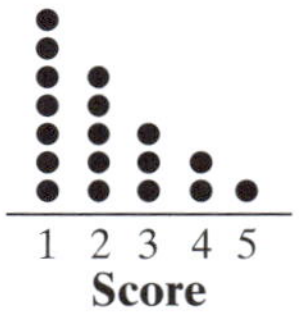

positively skewed

6 a bimodal
[The 2 modes are 1 and 5.]

b single mode
[The mode is 4.]

c bimodal
[The 2 modes are 2 and 6.]

d single mode
[The mode is 7.]

STEM-AND-LEAF PLOTS, HISTOGRAMS AND DESCRIBING DATA
INTERMEDIATE TEST PAGE 95

1 Options A, B and C are all symmetric. Option D is negatively skewed.
∴ [D] ✓ (1 mark)

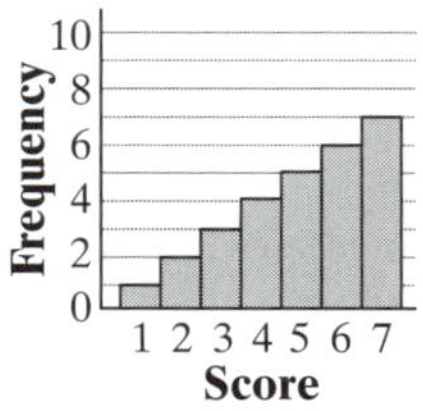

2

5	4
6	3 7
7	0 2 6 8 9
8	1 4 4
9	2 5

There are two scores of 84 and only one of every other score. The mode is 84.
∴ [D] ✓ (1 mark)

3 Option A is both skewed and bimodal.

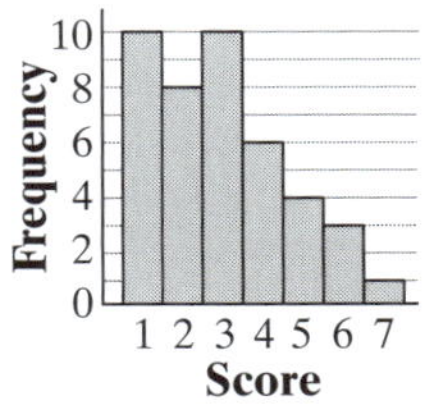

[A is positively skewed and bimodal, B is bimodal but symmetric and C and D have only single modes.]
∴ [A] ✓ (1 mark)

4 a

Class X		Class Y
8 7 6 5 5 4 3	6	4 8 9
9 7 4 4 2 0	7	1 3 3 5 6 8
6 2 1	8	0 1 2 4 5 7
4	9	0 3 7

✓✓✓ (3 marks)

b The data set for class X is positively skewed. ✓
The data set for class Y is symmetric. ✓ (2 marks)

5

Score	1	2	3	4	5	6
Frequency	6	5	8	6	8	7

a Total rolls
$= 6 + 5 + 8 + 6 + 8 + 7$
$= 40$ ✓ (1 mark)

b

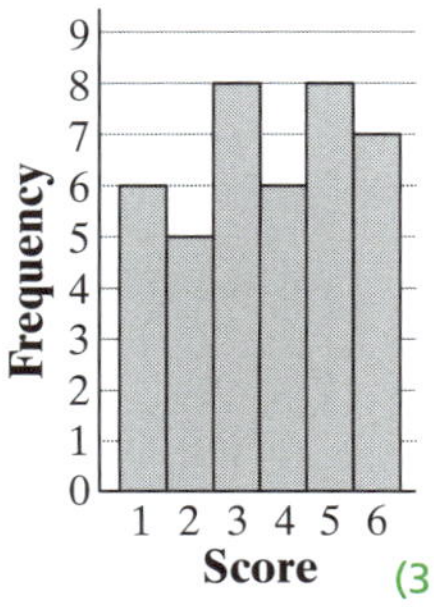

✓✓✓ (3 marks)

c

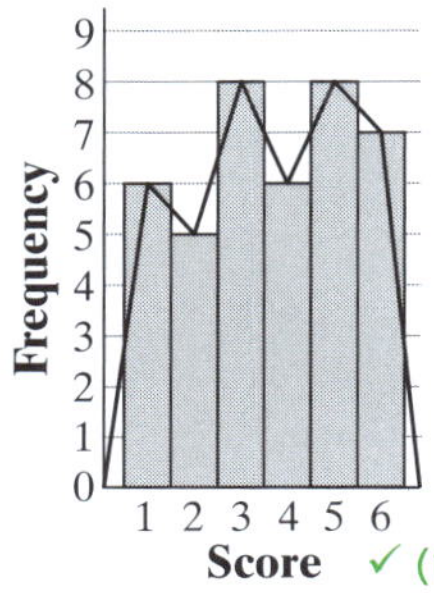

✓ (1 mark)

d The histogram is not smooth or symmetrical. It is slightly negatively skewed. ✓ (1 mark)

Possible answers:

e The histogram gives an instant picture of the shape of the distribution. ✓ (1 mark)

f The table gives all the data clearly. ✓ (1 mark)

6 a

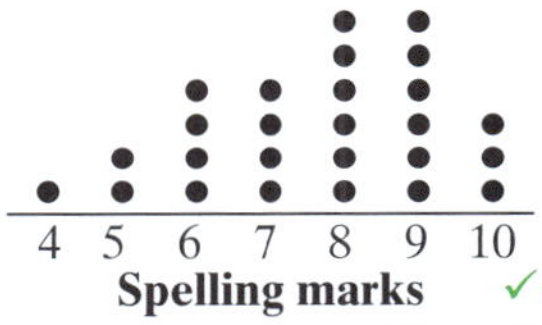

✓✓ (2 marks)

b The data is bimodal, the two modes are 8 and 9. ✓ The dot plot is negatively skewed. ✓ (2 marks)

(Total: 20 marks)

STEM-AND-LEAF PLOTS, HISTOGRAMS AND DESCRIBING DATA ADVANCED TEST PAGE 96

1

0	8
1	2 3 5 8 8 9
2	0 1 2 [redacted] 5 6 7
3	1 2 3 4 5 5 6
4	0 1 4 7
5	2

a The amount is between 22 and 25 litres. ✓ It could be 22, 23, 24 or 25 litres. ✓ (2 marks)

b There are 26 cows altogether.
% that produced > 40 L

$= \frac{4}{26} \times 100\%$

$= 15.3846\ldots\%$

Just over 15% of the cows produced more than 40 litres. ✓ (1 mark)

c Yes, it is bimodal. The two modes are 18 and 35. (But if Sweetpea produced 22 or 25 litres there would be a third mode.) ✓ (1 mark)

d The shape is fairly symmetrical. ✓ (1 mark)

e Possible answers:
A stem-and-leaf plot allows all the data to be seen. ✓
The shape of the display can give information about the data. ✓ (2 marks)

2 a Number of journeys in the morning

$= 1 + 2 + 2 + 3 + 4 + 5 + 6 + 5$

$= 28$

Alice made 28 journeys to work. ✓ (1 mark)

b Morning trip: mode is 11.

Morning drive to work

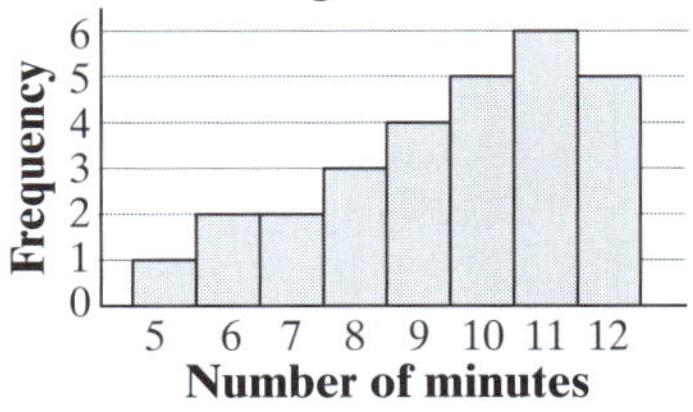

Afternoon trip: bimodal, two modes 8 and 9

Afternoon drive from work

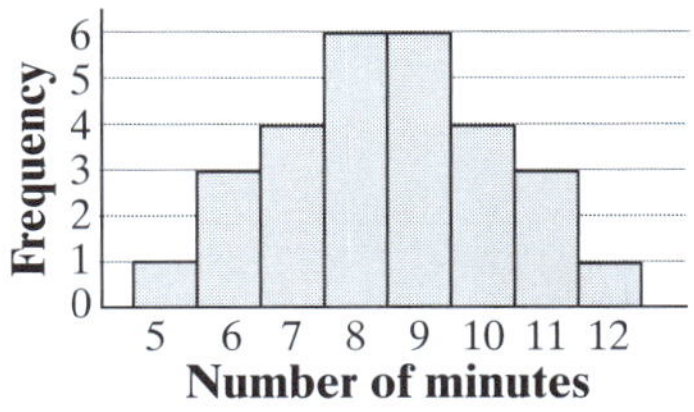

The most common morning trip is two to three minutes longer than the afternoon trip. ✓ (1 mark)

c The graph for the morning trip is negatively skewed and the graph for the afternoon trip is symmetric. ✓
In the morning there are more longer trips than shorter ones whereas in the afternoon there are the same number of longer and shorter trips. Alice can expect to take longer to get to work in the morning than to get home in the afternoon. ✓ (2 marks)

3 a

Score	Frequency
13	3
14	7
15	6
16	4
17	3
18	2
19	2
20	1

✓✓ (2 marks)

b

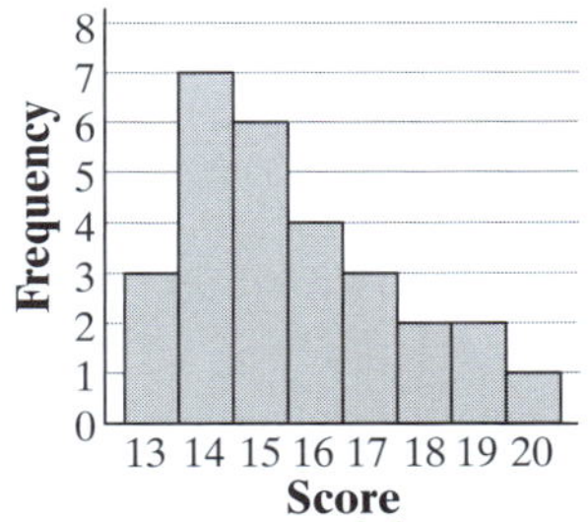

✓✓ (2 marks)

c The histogram is positively skewed. ✓
More students scored around 14 marks in the test and fewer students scored full marks. ✓ (2 marks)

d If every score decreased by two marks the shape of the display would still be exactly the same, but the location would be two marks fewer. The mode for example, would be 12 not 14. ✓ (1 mark)

4 a Ryan [He only played in 13 games while the others all played in 14.] ✓ (1 mark)

b Total of Jonty's goals
$= 2 \times 0 + 3 \times 1 + 4 \times 2 + 3 \times 3 + 2 \times 4$
$= 28$ ✓ (1 mark)

c Hunter ✓ (1 mark)

d Carlo ✓ (1 mark)

e It is not possible to tell who the weakest player is. Carlo scored the least goals, but he might be good in defence and not specifically a goal scorer. ✓ (1 mark)

5 **a**

Boys		Girls
	13	2 5 7
6 4	14	0 2 2 5 8
7 6 5 3	15	1 3 5 7 9 9
8 8 5 4 1	16	0 3 3 6 8
9 7 6 5 4 3 2	17	4 6
5 2 1	18	

✓✓✓ (3 marks)

b 168 cm ✓ (1 mark)

c Difference $= 185 - 132$
$= 53$ cm ✓ (1 mark)

d The heights of the boys on the whole are greater than the heights of the girls. ✓ The spread of scores for each group is fairly similar. The display for the girls is quite symmetrical while that for the boys is slightly negatively skewed. ✓ (2 marks)

(Total: 30 marks)

COMPARING DATA DISPLAYS AND USING MEAN, MEDIAN AND RANGE
SKILLS CHECK PAGE 98

1 **a** $5 + 7 + 8 + 16 = 36$
Mean $= 36 \div 4$
$= 9$

b $5 \times 2 + 6 \times 3 + 7 \times 6 + 8 \times 2 + 9 + 10 = 105$
Mean $= 105 \div 15$
$= 7$

2 **a** Range $= 32 - 3$
$= 29$

b Highest score = 18
Lowest score = 6
Range $= 18 - 6 = 12$

3 **a** 2, 5, 7, 9, 16
Median = 7 [middle score]

b 15, 19, 20, 23, 27, 31, 34 [in order]
Median = 23

c 3, 4, 8, 10
Median $= \frac{4 + 8}{2} = 6$

d 76, 79, 85, 86, 94, 97
Median $= \frac{85 + 86}{2} = 85.5$

4 **a** Mean
$= \frac{1+2+2+5+6+7+7+7+8}{9}$
$= \frac{45}{9}$
$= 5$

b Mode = 7 [There are more 7s than any other number.]

c Median = 6 [There are 9 scores. The middle score is the 5th score, which is 6.]

d Range $= 8 - 1$
$= 7$

5 **a** Mode = 58

b In order the scores are:
37, 58, 58, 62, 63, 67
Median $= \frac{58 + 62}{2}$
$= 60$
[the average of the two middle scores]

c Mean
$= \frac{37+58+58+62+63+67}{6}$
$= \frac{345}{6}$
$= 57.5$

d Range $= 67 - 37$
$= 30$

6

Score	Frequency
1	6
2	8
3	8
4	11
5	9
6	3
Total	45

a Mode = 4 [The highest frequency is 11. The score with the highest frequency is 4.]

b Range $= 6 - 1 = 5$

c $\Sigma fx = 1 \times 6 + 2 \times 8 + 3 \times 8 + 4 \times 11 + 5 \times 9 + 6 \times 3$
$= 153$
Mean $= \frac{153}{45}$
$= 3.4$

d Median = 4 [The middle score is the 23rd score. There are 6 ones, 8 twos and 8 threes which makes a total of 22, so the 23rd score is the first score of 4.]

7 **a** Range $= 20 - 15$
$= 5$

b Mode = 19

c $7 + 6 + 2 + 6 + 8 + 5 = 34$
$7 \times 15 + 6 \times 16 + 2 \times 17 + 6 \times 18 + 8 \times 19 + 5 \times 20$
$= 595$
Mean $= 595 \div 34$
$= 17.5$

d Median = 18 [The 17th and 18th scores are both 18.]

8 **a** Mode = 80 [There are 3 scores of 80.]

b Median = 78

c Range $= 97 - 52$
$= 45$

d Sum of scores = 2062
Mean $= 2062 \div 27$
$= 76.370\,370\,370\ldots$
$= 76.4$ (1 d.p.)

COMPARING DATA DISPLAYS AND USING MEAN, MEDIAN AND RANGE
INTERMEDIATE TEST PAGE 99

1 20, 13, 17, 15, 15, 18, 12, 16, 15
In order:
12, 13, 15, 15, 15, 16, 17, 18, 20
Mean
$= (12 + 13 + 3 \times 15 + 16 + 17 + 18 + 20) \div 9$
$= 15.666\ldots$
Median = 15 [the 5th score]
Mode = 15 [There are 3 scores of 15.]
Range $= 20 - 12$
$= 8$

So the mean is largest.
∴ [A] ✓ (1 mark)

2 7, 8, 10, 5, 3, 9, 7, 4, 7, 6
In order:
3, 4, 5, 6, 7, 7, 7, 8, 9, 10
Mode = 7
Range = 10 − 3
= 7
Mean
= (3 + 4 + 5 + 6 + 3 × 7 + 8 + 9 + 10) ÷ 10
= 6.6
Median = 7
The mean is not equal to 7.
∴ [C] ✓ (1 mark)

3 Mean of 5 scores = 8
Sum of the scores = 5 × 8
= 40
New sum = 40 + 17
= 57
New number of scores = 5 + 1
= 6
New mean = 57 ÷ 6
= 9.5
∴ [C] ✓ (1 mark)

4 One set of scores will be more consistent if the range is lower. [The scores will be less spread out.]
∴ [B] ✓ (1 mark)

5

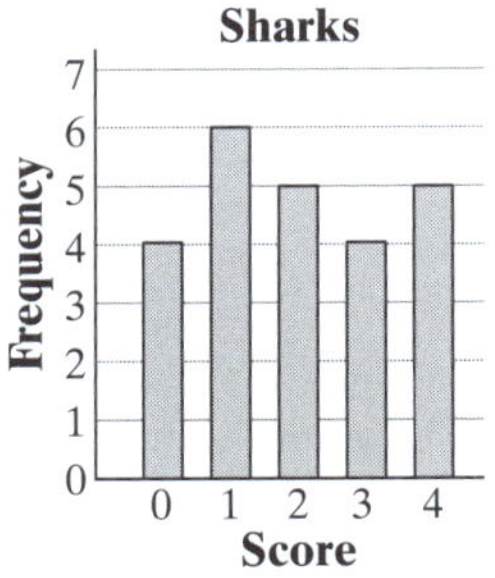

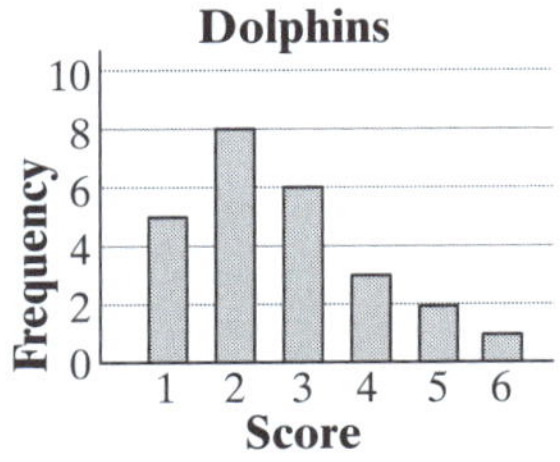

a range for Sharks
= 4 − 0
= 4 ✓ (1 mark)

b range for Dolphins
= 6 − 1
= 5 ✓ (1 mark)

c mode for Sharks = 1 ✓ (1 mark)

d mode for Dolphins = 2 ✓ (1 mark)

e total frequency for Sharks
= 4 + 6 + 5 + 4 + 5
= 24
So the median is the average of the 12th and 13th scores.
median for Sharks = 3 ✓ (1 mark)

f total frequency for Dolphins
= 5 + 8 + 6 + 3 + 2 + 1
= 25
The median is the 13th score.
median for Dolphins = 2 ✓ (1 mark)

g mean for Sharks
= (4 × 0 + 6 × 1 + 5 × 2 + 4 × 3 + 5 × 4) ÷ 24
= 2 ✓ (1 mark)

h mean for Dolphins
= (5 × 1 + 8 × 2 + 6 × 3 + 3 × 4 + 2 × 5 + 6) ÷ 25
= 2.68 ✓ (1 mark)

i The Dolphins have a greater range, mode and mean than the Sharks, although the median is the same for both. On the whole, the Dolphins appear to be the stronger team. ✓ (1 mark)

6 **a** Class 9P: mean = 70.625 [by calculator] ✓
Class 9S: mean = 69.6
Class 9P has the higher mean. ✓ (2 marks)

b The scores in order:
9P: 52, 59, 62, 63, 66, 68, 68, 70, 71, 74, 75, 76, 78, 81, 82, 85
9S: 55, 59, 60, 61, 64, 65, 67, 72, 73, 73, 75, 77, 79, 80, 84
9P: median = 70.5 ✓
9S: median = 72
Class 9S has the higher median. ✓ (2 marks)

c 9P: range = 85 − 52
= 33
9S: range = 84 − 55
= 29 ✓
Difference = 33 − 29
= 4
9P has the greater range by 4 marks. ✓ (2 marks)

d No, I don't agree.
While 9P has the higher mean, class 9S has the higher median. So there is not much difference in the location of the scores. ✓
9P has the higher range so the scores are a little less consistent. ✓ (2 marks)

7

Class T		Class K
7 6	5	6 6 9
6 5 3	6	0 1 1 4 7 7
9 8 4 2 1	7	2 3 5 6 6 8 8 9
8 8 6 5 5 3 1 0	8	1 4 5 6 7 9
4 4 3 2 1 0 0	9	3 4

a Class T: median = 83 ✓
Class K: median = 76 ✓
[The median is the 13th score for both classes.] (2 marks)

b Class T performed better on the test. Its median mark was higher. ✓
The range of marks was the same for both classes. In fact the upper and lower extremes were exactly the same.
The data for class K is symmetric while the data for class T is negatively skewed. ✓ (2 marks)
(Total: 25 marks)

COMPARING DATA DISPLAYS AND USING MEAN, MEDIAN AND RANGE
ADVANCED TEST PAGE 100

1

Males		Females
7 3	4	5
8 6 4 0	5	3 9
9 7 7 7 5 3 2 2 1	6	4 6 8 8 9
8 7 7 6 5 4 4 2 1 1 0	7	0 2 5 7 8 9
7 7 5 4 3 2 0	8	1 3 3 6 8
6	9	0 4 7

a Range for males = 96 – 43
= 53

Range for females = 97 – 45
= 52 ✓

The range is very slightly higher for males than for females, but it is very similar for both groups. ✓ (2 marks)

b Median for males = 71
Median for females = 76 ✓
So the median is 5 years higher for females than for males. ✓ (2 marks)

c Mean for males = 70.3 (1 d.p.; by calculator)
Mean for females = 74.8 (1 d.p.; by calculator) ✓
So the mean age for females is 4.5 years older than for males. ✓ (2 marks)

d The modal age for males = 67
The ages for females are bimodal.
The modes for females are 68 and 83. ✓
So the modal age for males is very similar to one of the modal ages for females, but there is a second mode for females that is much higher. As the modal age only occurs once more than most other ages it is not a particularly useful measure. ✓ (2 marks)

e From the stem-and-leaf plot we can immediately see that there are many more males than females. From the mean and median we can also see that males tend to have problems at a younger age than females. ✓ (1 mark)

f The range for males decreases by 9 to 44. ✓
The mean age would decrease to just under 70 (69.7), but the median and mode would not change. ✓
There is now a more obvious difference between the males and females, with the males tending to present at a younger age than females. ✓ (3 marks)

2

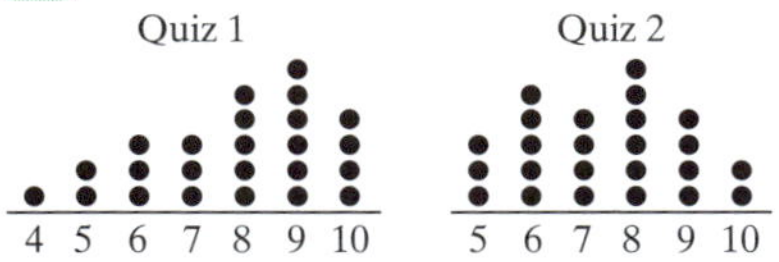

a There are 24 students in the class. ✓ (1 mark)

b Yes, Jesse is correct. ✓
The range for quiz 1 is 6 while the range for quiz 2 is 5. So the spread of scores is less in quiz 2, (but only slightly). ✓ (2 marks)

c Quiz 1:
mean = 7.8 (1 d.p.)
median = 8 ✓
Quiz 2:
mean = 7.375
median = 7.5 ✓
So, both the mean and median were higher for quiz 1, so that is the quiz in which the students performed best. ✓ (3 marks)

3 a Total frequency
$= 4 + 5 + 7 + 6 + 2$
$= 24$
There are 24 students in the class. ✓ (1 mark)

b First test:
Mean
$= (4 \times 3 + 5 \times 4 + 7 \times 5 + 6 \times 6 + 2 \times 7) \div 24$
$= 4.875$ ✓

Second test:
Mean
$= (3 \times 4 + 4 \times 5 + 5 \times 6 + 6 \times 7 + 4 \times 8 + 2 \times 9) \div 24$
$= 6.4166\ldots$
Increase
$= 6.4166\ldots - 4.875$
$= 1.541\,666\ldots$
$= 1.5$ (1 d.p.) ✓ (2 marks)

c There are 24 scores altogether. The median in each test is the average of the 12th and 13th score.
First test: Median = 5 ✓
Second test: median = 6.5
Difference = 1.5 ✓ (2 marks)

d First test: range = 7 – 3
= 4

Second test: range = 9 – 4
= 5

The range increased by one mark. ✓ (1 mark)

e On the whole, the students performed better in the second test than in the first test. The mean and median both increased by 1.5. ✓
The most common score in the first test was 5 and in the second test 7, so this also increased. The range was slightly larger for the second test so those marks were a little less consistent. ✓ (2 marks)

4

Number of messages	Frequency in morning	Frequency in afternoon
0	8	5
1	6	7
2	1	4
3	2	3
4	3	1

Morning: mean = 1.3
median = 1
mode = 0
range = 4 ✓
Afternoon: mean = 1.4
median = 1
mode = 1
range = 4 ✓
The median and range are the same for both times. The mode and mean are slightly higher in the afternoon. ✓
There is not a lot of difference statistically. ✓ (4 marks)
(Total: 30 marks)

RELATIVE FREQUENCIES AND PROBABILITIES
SKILLS CHECK

PAGE 102

1 a Relative frequency of heads
$= \frac{12}{20}$
$= \frac{3}{5}$

b Number of tails = 20 – 12
= 8

Relative frequency of tails
$= \frac{8}{20}$
$= \frac{2}{5}$

2 **a** $P(4) = \frac{1}{6}$

b $P(1) = \frac{1}{6}$

c $P(6) = \frac{1}{6}$

d $P(7) = 0$ [It is impossible.]

3 **a** There are 9 numbers.
$P(7) = \frac{1}{9}$

b There are 4 even numbers.
$P(\text{even}) = \frac{4}{9}$

c There are 5 odd numbers.
$P(\text{odd}) = \frac{5}{9}$

d All the numbers are less than 10.
$P(\text{number less than } 10) = 1$
[It is certain.]

e $P(12) = 0$ [It is impossible.]

f $P(\text{number greater than } 3)$
$= \frac{6}{9}$
$= \frac{2}{3}$

4 **a** relative frequency of 2 $= \frac{7}{50}$

b relative frequency of 5 $= \frac{9}{50}$

c relative frequency of 6 $= \frac{8}{50}$
$= \frac{4}{25}$

5 Total counters = 5 + 4 + 3
= 12

a 5 counters are black.
$P(\text{black}) = \frac{5}{12}$

b 4 counters are yellow.
$P(\text{yellow}) = \frac{4}{12}$
$= \frac{1}{3}$

c No counters are red.
$P(\text{red}) = 0$

d 7 counters are not black.
$P(\text{not black}) = \frac{7}{12}$

6

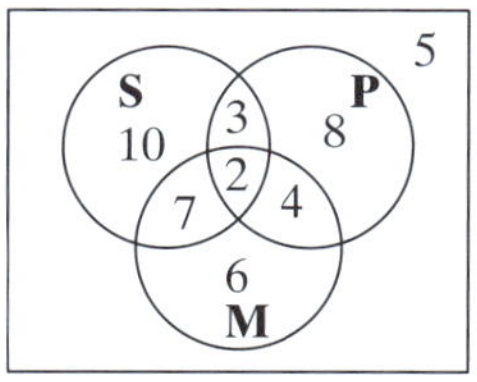

a Number surveyed
= 10+7+3+2+8+4+6+5
= 45

b Number in no activities = 5
[There are 5 not in any of the circles.]

c Number in all three activities = 2
[There are 2 in all three circles.]

d Number who went to movie and party, no sport = 4

e Number who played sport
= 10 + 7 + 3 + 2
= 22

7

	Overweight	Not overweight	Total
Men	73	56	129
Women	42	60	102
Total	115	116	231

a People surveyed = 231

b 73 out of 129 men were overweight; $\frac{73}{129}$

c 60 out of 102 women were not overweight; $\frac{60}{102} = \frac{10}{17}$

d 73 of the 115 overweight people were men; $\frac{73}{115}$

RELATIVE FREQUENCIES AND PROBABILITIES
INTERMEDIATE TEST PAGE 103

1 $P(\text{five}) = \frac{1}{6}$ [There are 6 outcomes, one of those six is the desired outcome.]
∴ [B] ✓ (1 mark)

2 $P(\text{yellow}) = \frac{3}{8}$
[There are 8 marbles to choose from; 3 of those marbles are yellow.]
∴ [B] ✓ (1 mark)

3 Number of traffic lights
= 37 + 28 + 9
= 74
Number of amber lights = 9
Relative frequency of amber
$= \frac{9}{74}$
∴ [D] ✓ (1 mark)

4 **a** Missing number = 63 – 32
= 31 ✓
(1 mark)

b **i** $P(X) = \frac{32}{63}$ ✓ (1 mark)
[32 out of the 63 who prefer brand A]

ii $P(B) = \frac{56}{87}$ ✓ (1 mark)
[56 out of the 87 who prefer brand Y]

iii Total people
= 63 + 75
= 138 ✓
$P(B \text{ and } X) = \frac{19}{138}$ ✓
(2 marks)

5 Number of pegs = 2 + 3 + 1
= 6
$P(\text{blue peg}) = \frac{2}{6}$ ✓
$= \frac{1}{3}$ ✓ (2 marks)

6 **a** $P(\text{grey}) = \frac{2}{5}$ ✓ (1 mark)
[There are five spaces, two of which are grey.]

b Lands on red or grey
37 + 41 = 78 times
Times it lands on blue
= 100 – 78
= 22 ✓
Relative frequency of blue
$= \frac{22}{100}$
$= \frac{11}{50}$ ✓ (2 marks)

c Relative frequency of red
$= \frac{37}{100}$ ✓
In 200 spins it should land on red.
$\frac{37}{100} \times 200 = 74$ times. ✓
(2 marks)

7

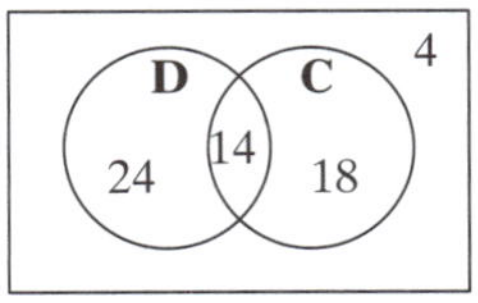

✓

4 households had neither a dog nor a cat.
[14 had both so 38 – 14 = 24 had a dog and not a cat. 32 – 14 = 18 had a cat and not a dog.
24 + 14 + 18 = 56 so 4 (60 – 56) had neither animal.]
$P(\text{neither animal}) = \frac{4}{60}$
$= \frac{1}{15}$ ✓
(2 marks)

8 **a** Relative frequency of pin landing point down
$= \frac{45}{100}$
$= 45\%$ ✓ (1 mark)

b Relative frequency of pin landing point up
$= \frac{55}{100}$
$= \frac{11}{20}$ ✓
In 120 trials the pin should land point up approximately
$\frac{11}{20} \times 120 = 66$ times. ✓
[Or: Relative frequency of pin landing point up = 55%. In 120 trials it should land point up 55% of 120 = 66 times.] (2 marks)
(Total: 20 marks)

RELATIVE FREQUENCIES AND PROBABILITIES
ADVANCED TEST PAGE 104

1

Number	1	2	3	4	5	6
Frequency	8	11	10	9	9	13

a **i** Relative frequency of $2 = \frac{11}{60}$ ✓ (1 mark)

ii Relative frequency of $5 = \frac{9}{60}$
$= \frac{3}{20}$ ✓ (1 mark)

iii Total even scores
$= 11 + 9 + 13$
$= 33$
Relative frequency of even $= \frac{33}{60}$
$= \frac{11}{20}$ ✓ (1 mark)

iv Total greater than 2
$= 10 + 9 + 9 + 13$
$= 41$
Relative frequency
$= \frac{41}{60}$ ✓ (1 mark)

v Total less than 3
$= 8 + 11$
$= 19$
Relative frequency
$= \frac{19}{60}$ ✓ (1 mark)

b $60 \div 6 = 10$
Each score would be expected to occur 10 times.
✓ (1 mark)

c The scores 2 and 6 occurred more times than would be expected. ✓ (1 mark)

d $240 \div 6 = 40$
Each score would be expected to occur 40 times. ✓
If the relative frequency for 1 was used that would give an expected result of 32, but if the die is fair each score should occur an equal number of times. ✓ (2 marks)

2 **a** Relative frequency of heads $= \frac{3}{8}$
So relative frequency of tails $= \frac{5}{8}$
Number of tails
$= \frac{5}{8} \times 24$
$= 15$ ✓ (1 mark)

b Tails occurred 15 times
Number of heads = 40 – 15
= 25
Relative frequency of heads
$= \frac{25}{40}$
$= \frac{5}{8}$ ✓ (1 mark)

c In first trial, number of heads
= 24 – 15
= 9
Total heads = 9 + 25
= 34
Total tosses = 24 + 40
= 64
Relative frequency of heads
$= \frac{34}{64}$
$= \frac{17}{32}$ ✓
The relative frequency is not $\frac{1}{2}$. So, Jonah is wrong.
✓ (2 marks)

3

a $P(3) = \frac{3}{12}$
$= \frac{1}{4}$ ✓ (1 mark)

b In 36 spins you would expect to spin 3 a quarter of the times. So you would expect to spin 3 nine times. In three different trials, 3 was spun less than this expected number, although not much less. ✓
It may be that the spinner is not fair, but we need many more trials to be certain of this. ✓ (2 marks)

c 8 is $\frac{1}{3}$ of the number of times spun.
Number of times = 3 × 8
= 24 ✓
Number of times 3 was spun
$= \frac{1}{4} \times 24$
$= 6$

Number of times 1 was spun
$= 24 - (8 + 6)$
$= 10$ ✓ (2 marks)

4 **a** There are 6 blue marbles.
$P(\text{blue}) = \frac{2}{9} = \frac{6}{27}$
So there are 27 marbles altogether. ✓
Number of red marbles
$= 27 - (8 + 6)$
$= 13$
$P(\text{red}) = \frac{13}{27}$ ✓ (2 marks)

b $P(\text{blue or red}) = \frac{1}{4} + \frac{2}{5}$
$= \frac{13}{20}$ ✓
$P(\text{green})$
$= 1 - P(\text{blue or red})$
$= 1 - \frac{13}{20}$
$= \frac{7}{20}$ ✓ (2 marks)

5

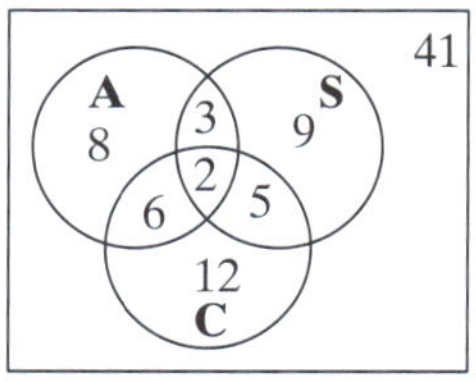

a Total Year 9 students = 86
Number representing in swimming
$= 3 + 9 + 2 + 5$
$= 19$
$P(\text{swimming}) = \frac{19}{86}$ ✓ (1 mark)

b $P(\text{all 3 sports}) = \frac{2}{86}$
$= \frac{1}{43}$ ✓ (1 mark)

c $P(\text{only cross-country})$
$= \frac{12}{86}$
$= \frac{6}{43}$ ✓ (1 mark)

d $P(\text{athletics, cross-country, no swimming})$
$= \frac{6}{86}$
$= \frac{3}{43}$ ✓ (1 mark)

e Number representing in exactly 2 sports
$= 3 + 5 + 6$
$= 14$
$P(\text{exactly 2 sports})$
$= \frac{14}{86}$
$= \frac{7}{43}$ ✓ (1 mark)

f Number in swimming or athletics but not both
$= 9 + 5 + 6 + 8$
$= 28$
$P(\text{swimming or athletics but not both})$
$= \frac{28}{86}$
$= \frac{14}{43}$ ✓ (1 mark)

6 14 had not seen either movie.
Number who had seen a movie
$= 50 - 14$
$= 36$
Now $19 + 23 = 42$
$42 - 36 = 6$
So 6 students must have seen both. ✓

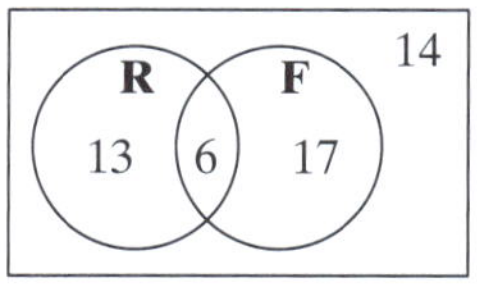

$P(\text{seen both}) = \frac{6}{50}$
$= \frac{3}{25}$ ✓ (2 marks)

(Total: 30 marks)

TWO-STEP CHANCE EXPERIMENTS
SKILLS CHECK PAGE 106

1 **a** head, tail
b 6, 7, 8, 9

2 **a** 1, 2, 3, 4, 5, 6
b A, E, G, N, R, S, T

3 $P(\text{not k}) = 1 - \frac{1}{26}$
$= \frac{25}{26}$

4

+	1	2	3	4	5	6
1	2	3	4	5	6	7
2	3	4	5	6	7	8
3	4	5	6	7	8	9
4	5	6	7	8	9	10
5	6	7	8	9	10	11
6	7	8	9	10	11	12

a $P(8) = \frac{5}{36}$

b $P(\text{less than 4}) = \frac{3}{36}$
$= \frac{1}{12}$

c $P(\text{greater than 6}) = \frac{21}{36}$
$= \frac{7}{12}$

d $P(\text{even}) = \frac{18}{36}$
$= \frac{1}{2}$

5 **a** If the first pen is replaced there are 10 pens to choose from, 7 of which are blue.
$P(\text{blue}) = \frac{7}{10}$

b If the first pen is not replaced, 9 pens remain, 6 of which are blue.
$P(\text{blue}) = \frac{6}{9}$
$= \frac{2}{3}$

6

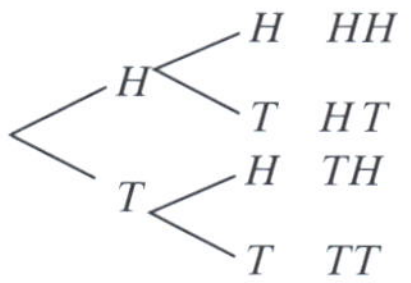

a $P(HH) = \frac{1}{4}$

b $P(TT) = \frac{1}{4}$

c $P(\text{one head and one tail})$
$= \frac{2}{4} = \frac{1}{2}$

d $P(\text{at least one head}) = \frac{3}{4}$

7 **a** Total tees $= 5 + 4$
$= 9$
$P(\text{white}) = \frac{4}{9}$

b 8 tees remain, 3 of those are white.
$P(\text{white}) = \frac{3}{8}$

c If the tee is replaced 4 of the 9 are white.
$P(\text{white}) = \frac{4}{9}$

8 **a**

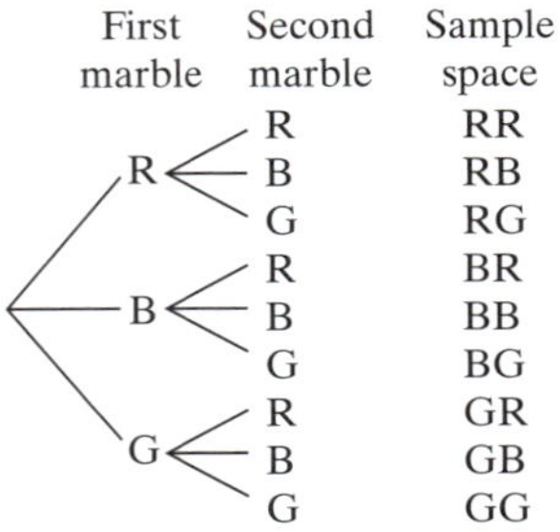

b $P(\text{one red, one blue}) = \frac{2}{9}$

9 **a**

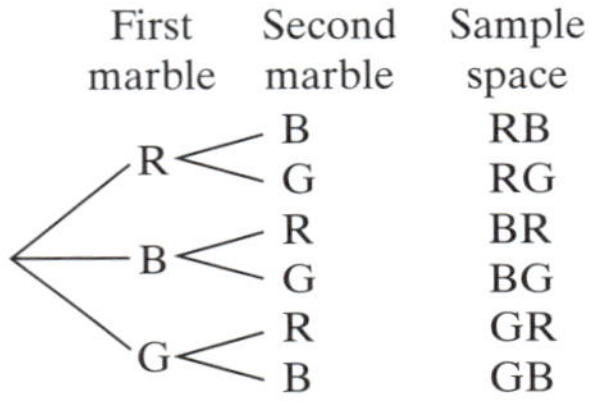

b $P(\text{one red, one blue})$
$= \frac{2}{6} = \frac{1}{3}$

TWO-STEP CHANCE EXPERIMENTS
INTERMEDIATE TEST PAGE 107

1 The probability cannot be $\frac{5}{2}$.
[The probability of any event must be between 0 and 1. $\frac{5}{2}$ is greater than one.]
∴ [C] ✓ (1 mark)

2 If the first peg is red, 4 blue, 2 red and 2 green pegs remain.
Total remaining pegs
$= 4 + 2 + 2$
$= 8$
$P(\text{red peg}) = \frac{2}{8} = \frac{1}{4}$
∴ [B] ✓ (1 mark)

3 Possible outcomes:
HH, HT, TH, TT
$P(TT) = \frac{1}{4}$
∴ [A] ✓ (1 mark)

4

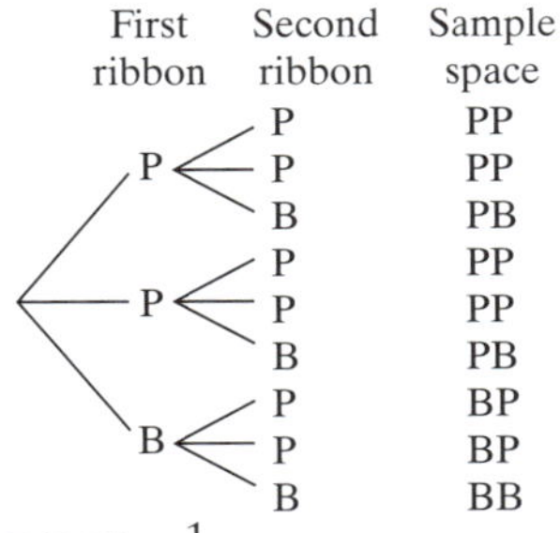

$P(BB) = \frac{1}{9}$
∴ [B] ✓ (1 mark)

5 **a**

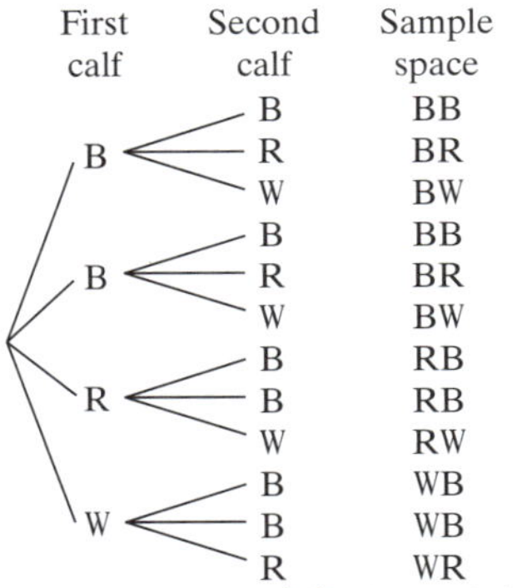

✓✓ (2 marks)

b $P(BB) = \frac{2}{12}$
$= \frac{1}{6}$ ✓ (1 mark)

c $P(\text{at least 1 is red}) = \frac{6}{12}$
$= \frac{1}{2}$ ✓
(1 mark)

d $P(\text{black}) = \frac{1}{3}$ ✓ (1 mark)
[There are three calves to choose from, 1 of these is black.]

6

	1	2	3	4	5	6
1	0	1	2	3	4	5
2	1	0	1	2	3	4
3	2	1	0	1	2	3
4	3	2	1	0	1	2
5	4	3	2	1	0	1
6	5	4	3	2	1	0

$P(1 \text{ or } 3) = \frac{16}{36} = \frac{4}{9}$ ✓

$P(2, 4 \text{ or } 5) = \frac{14}{36} = \frac{7}{18}$ ✓

Paris has a greater chance of winning because the probability of getting a 1 or a 3 is higher than the probability of getting a 2, 4 or 5. ✓ (3 marks)

7 **a**

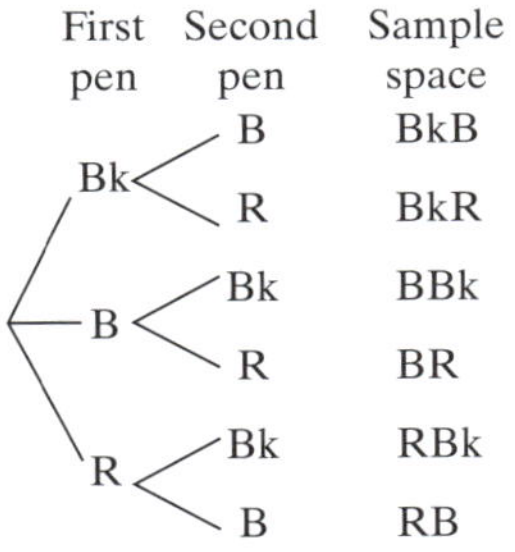

✓

$P(\text{Black and Blue}) = \frac{2}{6}$
$= \frac{1}{3}$ ✓

[BkB or BBk] (2 marks)

b $P(BkB) = \frac{1}{2}$ ✓ (1 mark)
[The first pen was black so only consider those elements of the sample space that begin with black.]

8 **a** $P(\text{B is first}) = \frac{1}{5}$ ✓ (1 mark)
[There are 5 cards that could be first and B is one of those.]

b $P(\text{C is second}) = \frac{1}{5}$ ✓
(1 mark)
[There are 5 cards that could be second and C is one of those.]

c
AB AC AD AE
BA BC BD BE
CA CB CD CE
DA DB DC DE
EA EB EC ED ✓
(1 mark)

d $P(\text{DE or ED}) = \frac{2}{20}$
$= \frac{1}{10}$ ✓
(1 mark)

e If the first card was replaced there would be 5 more possible pairs.
[AA, BB, CC, DD, EE]
$P(\text{DE or ED}) = \frac{2}{25}$ ✓
(1 mark)
(Total: 20 marks)

TWO-STEP CHANCE EXPERIMENTS
ADVANCED TEST PAGE 108

1 **a**

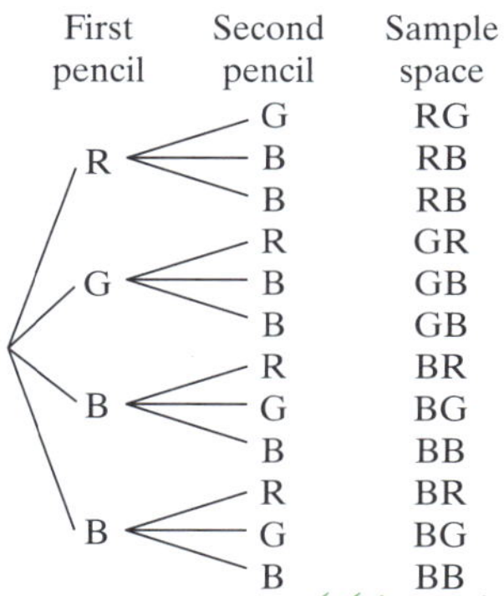

✓✓ (2 marks)

b **i** $P(BB) = \frac{2}{12}$

$= \frac{1}{6}$ ✓ (1 mark)

ii $P(\text{same colour})$

$= P(BB)$

$= \frac{1}{6}$ ✓ (1 mark)

iii $P(\text{first is red}) = \frac{1}{4}$ ✓ (1 mark)

iv $P(\text{exactly 1 blue})$

$= \frac{8}{12}$

$= \frac{2}{3}$ ✓ (1 mark)

v $P(\text{neither is blue})$

$= P(RG \text{ or } GR)$

$= \frac{2}{12}$

$= \frac{1}{6}$ ✓ (1 mark)

vi $P(\text{neither red})$

$= \frac{6}{12}$

$= \frac{1}{2}$ ✓ (1 mark)

vii $P(GB \text{ or } BG)$

$= \frac{4}{12}$

$= \frac{1}{3}$ ✓ (1 mark)

c If pencil is replaced:

$P(BB) = \frac{4}{16}$

$= \frac{1}{4}$ ✓

So, the probability of getting two blue pencils is greater if the first pencil is replaced. ✓ (2 marks)

2 **a** $P(\text{Cath's mug is blue})$

$= \frac{2}{6} = \frac{1}{3}$ ✓ (1 mark)

b $P(\text{Cath's mug now blue})$

$= \frac{2}{5}$ ✓ (1 mark)

c If the colour of Jack's mug is not known, then Cath has an equal chance of getting any of the mugs so she has two chances in six of getting a blue one. But, if we know that Jack's mug is white, Cath can't get that mug so there are now only five mugs from which she can choose. She has two chances in five of getting a blue one. ✓ (1 mark)

3

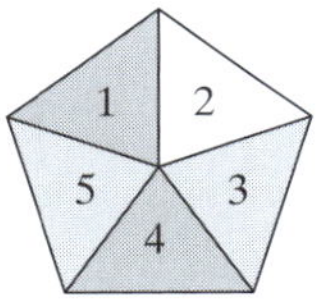

a It doesn't matter what number is spun first, there is one chance in five of the same number occurring again.

$P(\text{same number}) = \frac{1}{5}$ ✓ (1 mark)

b It doesn't matter which number is spun first, there is one chance in five of spinning a number that when added to the first number will make 6.

$P(\text{adding to 6}) = \frac{1}{5}$ ✓ (1 mark)

c $P(\text{not the same number})$

$= 1 - \frac{1}{5}$

$= \frac{4}{5}$

[Half will have the first number greater than the second.]

$P(\text{first number is greater})$

$= \frac{2}{5}$ ✓ (1 mark)

4 **a** $P(RR) = \frac{3}{10} \times \frac{2}{9}$

$= \frac{1}{15}$ ✓ (1 mark)

b $P(\text{neither is green})$

$= \frac{5}{10} \times \frac{4}{9}$

$= \frac{2}{9}$ ✓

$P(\text{at least one green})$

$= 1 - P(\text{neither is green})$

$= 1 - \frac{2}{9}$

$= \frac{7}{9}$ ✓ (2 marks)

c $P(\text{exactly one is green})$

$= P(G\tilde{G}) + P(\tilde{G}G)$

$= \frac{5}{10} \times \frac{5}{9} + \frac{5}{10} \times \frac{5}{9}$

$= \frac{5}{9}$ ✓ (1 mark)

5 **a** $P(\text{same song}) = \frac{1}{10}$

[It doesn't matter which song is played first.] ✓ (1 mark)

b $5\% = \frac{1}{20}$

So if there were 20 songs the probability would be 5%. There would need to be 21 songs. ✓ (1 mark)

6 **a** **i** If the card is replaced there are 10 possible cards that could be chosen first and 10 that could be chosen second.

Number of outcomes

$= 10 \times 10$

$= 100$

Outcomes that have a sum of 7:

$1 + 6, 2 + 5, 3 + 4, 4 + 3, 5 + 2$ and $6 + 1$

$P(\text{sum of 7})$

$= \frac{6}{100}$

$= \frac{3}{50}$ ✓ (1 mark)

ii Without replacement there are only 9 cards that could be chosen second.

Number of outcomes
$= 10 \times 9$
$= 90$

$P(\text{sum of } 7) = \frac{6}{90}$
$= \frac{1}{15}$ ✓ (1 mark)

b Odd card numbers are 1, 3, 5, 7 and 9.
Prime card numbers are 2, 3, 5 and 7.
Numbers both odd and prime are 3, 5 and 7.
$P(\text{odd and prime}) = \frac{3}{10}$ ✓
(1 mark)

c $P(\text{not odd nor prime})$
$= 1 - \frac{3}{10}$
$= \frac{7}{10}$

$P(\text{both not odd nor prime})$
$= \frac{7}{10} \times \frac{7}{10}$
$= \frac{49}{100}$ ✓

Taseefa will lose money. The probability of winning is less than one-half. ✓
(2 marks)

d If the first card is not replaced:
$P(\text{both not odd nor prime})$
$= \frac{7}{10} \times \frac{6}{9}$
$= \frac{7}{15}$ ✓

No, it will not make a difference. The probability of winning is still less than half. ✓ (2 marks)
(Total: 30 marks)

SAMPLE EXAM PAPER 1
Part A: Multiple Choice PAGE 110

1 [The total length of the right-hand side must be the same as the left-hand side, 10 m. So the missing measurement on the right side is 4 m.]

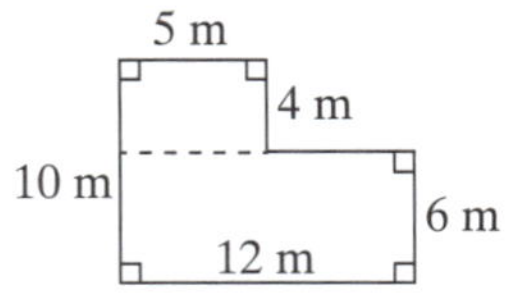

$A = 5 \times 4 + 12 \times 6$
$= 20 + 72$
$= 92$
Area is 92 m².
∴ [A] ✓ (1 mark)

2 If $k = -3$,
$2k^2 = 2 \times (-3)^2$
$= 2 \times 9$
$= 18$
∴ [A] ✓ (1 mark)

3 $x^2 = 7^2 + 4^2$
$= 65$

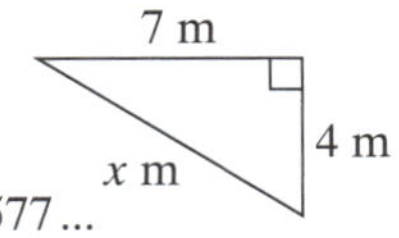

$x = \sqrt{65}$
$= 8.0622577\ldots$
Of the options the value of x is closest to 8.
∴ [A] ✓ (1 mark)

4 $0.000476 = 4.76 \times 10^{-4}$
[The decimal point moves 4 places.]
∴ [B] ✓ (1 mark)

5
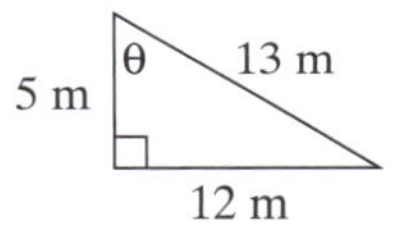

$\sin\theta = \frac{\text{opposite}}{\text{hypotenuse}}$
$= \frac{12}{13}$
∴ [D] ✓ (1 mark)

6 D is a pyramid not a prism.
∴ [D] ✓ (1 mark)

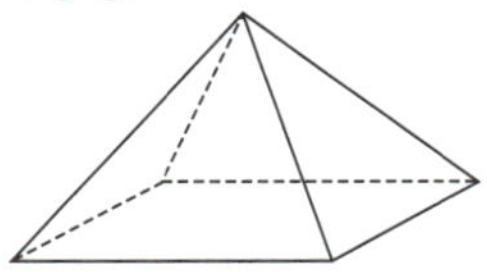

[A is a rectangular prism (or cube), B is an octagonal prism and C is a triangular prism.]

7 835 = 840 (2 significant figures)
∴ [D] ✓ (1 mark)
[The third figure is not required. Because it is 5 the second figure must be rounded up. The original number was close to 800 not 80 so the answer must be close to 800 not 80.]

8 [Melanie works 6 hours at 1½ times the normal rate of pay.]
Earnings $= 6 \times 1.5 \times \$18.60$
$= \$167.40$
∴ [C] ✓ (1 mark)

9 One microsecond is one-millionth of a second. $\frac{1}{1\,000\,000}$
∴ [D] ✓ (1 mark)

10 $7 - 2(x - 3) = 7 - 2x + 6$
$= 13 - 2x$
∴ [C] ✓ (1 mark)

11
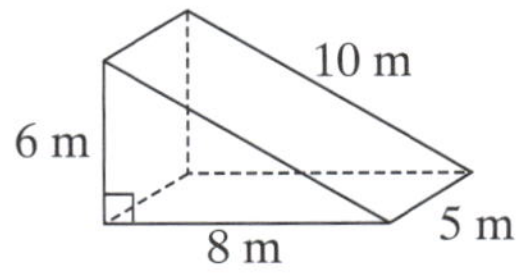

[The base of the prism is a triangle.]
Base: $A = \frac{1}{2} \times 8 \times 6$
$= 24$
area of base = 24 m²
$V = Ah$
$= 24 \times 5$
$= 120$
volume = 120 m³
∴ [A] ✓ (1 mark)

12 $A = 6s^2$
$= 6 \times 5^2$
$= 6 \times 25$
$= 150$
Surface area = 150 m²
∴ [C] ✓ (1 mark)

13 $\frac{392}{21.3 \times 4.8} = \frac{392}{102.24}$
$= 3.8341158\ldots$
$= 3.8$ (1 d.p.)
∴ [A] ✓ (1 mark)

14 $y = 4 - x^2$ is a parabola.
It is 'upside down'.
When $x = 0$, $y = 4$

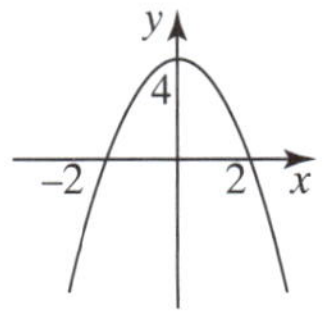

∴ [B] ✓ (1 mark)

15 $6^0 + 2x^0 = 1 + 2 \times 1$
$= 1 + 2$
$= 3$
∴ **[D]** ✓ (1 mark)

16 100 m in 15 seconds
= 400 m in 1 minute
[multiplying by 4]
= 24 000 m in 1 hour
[multiplying by 60]
= 24 km per hour
∴ **[A]** ✓ (1 mark)

17

11	3 5 5
12	0 4 6 8 9
13	1 2 3 5 7 9
14	0 1 4 7 8

There are 19 scores altogether. So, the tenth score is the median.
Median = 132
∴ **[C]** ✓ (1 mark)

18

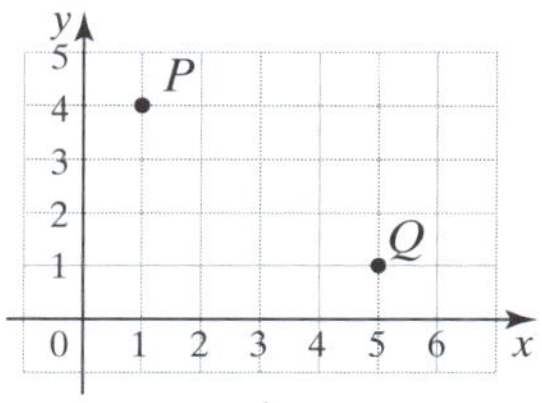

Gradient $= \frac{\text{rise}}{\text{run}}$
$= -\frac{3}{4}$
∴ **[B]** ✓ (1 mark)

19 $P(1, 4)$, $Q(5, 1)$
$d^2 = 3^2 + 4^2$
$= 25$
$d = \sqrt{25}$
$= 5$
[Or use the distance formula:
$d = \sqrt{(x_2 - x_1)^2 + (y_2 - y_1)^2}$
$= \sqrt{(5-1)^2 + (1-4)^2}$
$= \sqrt{16 + 9}$
$= \sqrt{25}$
$= 5$]
The distance is 5 units.
∴ **[C]** ✓ (1 mark)

20 8, 7, 9, 9, 6
Mean $= \frac{8 + 7 + 9 + 9 + 6}{5}$
$= 7.8$
Mode = 9
Range = 9 – 6
= 3
6, 7, 8, 9, 9
Median = 8
∴ **[B]** ✓ (1 mark)

21 $\cos\theta = 0.758$
$\theta = 40.711\,80\ldots°$
$= 41°$ (nearest degree)
∴ **[C]** ✓ (1 mark)
[using $\cos^{-1}$ on a calculator]

22 $P(\text{head}) = \frac{1}{2}$

$P(\text{tail}) = \frac{1}{2}\ (= 0.5)$

Both are correct
∴ **[D]** ✓ (1 mark)

23

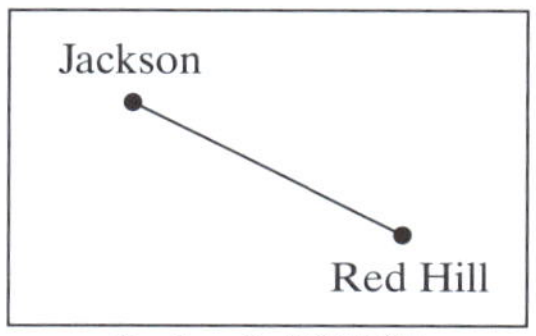

Scale 1 : 200 000

On the map:
1 cm = 200 000 cm
= 2000 m
= 2 km
6 cm = 6 × 2 km
= 12 km
∴ **[B]** ✓ (1 mark)

24 $P(-2, 8)$, $Q(4, 2)$
$x = \frac{-2 + 4}{2}$ $\quad y = \frac{8 + 2}{2}$
$= 1$ $\quad = 5$
The midpoint is (1, 5).
∴ **[B]** ✓ (1 mark)

25 $(2x^3)^2 = 4x^6$
$[2^2 = 4,\ (x^3)^2 = x^{3 \times 2} = x^6$
Or $(2x^3)^2 = 2x^3 \times 2x^3$
$= 4x^6]$
∴ **[D]** ✓ (1 mark)

26 Number of pupils = 14 + 16
= 30
$P(\text{girl}) = \frac{14}{30}$
$= \frac{7}{15}$
∴ **[A]** ✓ (1 mark)

27 Number of tosses = 20 + 30
= 50
Relative frequency of heads
$= \frac{20}{50}$
$= \frac{2}{5}$
∴ **[B]** ✓ (1 mark)

28 $r = 4,\ h = 7$
$V = \pi r^2 h$
$= \pi \times 4^2 \times 7$
$= 351.858\,3772\ldots$
Volume = 352 cm^3 to the nearest cm^3.
∴ **[B]** ✓ (1 mark)

29 The displays in options A, B and D are all reasonably symmetric. Option C is negatively skewed.
∴ **[C]** ✓ (1 mark)

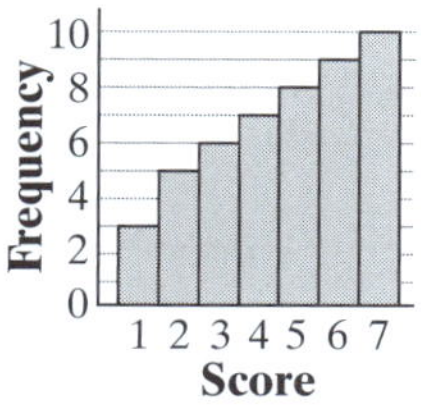

30

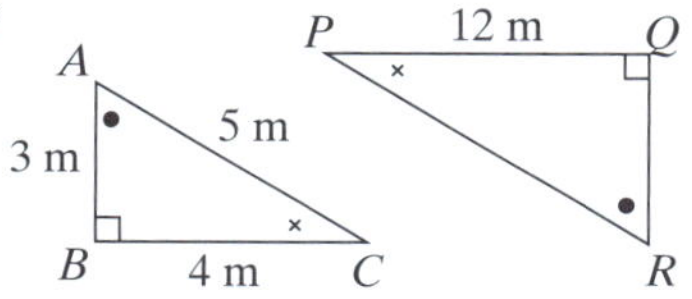

$\frac{QR}{AB} = \frac{PQ}{BC}$
$\frac{QR}{3} = \frac{12}{4}$
$= 3$
$QR = 3 \times 3$
$= 9$
Length of QR is 9 m.
∴ **[A]** ✓ (1 mark)
(Total Part A: 30 marks)

SAMPLE EXAM PAPER 1
Part B: Short Answer PAGE 113

31 Cash price = $3600
a Deposit = 30% of $3600
$= \frac{30}{100} \times \$3600$
= $1080 ✓ (1 mark)

b Balance = $3600 – $1080
= $2520 ✓ (1 mark)

c $I = PRN$
$= \$2520 \times \frac{5}{100} \times 2$
= $252
Nico will pay $252 interest. ✓ (1 mark)

d Total to be repaid
$= \$2520 + \$252$
$= \$2772$ ✓
2 years = 24 months
Each monthly instalment
$= \$2772 \div 24$
$= \$115.50$ ✓ (2 marks)

32

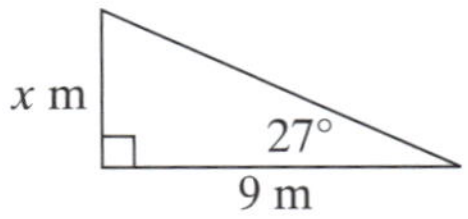

$\tan 27° = \dfrac{x}{9}$

$x = 9 \times \tan 27°$ ✓
$= 4.585\,729\ldots$
$= 4.6$ (1 d.p.) ✓
(2 marks)

33

No. of computers	0	1	2	3	4	5
No. of students	13	94	215	76	38	4

a Total students
$= 13 + 94 + 215 + 76 + 38 + 4$
$= 440$ ✓ (1 mark)

b Total computers
$= 94 + 2 \times 215 + 3 \times 76 + 4 \times 38 + 5 \times 4$
$= 924$ ✓ (1 mark)

c Mean $= 924 \div 440$
$= 2.1$ ✓ (1 mark)

34 **a** $A = kh$
When $h = 38$, $A = 1425$
$1425 = k \times 38$
$k = 1425 \div 38$
$= 37.5$ ✓ (1 mark)

b It is the amount that Jim earns every hour. He earns \$37.50 per hour. ✓(1 mark)

c $A = 37.5h$
When $h = 35$,
$A = 37.5 \times 35$
$= 1312.5$
Jim earns \$1312.50. ✓
(1 mark)

d $A = 37.5h$
$1575 = 37.5h$
$h = 1575 \div 37.5$
$= 42$ ✓
Jim worked 42 h that week.
Extra hours = 42 – 38 = 4
Jim worked 4 extra hours. ✓ (2 marks)

35

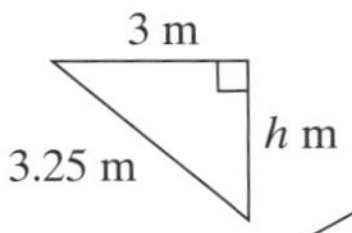

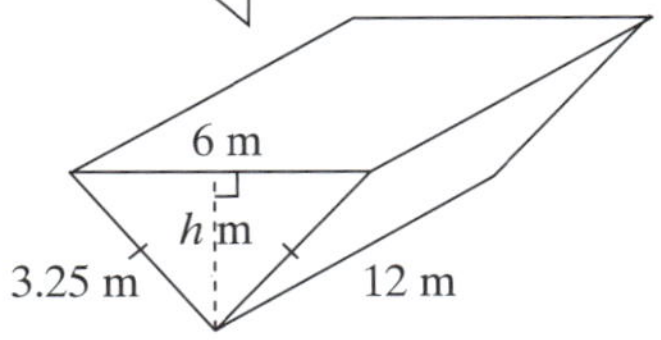

a $h^2 = 3.25^2 - 3^2$
$= 1.5625$
$h = \sqrt{1.5625}$ $(h > 0)$
$= 1.25$ ✓ (1 mark)

b $A = \dfrac{1}{2}bh$
$= \dfrac{1}{2} \times 6 \times 1.25$
$= 3.75$
The area is 3.75 m^2. ✓
(1 mark)

c There are two triangular faces, each with area 3.75 m^2, two rectangular faces that are 3.25 m by 12 m and one rectangular face that is 6 m by 12 m.
$A = 2 \times 3.75 + 2 \times 3.25 \times 12 + 6 \times 12$
$= 157.5\ m^2$ ✓ (1 mark)

d $V = AH$
$= 3.75 \times 12$
$= 45$ ✓
The volume is 45 m^3.
The capacity is 45 000 L. ✓
(2 marks)
(Total Part B: 20 marks)

SAMPLE EXAM PAPER 2
Part A: Multiple Choice PAGE 114

1 $a = 3, b = -5$
$ab^2 = 3 \times (-5)^2$
$= 3 \times 25$
$= 75$
∴ **[D]** ✓ (1 mark)

2

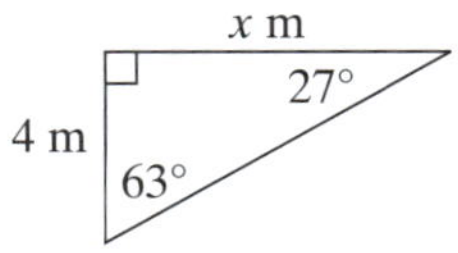

$\tan 63° = \dfrac{x}{4}$

$x = 4 \tan 63°$
∴ **[D]** ✓ (1 mark)

3 $3^{-4} = \dfrac{1}{3^4} = \dfrac{1}{81}$
∴ **[A]** ✓ (1 mark)

4

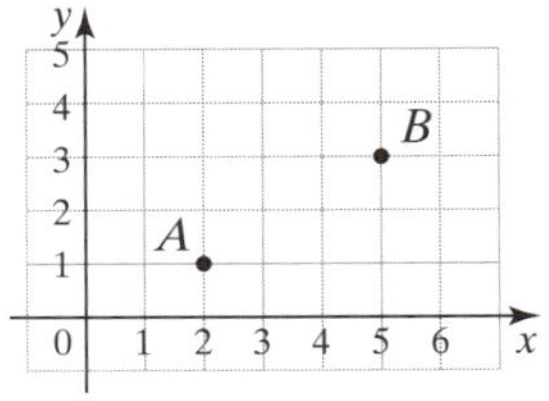

Gradient $= \dfrac{\text{rise}}{\text{run}}$
$= \dfrac{2}{3}$

[Or: $A(2, 1), B(5, 3)$
$m = \dfrac{y_2 - y_1}{x_2 - x_1}$
$= \dfrac{3 - 1}{5 - 2}$
$= \dfrac{2}{3}$]
∴ **[A]** ✓ (1 mark)

5 $y = -\dfrac{x}{6}$

$y = -\dfrac{1}{6}x$

This is the equation of a straight line.
[gradient $= -\dfrac{1}{6}$, y-intercept = 0]
∴ **[C]** ✓ (1 mark)

6 6.2037 = 6.20 (3 significant figures)
[Three figures are required. The fourth figure is 3, which is less than 5 so there is no need to round up. In the answer the 0 is significant.]
∴ **[C]** ✓ (1 mark)

7 72 km per hour
= 72 000 m per hour
= 1200 m per minute
= 20 m per second
∴ **[A]** ✓ (1 mark)

8 Number of red supporters
$= 12 + 18$
$= 30$
P(red supporter is under 16)
$= \dfrac{12}{30}$
$= \dfrac{2}{5}$
∴ **[B]** ✓ (1 mark)

9 $y = 4 - 3x$
$y = -3x + 4$
Gradient is -3
[The gradient is the coefficient of x (the number by which x is multiplied).]
$\therefore$ **[D]** ✓ (1 mark)

10

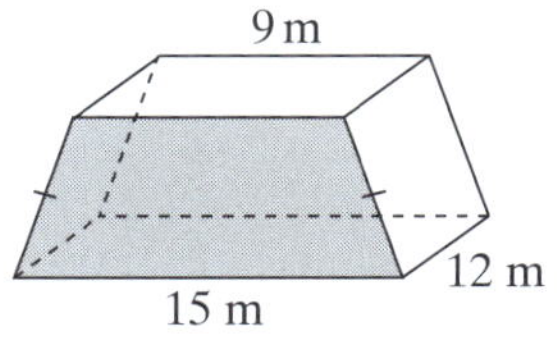

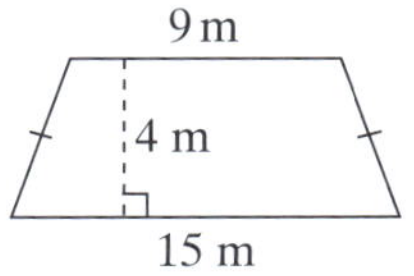

Area of trapezium

$A = \frac{1}{2}h(a + b)$
$= \frac{1}{2} \times 4 \times (9 + 15)$
$= 2 \times 24$
$= 48$

Area is 48 m^2.

$V = Ah$
$= 48 \times 12$
$= 576$

Volume is 576 m^3.
$\therefore$ **[B]** ✓ (1 mark)

11 $P(3, 5)$, $Q(-2, 5)$
$PQ = (3 - (-2))$ units
$= 5$ units
$\therefore$ **[B]** ✓ (1 mark)

12 $2(2a + 3) + 4(a - 2)$
$= 4a + 6 + 4a - 8$
$= 8a - 2$
$\therefore$ **[D]** ✓ (1 mark)

13

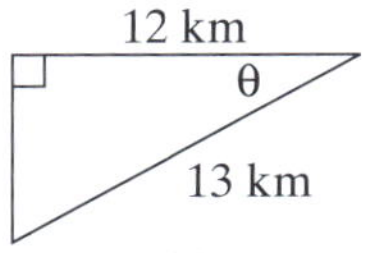

$\cos\theta = \frac{12}{13}$
$\theta = 22.619\,864\,95\ldots^\circ$
$= 23^\circ$ (nearest degree)
$\therefore$ **[A]** ✓ (1 mark)

14 $16m^{-6} \div 8m^{-2} = 2m^{-6-(-2)}$
$= 2m^{-4}$
$\therefore$**[A]** ✓ (1 mark)

15

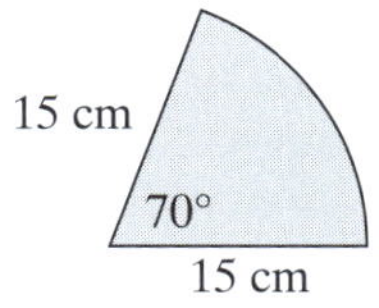

$A = \frac{\theta}{360} \times \pi r^2$
$= \frac{70}{360} \times \pi \times 15^2$
$= 137.444\,6786\ldots$

Area = 137 cm^2 (nearest cm^2)
$\therefore$ **[B]** ✓ (1 mark)

16 $\frac{n}{4} = \frac{2}{3}$
$n = \frac{2}{3} \times 4$
$= 2\frac{2}{3}$
$\therefore$ **[C]** ✓ (1 mark)

17 $A = 6s^2$
$541.5 = 6s^2$
$s^2 = 541.5 \div 6$
$= 90.25$
$s = \sqrt{90.25}$ $(s > 0)$
$= 9.5$

The length of each side is 9.5 cm.
$\therefore$ **[C]** ✓ (1 mark)

18 8, 7, 2, 5, 8, 11
mean $= \frac{8 + 7 + 2 + 5 + 8 + 11}{6}$
$= \frac{41}{6}$
$= 6\frac{5}{6}$

2, 5, 7, 8, 8, 11
median $= \frac{7 + 8}{2}$
$= 7.5$
mode = 8
range = 11 − 2
= 9

The range is the greatest
$\therefore$ **[D]** ✓ (1 mark)

19 $r = 9, h = 7$
$A = 2\pi r^2 + 2\pi rh$
$= 2 \times \pi \times 9^2 + 2 \times \pi \times 9 \times 7$
$= 904.778\,684\ldots$
$= 905$ (nearest whole number)
$\therefore$ **[B]** ✓ (1 mark)

20 Commission = 3.5% of \$7000
$= \frac{3.5}{100} \times \$7000$
$= \$245$
Pay = \$800 + \$245
= \$1045
$\therefore$ **[C]** ✓ (1 mark)

21

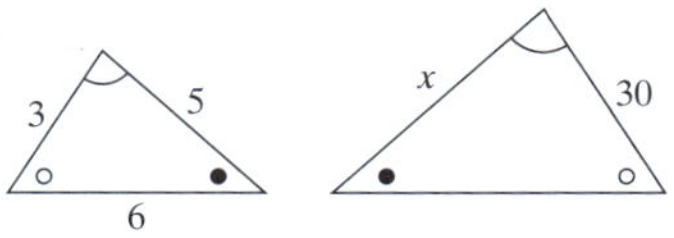

$\frac{x}{5} = \frac{30}{3}$
$= 10$
(corresponding sides of similar triangles)
$x = 5 \times 10$
$= 50$
$\therefore$ **[C]** ✓ (1 mark)

[The side marked x corresponds to the side marked 5 because they are in the same position in relation to the equal angles. Similarly the side marked 30 corresponds to the side marked 3.]

22 $x - y = 4$
[Substitute each point into the equation to see if it is satisfied.]

(5, 1)	5 − 1 = 4	(correct)
(−1, 5)	−1 − 5 = −6	(incorrect)
(4, 0)	4 − 0 = 4	(correct)
(2, −2)	2 − (−2) = 4	(correct)

(−1, 5) does not lie on the line.
$\therefore$ **[B]** ✓ (1 mark)

23

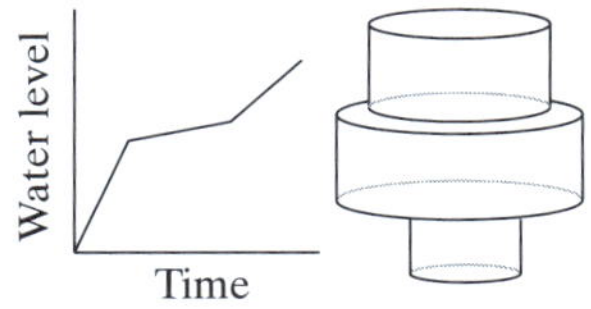

The water level in the container rises quickly, then slower, then more quickly but not as fast as it first did.
$\therefore$ **[B]** ✓ (1 mark)

24 $V = 36$, $h = 8$
$V = Ah$
$36 = A \times 8$
$A = 36 \div 8$
$= 4.5$

The area is 4.5 cm^2.
$\therefore$ **[B]** ✓ (1 mark)

[It does not matter what type of prism it is.]

25 [Check each triangle to see whether Pythagoras' Theorem holds.]

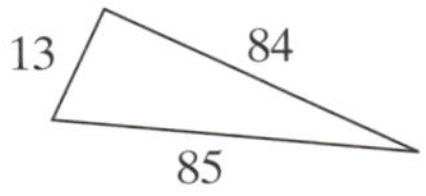

$13^2 + 84^2 = 169 + 7056$
$= 7225$
$85^2 = 7225$
$13^2 + 84^2 = 85^2$
The triangle is right-angled.

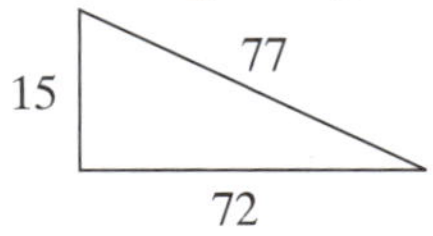

$15^2 + 72^2 = 225 + 5184$
$= 5409$
$77^2 = 5929$
$15^2 + 72^2 \neq 77^2$
The triangle is not right-angled.

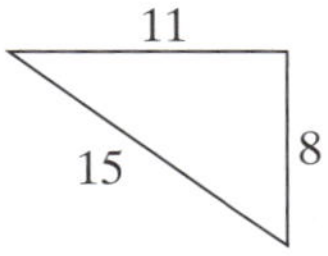

$11^2 + 8^2 = 121 + 64$
$= 185$
$15^2 = 225$
$11^2 + 8^2 \neq 15^2$
The triangle is not right-angled.

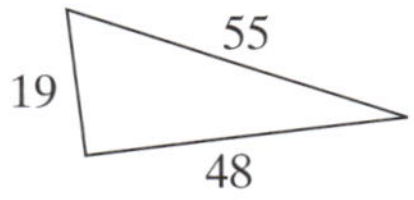

$19^2 + 48^2 = 361 + 2304$
$= 2665$
$55^2 = 3025$
$19^2 + 48^2 \neq 55^2$
The triangle is not right-angled.
Only the first triangle is right-angled.
$\therefore$ **[A]** ✓ (1 mark)

26 $0.7\% = 0.7 \div 100 = 0.007$
$6 \times 10^{-2} = 0.06$
$\frac{1}{50} = 0.02$
Numbers are:
0.007, 0.06, 0.02, 0.04
Smallest is 0.007
$\therefore$ **[A]** ✓ (1 mark)

27 $x^2 + y^2 = 100$
Circle, centre (0, 0), radius 10 units
Distance from (0, 0) to (6, 8):

$$d = \sqrt{(x_2 - x_1)^2 + (y_2 - y_1)^2}$$
$$= \sqrt{(6-0)^2 + (8-0)^2}$$
$$= \sqrt{100}$$
$$= 10$$

The distance from the centre to (6, 8) is 10 units.
$\therefore$ the point lies on the circle
$\therefore$ **[B]** ✓ (1 mark)

28 A nanosecond is one thousand millionth of a second.

$$\frac{1}{1\,000\,000\,000}$$

$\therefore$ **[D]** ✓ (1 mark)

29 The graph has its vertex at the origin. Its equation is of the form $y = ax^2$ and it passes through the point (2, 2).

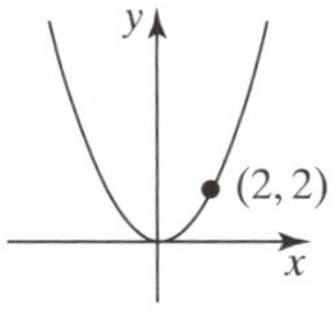

$y = ax^2$
$2 = a \times 2^2$
$= 4a$
$a = \frac{1}{2}$
$y = \frac{1}{2}x^2$
$= \frac{x^2}{2}$
$\therefore$ **[D]** ✓ (1 mark)

30 $r = 2.5, \quad h = 3$
$V = \pi r^2 h$
$= \pi \times 2.5^2 \times 3$
$= 58.904\,86\ldots$
$= 60$ (nearest ten)
$\therefore$ the volume is 60 m^3
Capacity = 60 000 L
= 60 kL
$\therefore$ **[C]** ✓ (1 mark)
(Total Part A: 30 marks)

SAMPLE EXAM PAPER 2
Part B: Short Answer PAGE 117

31 **a** $3x(2x - 5) - 4(2x - 7)$
$= 6x^2 - 15x - 8x + 28$
$= 6x^2 - 23x + 28$ ✓ (1 mark)

b $\frac{36a^5bc^2}{(3abc)^2} = \frac{36a^5bc^2}{9a^2b^2c^2}$
$= \frac{4a^3}{b}$ ✓ (1 mark)

32 **a** Distance
= 150 000 000 km
= 150 000 000 × 1000 m
= 150 000 000 000 m
= 1.5×10^{11} m ✓ (1 mark)

b Time
= $((1.5 \times 10^{11}) \div (3 \times 10^8))$ s
= 500 s ✓
= 8 min 20 s ✓ (2 marks)

33 **a** 3 years = 3 × 12 months
= 36 months
Total repaid
= 36 × \$298.75
= \$10 755 ✓ (1 mark)

b Interest = \$10 755 – \$9000
= \$1755 ✓ (1 mark)

c $I = PRN$
$1755 = 9000 \times R \times 3$
$= 27\,000R$
$R = 1755 \div 27\,000$
$= 0.065$ ✓
Interest rate
$= 0.065 \times 100\%$
$= 6.5\%$ ✓ (2 marks)

34

Class 9H		Class 9R
7 4 3	5	2 8
9 6 2 2 1 0	6	1 1 3 5
9 8 7 5 4 2 1	7	0 2 6 7 9
8 6 5 3 1 0	8	1 2 4 5 7 7 7 9
6 5 4 1	9	2 3 3 4 5 5

a Class 9H: median $= \frac{75 + 77}{2}$
$= 76$
Class 9R: median = 82
Difference = 82 – 76
= 6
The median of class 9R is 6 marks higher than that of class 9H. ✓ (1 mark)

b 9H: range = 96 – 53
= 43
9R: range = 95 – 52
= 43
The range for both classes is the same. ✓ (1 mark)

c The median was higher for 9R so that class performed better than 9H.
The range was the same for both classes so they were equally consistent. ✓
The display for 9H is reasonably symmetric but the display for 9R is negatively skewed. Because it has more higher marks than 9H, this is another indication that 9R did better. ✓ (2 marks)

35 **a**

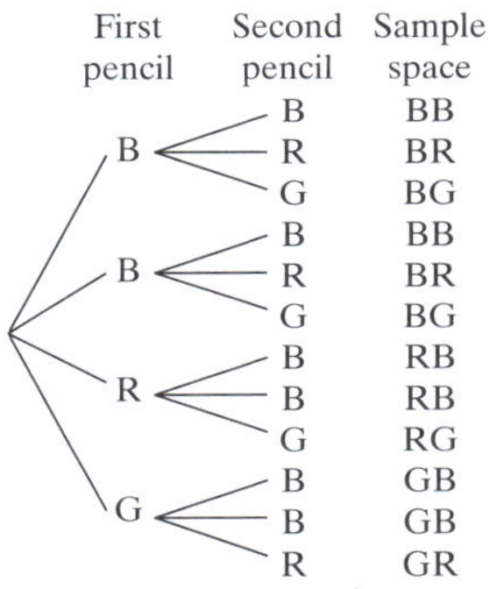

✓ (1 mark)

b P(one blue and one red)
$= P(\text{RB or BR})$
$= \frac{4}{12}$
$= \frac{1}{3}$ ✓ (1 mark)

36 **a** $r = 1.5$, $h = 2.2$
$V = \pi r^2 h$
$= \pi \times 1.5^2 \times 2.2$
$= 15.5508836\ldots$
$= 15.6\text{ m}^3$ (1 d.p.) ✓ (1 mark)

b Capacity = 15.6 × 1000 L
= 15 600 L ✓ (1 mark)

37 **a** $\cos 32° = \frac{600}{AB}$
$AB = \frac{600}{\cos 32°}$ ✓
$= 707.50704\ldots$
$= 710\text{ m}$
(nearest 10 m) ✓ (2 marks)

b 5 km/h = 5000 m/h
Time taken = (710 ÷ 5000) h
= 0.142 h
= 8.52 min
It will take Cressida about 8½ minutes to walk from A to B. ✓ (1 mark)

(Total Part B: 20 marks)

SAMPLE EXAM PAPER 3
Part A: Multiple Choice PAGE 119

1 0.395 ÷ 48.95
= 0.008 069 4586 …
= 0.008 07 (3 sig. figs.)
∴ [D] ✓ (1 mark)

2 $-3(2x - 9) - 5(4x - 1)$
$= -6x + 27 - 20x + 5$
$= -26x + 32$
∴ [B] ✓ (1 mark)

3 $m = 3$
So $y = 3x + b$
The line passes through (–2, 5).
So $5 = 3 \times -2 + b$
$5 = -6 + b$
$b = 11$
The equation is $y = 3x + 11$.
∴ [D] ✓ (1 mark)

4 Options A and B have curves that are increasing.
Option B is increasing at a decreasing rate.
∴ [B] ✓ (1 mark)

5 $P(\text{first pen is blue}) = \frac{4}{9}$
If the first pen is blue
$P(\text{second pen is blue}) = \frac{3}{8}$
P(both pens are blue)
$= \frac{4}{9} \times \frac{3}{8}$
$= \frac{1}{6}$
∴ [C] ✓ (1 mark)

6

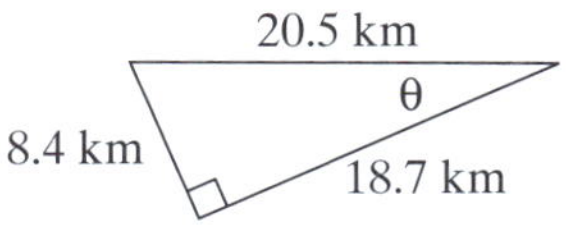

$\tan\theta = \frac{8.4}{18.7}$
$\theta = \tan^{-1}\left(\frac{8.4}{18.7}\right)$
= 24°11′22.25″
= 24°11′ (nearest minute)
[Or use sin or cos]
∴ [A] ✓ (1 mark)

7 $(6x^8 \div 3x^3)^2 = (2x^5)^2$
$= 4x^{10}$
∴ [D] ✓ (1 mark)

8 Diameter = 12 cm, so $r = 6$
Area of circle
$= \pi r^2$
$= \pi \times 6^2$
$= 113.097335\ldots\text{ cm}^2$
Diagonals of square are 12 cm.
[A square is also a rhombus.]
Area of square $= \frac{1}{2}xy$
$= \frac{1}{2} \times 12 \times 12$
$= 72\text{ cm}^2$
Shaded area
$= (113.097\ldots - 72)\text{ cm}^2$
$\approx 41\text{ cm}^2$
∴ [A] ✓ (1 mark)

9 $A(-2, 6)$, $B(6, 2)$
Midpoint $= \left(\frac{-2+6}{2}, \frac{6+2}{2}\right)$
$= (2, 4)$
Of the options the coordinates $x = 2$ and $y = 4$ only satisfy $y = x + 2$.
∴ [C] ✓ (1 mark)

10

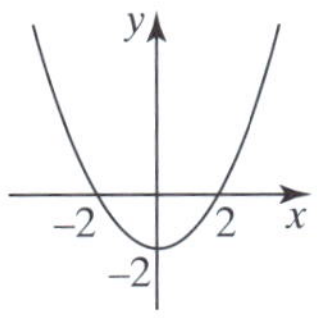

The parabola is of the form
$y = ax^2 + c$
When $x = 0$, $y = -2$
So $c = -2$ and $y = ax^2 - 2$
When $x = 2$, $y = 0$
$0 = a \times 2^2 - 2$
$4a - 2 = 0$
$4a = 2$
$a = \frac{1}{2}$
So $y = \frac{1}{2}x^2 - 2$
∴ [D] ✓ (1 mark)

11 $0.06 \text{ GL} = (0.06 \times 1000) \text{ ML}$
$= 60 \text{ ML}$
Option A is not correct.
$\therefore$ **[A]** ✓ (1 mark)

12

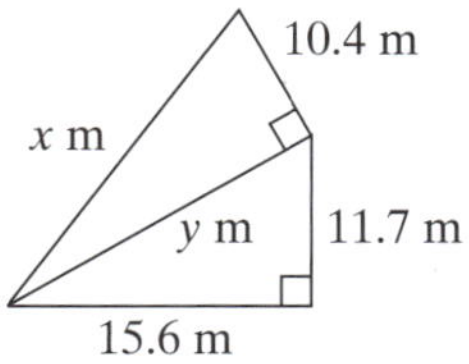

$y^2 = 11.7^2 + 15.6^2$
$= 380.25$
$x^2 = y^2 + 10.4^2$
$= 380.25 + 108.16$
$= 488.41$
$x = \sqrt{488.41} \quad (x > 0)$
$= 22.1$
$\therefore$ **[B]** ✓ (1 mark)

13 Total repaid $= 60 \times \$530$
$= \$31\,800$
$I = \$31\,800 - \$24\,000$
$= \$7800$
60 months = 5 years
Now $I = Prn$
$\$7800 = \$24\,000 \times r \times 5$
$r = \dfrac{\$7800}{\$24\,000 \times 5}$
$= 0.065$
$= 6.5\%$
$\therefore$ **[C]** ✓ (1 mark)

14 $2^{-5} \div 4^{-8} = 2^{-5} \div (2^2)^{-8}$
$= 2^{-5} \div 2^{-16}$
$= 2^{11}$
$\therefore$ **[D]** ✓ (1 mark)

15 Lowest score = 10
Range = 14
So highest score = 10 + 14 = 24
The mode is 13 so at least two scores are 13.
The scores are 10, 13, 13, 24.
Mean $= \dfrac{10 + 13 + 13 + 24}{4}$
$= 15$
$\therefore$ **[C]** ✓ (1 mark)

16 The triangles are similar because they are equiangular. (Alternate $\angle$s (|| sides) are equal, vertically opposite $\angle$s are equal.)

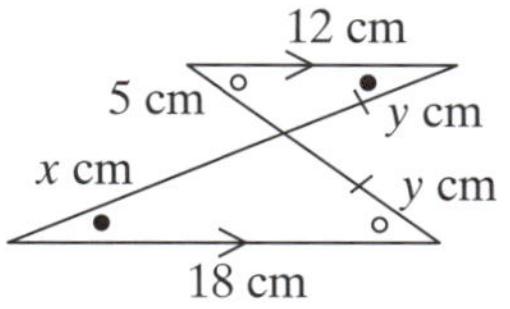

Let the unknown equal sides be y cm.
So $\dfrac{y}{5} = \dfrac{18}{12}$
$y = \dfrac{18}{12} \times 5$
$= 7.5$
$\dfrac{x}{7.5} = \dfrac{18}{12}$
$x = \dfrac{18}{12} \times 7.5$
$x = 11.25$
$\therefore$ **[A]** ✓ (1 mark)

17 $(-5, 4), (7, -2)$
$m = \dfrac{y_2 - y_1}{x_2 - x_1}$
$= \dfrac{-2 - 4}{7 - (-5)}$
$= -\dfrac{1}{2}$
$\therefore$ **[B]** ✓ (1 mark)

18 80 km/h
$= 80\,000$ m/h
$= 1333.333\ldots$ m/min
$= 22.222\ldots$ m/s
$= 111.111\ldots$ m in 5 seconds
$\therefore$ **[C]** ✓ (1 mark)

19 $A = 12\,755 \text{ cm}^2$
$2\pi rh = 12\,755$
$2 \times \pi \times 35 \times h = 12\,755$
$h = \dfrac{12\,755}{2 \times \pi \times 35}$
$= 58.000\,6085\ldots$
So height $\approx$ 58 cm
Now $V = \pi r^2 h$
$= \pi \times 35^2 \times 58$
$= 223\,210.15\ldots$
The volume is approximately $223\,210 \text{ cm}^3$.
$\therefore$ **[D]** ✓ (1 mark)

20 Let x mm be the length of the hypotenuse.
$x^2 = 119^2 + 120^2$
$= 28\,561$
$x = \sqrt{28\,561} \quad (x > 0)$
$= 169$
$A = 2 \times \dfrac{1}{2} \times 120 \times 119$
$+ (119 + 120 + 169) \times 50$
$= 34\,680 \text{ mm}^2$
$\therefore$ **[A]** ✓ (1 mark)
(Total Part A: 20 marks)

SAMPLE EXAM PAPER 3
Part B: Short Answer PAGE 121

21 $0.08 \div 126\,000$
$= 0.000\,000\,6349\ldots$
$= 6.35 \times 10^{-7}$ (3 sig. fig.) ✓
(1 mark)

22 $2xy^2(3x^2 - 2y) - 9x^2y(2 - xy)$
$= 6x^3y^2 - 4xy^3 - 18x^2y + 9x^3y^2$
$= 15x^3y^2 - 4xy^3 - 18x^2y$ ✓
(1 mark)

23 **a** Upon measuring, the longest boundary is 9 cm.
So 9 cm represents 900 m or 90 000 cm.
1 cm represents 10 000 cm
The scale is 1:10 000 ✓
(1 mark)

b On the map, the top boundary is 6 cm so represents a real distance of 600 m. ✓
The distance between the top and bottom is 3.5 cm, so a real distance of 350 m. ✓
$A = \dfrac{h}{2}(a + b)$
$= \dfrac{350}{2}(600 + 900)$
$= 262\,500 \text{ m}^2$
$= 26.25$ ha ✓ (3 marks)

24

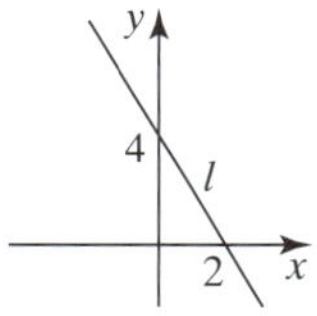

a $m = \dfrac{\text{rise}}{\text{run}}$
$= -\dfrac{4}{2}$
$= -2$ ✓ (1 mark)

b $l: y = -2x + 4$ ✓ (1 mark)

c At Q, $y = -6$
Now Q lies on l.
So, $-2x + 4 = -6$
$-2x = -10$
$x = 5$
So Q is the point $(5, -6)$.
✓ (1 mark)

d $(0, 0), (5, -6)$
$d^2 = 5^2 + (-6)^2$
$= 61$
The distance is $\sqrt{61}$ units. ✓ (1 mark)

e $x^2 + y^2 = 61$ ✓ (1 mark)

25 **a** 8 years = 8 × 12 months
= 96 months
Total repaid
= 96 × \$131.25
= \$12 600 ✓ (1 mark)

b $I = PRN$
$= P \times 0.05 \times 8$
$= 0.4P$ ✓
Now $P + I = 12\,600$
$P + 0.4P = 12\,600$
$1.4P = 12\,600$
$P = 12\,600 \div 1.4$
$= 9000$
The amount borrowed was \$9000. ✓ (2 marks)

26

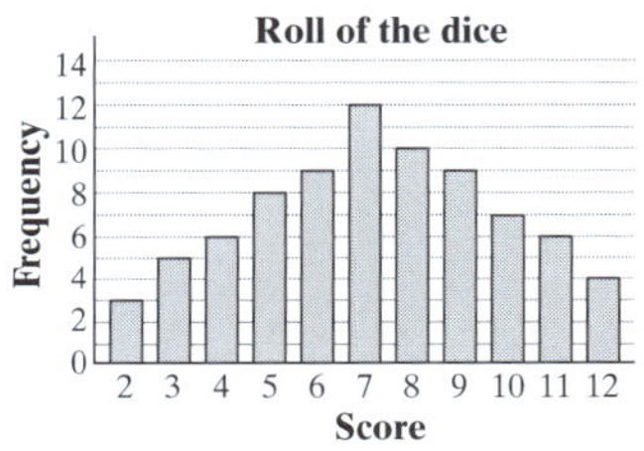

a Total frequency
= 3 + 5 + 6 + 8 + 9 + 12 + 10 + 9 + 7 + 6 + 4
= 79
Joe rolled the dice 79 times. ✓ (1 mark)

b 7 occurred 12 times
Relative frequency of 7
$= \frac{12}{79}$ ✓ (1 mark)

c Sum
= 3×2+5×3+6×4+…
= 568
mean = 568 ÷ 79
= 7.189 87…
= 7.2 (1 d.p.) ✓
(1 mark)

d The middle score is the 40th score so the median is 7. ✓ (1 mark)

e No, the skewness is negative not positive, but the display is actually fairly symmetrical. ✓ (1 mark)

f There are 6 × 6 or 36 different outcomes when two dice are rolled.
A score of 5 could be 1 + 4, 2 + 3, 3 + 2 or 4 + 1.
So there are 4 different outcomes that produce 5.

$P(\text{five}) = \frac{4}{36} = \frac{1}{9}$

You would expect to get 5 one-ninth of the times.
$\frac{1}{9}$ of 79 = 8.7777…
But 5 occurred 8 times, which is slightly less than you would expect. ✓ 1 mark)

g $P(\text{double}) = \frac{1}{6}$ ✓ (1 mark)

[It doesn't matter which number is rolled first, there is one chance in six of rolling the same number with the second die.]

27

$y = x^2 + 3$
When $y = 7$,
$x^2 + 3 = 7$
$x^2 = 4$
$x = \pm 2$ ✓

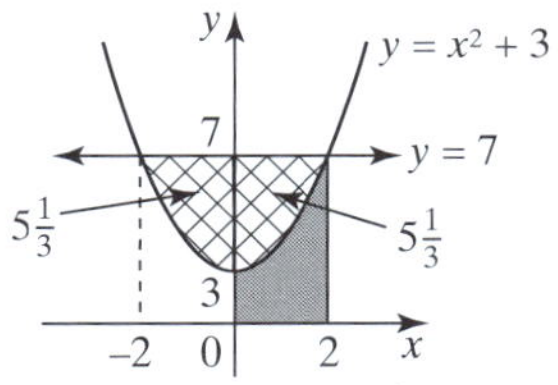

Hatched area = $10\frac{2}{3}$ units²

The parabola is symmetrical. So, the area bounded by the y-axis, the parabola and the line $y = 7$ will be $5\frac{1}{3}$ units². ✓

Area of rectangle bounded by the axes and the lines $x = 2$ and $y = 7$ is 2 × 7 units² or 14 units².
Shaded area
$= (14 - 5\frac{1}{3})$ units²
$= 8\frac{2}{3}$ units² ✓ (3 marks)

28 **a** Let the diameter be x cm.

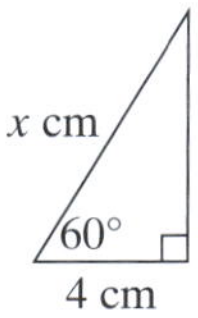

$\cos 60° = \frac{4}{x}$

$x = \frac{4}{\cos 60°}$

$= 8$
The diameter is 8 cm ✓
(1 mark)

b Semicircle:
Arc length $= \frac{1}{2}\pi d$
$= \frac{1}{2} \times \pi \times 8$
$= 12.566\,370\,61\ldots$
$= 12.6$ (1 d.p.) ✓
Let h cm be the height of the triangle.

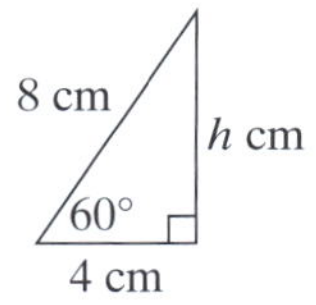

$h^2 = 8^2 - 4^2$
$= 64 - 16$
$= 48$
$h = \sqrt{48}$
$= 6.928\,203\,23\ldots$
$= 6.9$ (1 d.p.) ✓
[Or: let h cm be the height of the triangle
$\tan 60° = \frac{h}{4}$
$h = 4 \times \tan 60°$
$= 6.928\,203\,23\ldots$]
$P = 12.566\ldots + 6.928\ldots + 4$
$= 23.5$ (1 d.p.)
The perimeter is 23.5 cm correct to one decimal place. ✓ (3 marks)

c Semicircle:
$A = \frac{1}{2}\pi r^2$
$= \frac{1}{2} \times \pi \times 4^2$
$= 25.132\,74\ldots$ ✓
Triangle:
$A = \frac{1}{2}bh$
$= \frac{1}{2} \times 4 \times 6.9282\ldots$
$= 13.8564\ldots$
Total
$= 25.132\,74\ldots + 13.8564\ldots$
$= 38.989\,147\ldots$
$= 39$ (nearest unit)
The area is 39 cm² to the nearest square centimetre. ✓ (2 mark)
(Total Part B: 30 marks)

TEST & EXAM RESULTS

Transfer your **percentage score** that you calculated in the **Your Feedback** box at the end of each **test** and **exam** to the table below. This will help you work out your areas of strength and weakness.

Test Topic	Intermediate Test Score	Advanced Test Score
Review: Real Numbers	%	%
Review: Algebraic Techniques	%	%
Review: Indices and Scientific Notation	%	%
Review: Pythagoras' Theorem	%	%
Review: Units of Measurement and Time Scales	%	%
Number and Algebra: Earning Money	%	%
Number and Algebra: Simple Interest	%	%
Number and Algebra: Rates and Proportion	%	%
Number and Algebra: The Index Laws	%	%
Number and Algebra: The Distributive Law	%	%
Number and Algebra: Distance between Two Points	%	%
Number and Algebra: Midpoint and Gradient	%	%
Number and Algebra: Sketching Linear Graphs	%	%
Number and Algebra: The Equation of a Line	%	%
Number and Algebra: Non-linear Graphs	%	%
Measurement and Geometry: Area of Composite Shapes	%	%
Measurement and Geometry: Surface Area of Prisms	%	%
Measurement and Geometry: Volume of Prisms	%	%
Measurement and Geometry: Surface Area and Volume of Cylinders	%	%
Measurement and Geometry: Enlargement and Similarity	%	%
Measurement and Geometry: Ratio and Scale Factors	%	%
Measurement and Geometry: Trigonometry of Right-angled Triangles	%	%
Statistics and Probability: Collecting and Evaluating Data	%	%
Statistics and Probability: Stem-and-leaf plots, Histograms and Describing Data	%	%
Statistics and Probability: Comparing Data Displays and Using Mean, Median and Range	%	%
Statistics and Probability: Relative Frequencies and Probabilities	%	%
Statistics and Probability: Two-step Chance Experiments	%	%
Sample Exam Papers		
Sample Exam Paper 1		%
Sample Exam Paper 2		%
Sample Exam Paper 3		%

FEEDBACK CHECKLIST TO IMPROVE YOUR TEST & EXAM RESULTS

Do you want to improve your scores in the Sample Exams? Check that:

You are ready.

This is a revision workbook designed specifically for the revision of work already done. It is not a replacement for your textbook or class notes. Cover the topic in class or in your own time by reading your textbook or the ***Excel*** *Mathematics Study Guide Years 9–10*, before using this book for further practice on the topic.

You are revising in the right order.

Maths is a subject that builds on previous knowledge. Often you will need to have a good grasp of an early topic in order to fully understand a later topic.

Your standards are realistic.

You cannot expect to score 100% all the time. Remember that this book has been designed to help you identify your strengths as well as your weaknesses, so it is OK to make mistakes. The key to success is learning from those mistakes.

You are not rushing through the questions.

Many Maths students fail to read questions properly. This accounts for a large number of the mistakes made in tests and exams. Although your time is limited, you must still take enough time to read each question carefully. If necessary, read the question a second (or even third) time. Don't start your answer until you are sure that you understand the question. For the longer questions, take a moment or two to plan your answer.

You are allowing yourself enough time to study.

Preparation for any test or exam requires a sufficient investment of time to be undertaken successfully. This means that you have to plan ahead. Don't wait until a week or so before an exam, for example, to ask yourself the question: How much time will I need to devote to this subject to revise it thoroughly? Don't forget to allow sufficient time for your other subjects—Maths isn't the only subject you will have to sit an exam for!

You are OK.

You may be having some problems unrelated to your schoolwork. It is wise to sort out such problems as quickly as possible. Speak to your teacher, a parent or your school counsellor if you think you need help.

INDEX